Advances in Organic Synthesis

(*Volume 18*)

Edited by

Shazia Anjum
Institute of Chemistry,
The Islamia University of Bahawalpur,
Pakistan

Advances in Organic Synthesis

Volume # 18

Editor: Shazia Anjum

ISSN (Online): 2212-408X

ISSN (Print): 1574-0870

ISBN (Online): 978-981-5040-79-1

ISBN (Print): 978-981-5040-80-7

ISBN (Paperback): 978-981-5040-81-4

Published by Bentham Science Publishers Pte. Ltd. Singapore. All Rights Reserved.

need for a court order if at any point you breach any terms of this License Agreement. In no event will any delay or failure by Bentham Science Publishers in enforcing your compliance with this License Agreement constitute a waiver of any of its rights.

3. You acknowledge that you have read this License Agreement, and agree to be bound by its terms and conditions. To the extent that any other terms and conditions presented on any website of Bentham Science Publishers conflict with, or are inconsistent with, the terms and conditions set out in this License Agreement, you acknowledge that the terms and conditions set out in this License Agreement shall prevail.

Bentham Science Publishers Pte. Ltd.
80 Robinson Road #02-00
Singapore 068898
Singapore
Email: subscriptions@benthamscience.net

BENTHAM SCIENCE

CONTENTS

PREFACE

The series 'Advances in Organic Synthesis' are a dedicated set of different volumes covering all contemporary developments in organic synthesis. The contents of each volume are thoughtfully picked to capture the dire need of the readers and at the same time, it is rewarding to quench the thirst of researchers. The first chapter of volume 18 describes the latest syntheses and biological activities of anti-cancer Ferrocenes. Mehra and Lumb gave a brief account of the stable aromatic nature of ferrocene and bioferrocene compounds that possess reversible redox properties, and low toxicity that has also revolutionized the area of medicinal organometallic chemistry.

Masdeu *et al.* compiled a review on the synthesis of fused nitrogenated heterocycles using intramolecular Povarov Reaction. This strategy helped in synthesizing tri-, tetra-, penta-, hexa-, hepta-and octacyclic-fused heterocyclic compounds of diverse biological activities. While Singh and Kumari in their chapter 03 described the recent applications of barbituric acid as a synthetic precursor for the synthesis of various bioactive compounds. Barbituric acid can be employed in the condensation reaction as well as 5-/6- membered oxygen/nitrogen-containing heterocycles.

The use of ionic liquids (ILs) in organic syntheses has made a renaissance. It has revolutionized the entire trajectory of the synthetic pathway. In Chapter 4, Khokhar *et al.* described emerging applications of ILs as green solvents, green catalysts, and in coupling reactions. It depicts their remarkable use in environmentally benign organic synthesis which is the main focus of discussion in this chapter.

The last three chapters of these volumes are exclusively dedicated to the newly emerging field of nanotechnology. For example, chapter 05 of this volume illustrates the advanced use of Zinc oxide (ZnO) nanomaterials for biomedical applications. In this chapter, Bhuiyan *et al.* showed that ZnO nanomaterials could be developed by the organic synthesis process for excellent biocompatibility, selectivity, sensitivity, good chemical stability, non-toxicity, and fast electron transfer properties in a cost-effective manner. While Higazy *et al.*, in their chapter 06 covered the contemporary development of nanocomposite hydrophobic coatings for effective corrosion protection. Organic-inorganic nanocomposites can be employed as outstanding anti-corrosive coatings to provide longevity of steel construction service. Overall, they gave an overview of present and advanced research developments, such as graphene nanocomposite surfaces, etc. Interestingly, the last chapter of this volume by Zdiri *et al.* covers the morphologies and properties of virgin and waste polypropylene nanocomposites. The article gives information on different types of nanoparticles used for the enhancement of thermo-mechanical and physical behaviors of PP nanocomposites. Moreover, it also discusses the improvements in the properties of waste PP by nanoparticles' incorporation and the influence of clay nanoparticles on waste PP-based nanocomposites.

I hope that the readers will enjoy reading this volume which covers contemporary developments in organic syntheses. I would like to express my sincere thanks to the editorial staff of Bentham Science Publishers, particularly Ms. Asma Ahmed and Mr. Mahmood Alam for their constant help and support.

Shazia Anjum
Institute of Chemistry
The Islamia University of Bahawalpur
Pakistan

List of Contributors

A. Elamri	Université de Monastir, ENIM, Unité de Recherche Matériaux et Procédés Textiles, UR17ES33, 5019, Monastir, Tunisie
Carme Masdeu	Departamento de Química Orgánica I, Facultad de Farmacia and Centro de Investigación Lascaray (Lascaray Research Center), Universidad del País Vasco/Euskal Herriko Unibertsitatea (UPV/EHU), Paseo de la Universidad 7, 01006 Vitoria-Gasteiz, Spain
Concepcion Alonso	Departamento de Química Orgánica I, Facultad de Farmacia and Centro de Investigación Lascaray (Lascaray Research Center), Universidad del País Vasco/Euskal Herriko Unibertsitatea (UPV/EHU), Paseo de la Universidad 7, 01006 Vitoria-Gasteiz, Spain
Francisco Palacios	Departamento de Química Orgánica I, Facultad de Farmacia and Centro de Investigación Lascaray (Lascaray Research Center), Universidad del País Vasco/Euskal Herriko Unibertsitatea (UPV/EHU), Paseo de la Universidad 7, 01006 Vitoria-Gasteiz, Spain
Gyandshwar K. Rao	Department of Chemistry, Biochemistry and Forensic Science, Amity School of Applied Sciences, Amity University Haryana, Gurgaon 122413, India
Hayati Mamur	Department of Electrical and Electronics Engineering, Manisa Celal Bayar University, Turkey
Isha Lumb	Department of Chemistry, Baring Union Christian College, Batala-143505, India
Jesús M. de los Santos	Departamento de Química Orgánica I, Facultad de Farmacia and Centro de Investigación Lascaray (Lascaray Research Center), Universidad del País Vasco/Euskal Herriko Unibertsitatea (UPV/EHU), Paseo de la Universidad 7, 01006 Vitoria-Gasteiz, Spain
Kamalakanta Behera	Department of Chemistry, Faculty of Science, University of Allahabad, Prayagraj - 211002, India
Kamal Nayan Sharma	Department of Chemistry, Biochemistry and Forensic Science, Amity School of Applied Sciences, Amity University Haryana, Gurgaon 122413, India
Komal	Department of Chemistry, Biochemistry and Forensic Science, Amity School of Applied Sciences, Amity University Haryana, Gurgaon 122413, India
Khmais Zdiri	Université de Haute Alsace, ENSISA, Laboratoire de Physique et Mécanique Textiles, EA 4365, 68100, Mulhouse, France Université de Monastir, ENIM, Unité de Recherche Matériaux et Procédés Textiles, UR17ES33, 5019, Monastir, Tunisie Université de Lille, ENSAIT, Laboratoire de Génie et Matériaux Textiles, EA 2461, 59056, Roubaix, France
Mohamed S. Selim	Petroleum Application Department, Egyptian Petroleum Research Institute, Nasr City 11727, Cairo, Egypt Key Laboratory of Clean Chemistry Technology of Guangdong Regular Higher Education Institutions, School of Chemical Engineering and Light Industry, Guangdong University of Technology, Guangzhou, 510006, PR China
Mohammad Ruhul Amin Bhuiyan	Department of Electrical and Electronic Engineering, Islamic University, Bangladesh

M. Hamdaoui	Université de Haute Alsace, ENSISA, Laboratoire de Physique et Mécanique Textiles, EA 4365, 68100, Mulhouse, France
Olfat E. El-Azabawy	Key Laboratory of Clean Chemistry Technology of Guangdong Regular Higher Education Institutions, School of Chemical Engineering and Light Industry, Guangdong University of Technology, Guangzhou, 510006, PR China
O. Harzallah	Université de Haute Alsace, ENSISA, Laboratoire de Physique et Mécanique Textiles, EA 4365, 68100, Mulhouse, France
Ömer Faruk Dilmaç	Department of Chemical Engineering, Çankiri Karatekin University, Turkey
Sundaram Singh	Department of Chemistry, Indian Institute of Technology (BHU), Varanasi 221005, Uttar Pradesh, India
Savita Kumari	Department of Chemistry, Indian Institute of Technology (BHU), Varanasi 221005, Uttar Pradesh, India
Shruti Trivedi	Department of Chemistry, Institute of Science, Banaras Hindu University Varanasi - 221 005 (U.P.), India
Shreya Juneja	Department of Chemistry, Indian Institute of Technology Delhi, New Delhi - 110016, India
Siddharth Pandey	Department of Chemistry, Indian Institute of Technology Delhi, New Delhi - 110016, India
Shimaa A. Higazy	Petroleum Application Department, Egyptian Petroleum Research Institute, Nasr City 11727, Cairo, Egypt
Vishu Mehra	Department of Chemistry, Hindu College, Amritsar-143005, India
Vaishali Khokhar	Department of Chemistry, Indian Institute of Technology Delhi, New Delhi - 110016, India

CHAPTER 1

Recent Synthetic and Biological Advances in Anti-cancer Ferrocene-Analogues and Hybrids

Vishu Mehra[2,*] and **Isha Lumb**[1]

[1] *Department of Chemistry, Baring Union Christian College, Batala-143505, India*

[2] *Department of Chemistry, Hindu College, Amritsar-143005, India*

Abstract: Cancer is among the most severe risks to the global human population. The enduring crisis of drug-resistant cancer and the limited selectivity of anticancer drugs are significant roadblocks to its control and eradication, requiring the identification of new anticancer entities. The stable aromatic nature, reversible redox properties, and low toxicity of ferrocene revolutionized medicinal organometallic chemistry, providing us with bioferrocene compounds with excellent antiproliferative potential, which has been the focus of persistent efforts in recent years. Substituting the aryl/heteroaryl core for ferrocene in an organic molecule alters its molecular characteristics, including solubility, hydro-/lipophilicity, as well as bioactivities. Ferrocifen (ferrocene analogues of hydroxytamoxifen) has shown antiproliferative potential in both hormone-dependent (MCF-7) and hormone-independent (MDA-MB-231) breast cancer cells. It is now in pre-clinical trials against malignancies. These entities operate through various targets, some of which have been revealed and activated in response to product concentrations. They also react to the cancer cells by diverse mechanisms that can work in concert or in isolation, depending on signaling pathways that promote senescence or death. The behavior of ferrocene-containing hybrids with a range of anticancer targets is explained in this chapter.

Keywords: Anti-proliferative Potential, Azide-alkyne Cycloaddition, Biological Activities, Bio-organometallic, Bioferrocene Compounds, Cancer, Cytotoxicity, Ferrocene Compounds, Ferrocifen, Ferrociphenols.

1. INTRODUCTION

Organometallic chemistry and biochemistry have recently been combined to form a new subject known as bioorganometallic chemistry. This new research topic has piqued scientists' interest because of the unusual chemical structure and biological activity of organometallic compounds. These carbon-metal linkage compounds

---

* **Corresponding author Vishu Mehra:** Department of Chemistry, Hindu College, Amritsar-143005, India; Tel: +91-183-2547147; Email: vishu3984@gmail.com

Shazia Anjum (Ed.)

offer a potentially rich sector for the discovery of new pharmacological medicines with novel mechanisms of action, and the field is rapidly increasing. In recent years, there has been an increased interest in developing organometallic compounds as structural variants of existing drugs for treating drug resistance cancer [1]. Among the various organometallics, ferrocene [2], the archetypal organometallic compound, serves as a useful platform in bio-organometallic chemistry because of its important role in various fields, including stereoselective, stereospecific, and asymmetric transformations, electrochemistry, polymer chemistry, material science, biochemistry, crystal engineering, and drug design and development [3]. Ferrocene compounds are particularly appealing candidates for biological applications because of their durability in aqueous and aerobic settings, as well as the availability of a wide range of derivatives and outstanding electrochemical characteristics [4]. Ferrocene [5, 6] is a compelling target in fields like drug design mediators of protein redox processes, internal standards in electrochemistry, and organic synthesis, such as functionalization of cyclopentadienyl ligands, due to its sandwich-like structure and chemical representation $(\eta^5\text{-}C_5H_5)_2Fe$. In many ways, ferrocene is similar to benzene in that it behaves like an aromatic ring and conducts electrophilic reactions, including Friedel-alkylation, Craft's acylation, Vilsmeyer formulation, and mercuration reactions, which are all phenyl ring properties. Ferrocene derivatives with asymmetric substituents are extensively used as asymmetric hydrogenation catalysts [7]. Among organometallic compounds, ferrocene has a remarkable range of chemistry. Numerous studies have demonstrated ferrocene's efficacy *in vivo* and *in vitro*, as well as its potential as an anticancer, antimalarial, and antifungal agent [8 - 11]. The anticancer action of ferrocene-based compounds is linked to the oxidation state of the central iron atom. Only the ferrocenium salts with the central iron atom in the oxidation state of +3 have been found to exhibit anticancer activity. Incorporating ferrocene into bioactive compounds is a common technique in this field, with the most successful example being its incorporation into tamoxifen, resulting in the potential therapeutic candidate ferrocifen, which has the unique property of being antiproliferative against both the MCF-7 (hormone-dependent) and MDA-MB-231 (hormone-independent) breast cancer cell lines [12]. Many studies have also shown that ferrocene analogues have the potential to treat a wide range of illnesses, including fungal/bacterial infections, malaria, HIV, and cancer. This chapter aims to keep researchers informed about recent advances in the synthesis and evaluation of ferrocene-containing bioactive pharmacophores, focusing on the structure-activity relationship (2015-2020).

FERROCENE-BASED CONJUGATES HAVE ANTIPROLIFERATIVE POTENTIAL

Schobert and co-workers synthesized and analyzed the antiproliferative potential of ferrocene-derived *N*-heterocyclic carbene complexes of Gold (I) [13]. Ferrocene-carboxaldehyde **1** was initially treated with toluene sulfonyl methyl isocyanide **2** and methylamine, resulting in imidazoles **3**. Alkylation reaction of **3** with iodomethane or iodoethane provided the corresponding imidazolium salts **4**, which upon subsequent reaction with silver (I) oxide yielded the complexes **5**. Complex **5** was transmetallated with chloro (dimethyl sulfide)gold (I), resulting in the derivative **6**. The subsequent reaction of **6** with triphenylphosphine and sodium tetrafluoroborate afforded the target complex *viz.* cationic phosphano gold complex **7** (Scheme **1**).

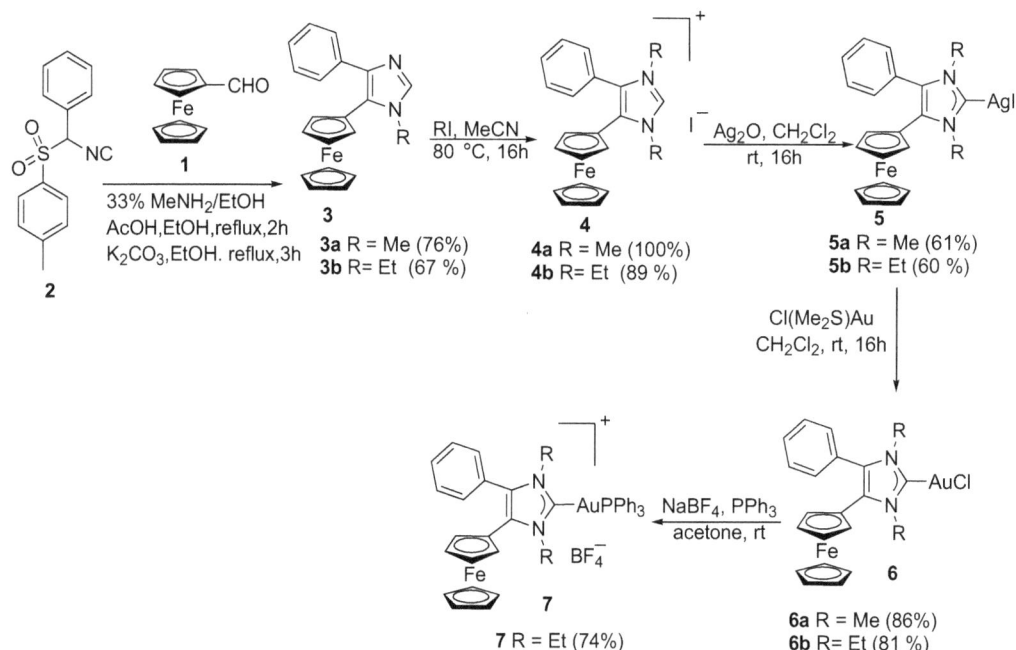

Scheme (1). Synthetic route to ferrocenyl substituted *N*-heterocyclic carbene complexes of Gold (I) **7**.

Ferrocene-substituted biscarbene complex **8** was prepared by substituting BF_4 for the counter anion in imidazolium iodide **4b**. The desired silver carbene complex **9** was synthesized by transmetallation of the complex **8** with chloro (dimethyl sulfide)gold (I) (0.5 equivalents) (Scheme **2**).

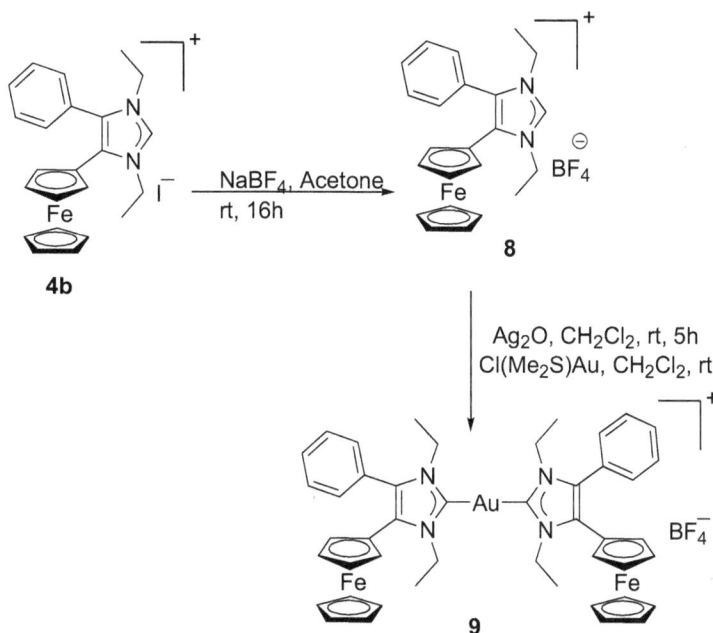

Scheme (2). The synthetic pathway to obtain ferrocenyl substituted *N*-heterocyclic carbene complexes of Gold (I) **9**.

When these complexes (**6**, **7**, and **9**) were tested for their antiproliferative potential, the *N*-methylated monocarbene complex **6a** showed good antiproliferative potential with IC_{50s} 7-20 μM against human cancer and non-malignant cells. Whereas, its ethyl counterpart **6b** had IC_{50s} between 0.2 to 4.0 μM. Complex **9** demonstrated a lower IC_{50} value and was more active than **7**. Furthermore, these complexes **6**, **7**, and **9** were tested against HT-29, a multidrug-resistant tumor cell line.

Ruan *et al.* [14] developed a library of (*E*)-2-methyl-3-ferrocenyl-*N*-acrylamides, and tested them *in vitro* for antiproliferative potential towards B16-F10 and A549 cell lines. Ferrocene carboxaldehyde **1** upon witting reaction with triphenyl-phosphorane-derivative **10** yielded **11**, which was hydrolysed to produce (*E*)- 2-methyl-3-ferrocenylacrylic acid **12**. Next step involving the synthesis of the desired (*E*)-2-methyl-3-ferrocenyl-*N*-acrylamides derivatives **13** proceeded with amide coupling of precursor **12** in the presence of standard coupling agents *viz.* EDCI and HOBt, with various substituted anilines, as shown in Scheme **3**.

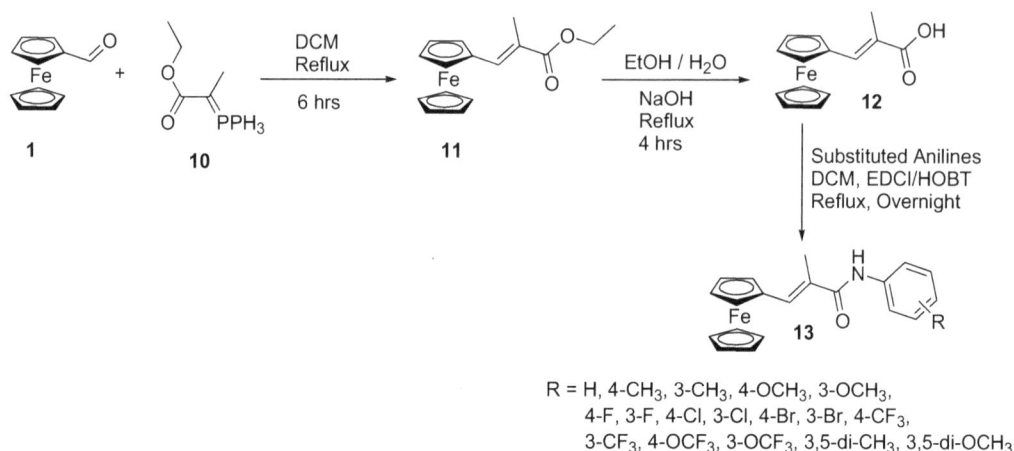

R = H, 4-CH$_3$, 3-CH$_3$, 4-OCH$_3$, 3-OCH$_3$,
4-F, 3-F, 4-Cl, 3-Cl, 4-Br, 3-Br, 4-CF$_3$,
3-CF$_3$, 4-OCF$_3$, 3-OCF$_3$, 3,5-di-CH$_3$, 3,5-di-OCH$_3$

Scheme (3). Synthesis of (*E*)-2-methyl-3-ferrocenyl-*N*-acrylamides derivatives **13**.

Further, the anticancer effectiveness of the synthesized derivatives was examined against two tumour cell lines, B16-F10 and A549 (with celecoxib as a control) by the MTT assay. The examination of substituents influence on the phenyl ring revealed that unsubstituted derivative, with an IC$_{50}$ value of 0.17 μM, had the strongest anticancer activity. The 4th and 3rd positions of the phenyl ring were probed using various electron-donating and electron-withdrawing groups. Except for Br at the third position, other electron-withdrawing substituents such as CF$_3$, OCF$_3$, F, and Cl, diminished the antitumor activity. On the contrary, the addition of CH$_3$ preserved the anti-tumor effect, but 4-OCH$_3$ was comparable to the unsubstituted derivative. Interestingly, the compound with the 3-OCH$_3$ substitution was virtually inert. The reason for this might be that these substituents have a lower binding affinity for the target protein. The inclusion of substituents such as 3-trifluoromethyl, 3-methyl, 3-trifluoromethoxy, 3-fluoro, 4-chloro, and 4-trifluoromethoxy groups on the phenyl core greatly improved the anticancer activity as compared to the unsubstituted derivatives. These results were quite different when tested against B16-F10. The most potent derivatives **13a** and **13b** are depicted in Fig. **1**.

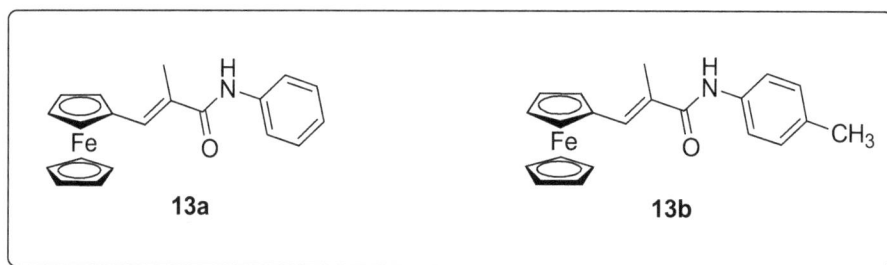

Fig. (1). Most Potent (*E*)-2-methyl-3-ferrocenyl-*N*-acrylamides **13a** and **13b**.

Ferrocenyl substituted Schiff base, and corresponding metal complexes were recently synthesized by Deghadi *et al.* [15]. The reaction between equimolar concentration (1:1) of 2-acetylferrocene **15** and 1,8-naphthalenediamine **14** under refluxing conditions, with methanol as a solvent, afforded the ferrocene-based ligand **16** (Scheme 4).

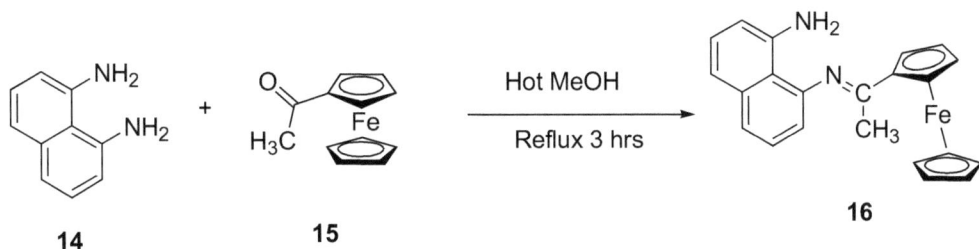

Scheme (4). Synthesis of Ferrocenyl substituted Schiff base ligand **16**.

The metal complexes were made by reacting equimolar amounts of various transition metals with ferrocenyl substituted Schiff base ligand **16** in suitable solvents, and their structures were studied using various spectroscopic techniques. The proposed structure of ferrocene derivatized Schiff base metal complexes is depicted in Fig. **2**.

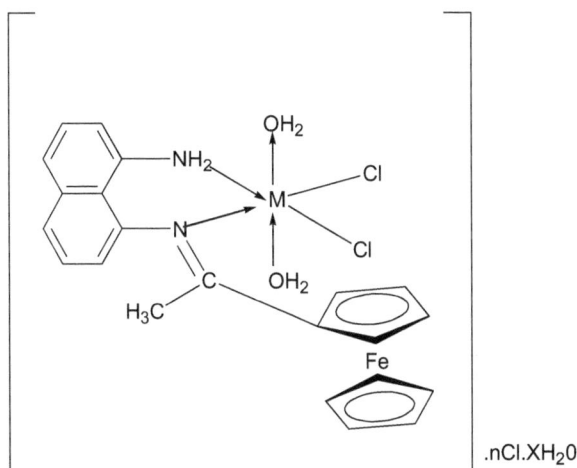

Fig. (2). Structure of ferrocene-derivatized Schiff base metal complex.

The synthesized ferrocenyl Schiff bases and their complexes were tested for anticancer activities against the MCF-7 (breast carcinoma) cells, at a 100 μM concentration. The inhibition fraction of the ligand and corresponding metal complexes (Fe (III), Cu (II), Cr(III), and Cd (II)) was found to be greater than

70%. Further, at various concentrations, the IC_{50} values of the complexes, Cr-(III--L, Cd-(II)-L, Fe-(III)-L, Cu-(II)-L, and ferrocenyl substituted Schiff base ligand **16** were calculated as 11.3, 16.6, 35, 37.7, and 20.2 μM. Results revealed that Cd (II) and Fe (III) complexes were more active than Cu (II)/Cr (III) complexes and Schiff base ligands.

Tucker and co-workers [16] synthesized bis-substituted ferrocene carrying either hydroxy/methoxy-alkyl or thyminyl/methylthyminyl group, and examined their anticancer potential in osteosarcoma (bone cancer) cells. Further, the results were compared to the anticancer activities of the known lead compound, **1-(*S,Rp*)**, a nucleoside analogue with high toxicity towards cancer cells. The synthetic methodology included the initial preparation of synthon **17** [17], which was needed for the synthesis of the compounds **1-(*S,Rp*)-Me$_2$19**, **1-(*S,Rp*)-OMe 22** and **1-(*S,Rp*)-*N*Me 25**. The double de-protection of **17** was carried out in the presence of TBAF and methyl amine to form **1-(*S,Rp*) 18**. Further treatment of **18** with K^{t}OBu and MeI resulted in the synthesis of bismethylated target **1-(*S,Rp*)-Me$_2$ 19**. The desired compound **1-(*S,Rp*)-OMe 22** was synthesized by deprotection, then *O*-methylation of **20** and finally benzoyl group removal from precursor **21**. Further, debenzoylation of **17** was done in the presence of ammonia and methanol to yield **23**, which underwent *N*-methylation resulting in the formation of **24**. Alcohol deprotection of **24** yielded the desired product **1-(*S,Rp*)-*N*Me 25** (Scheme 5).

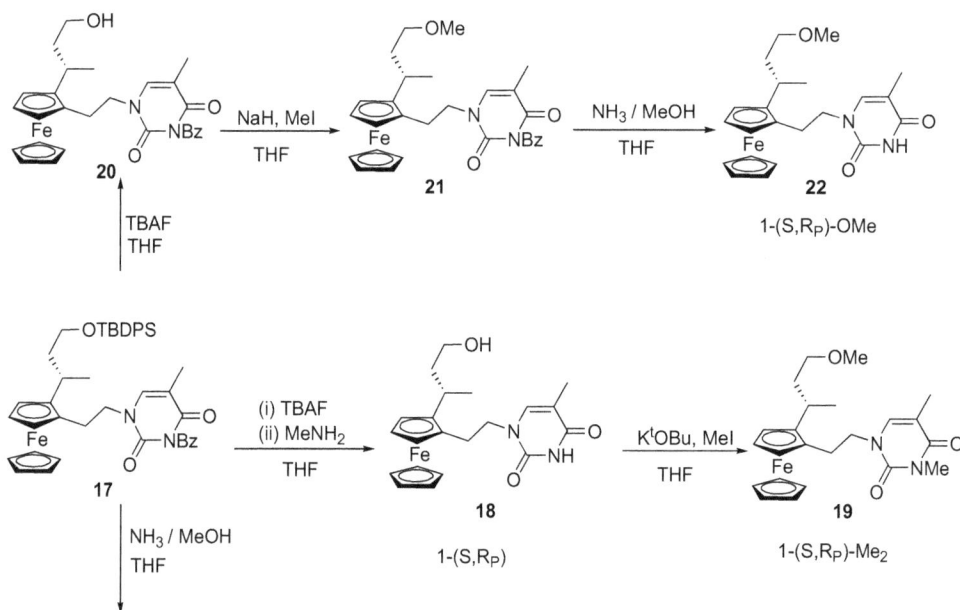

(Scheme 5) contd.....

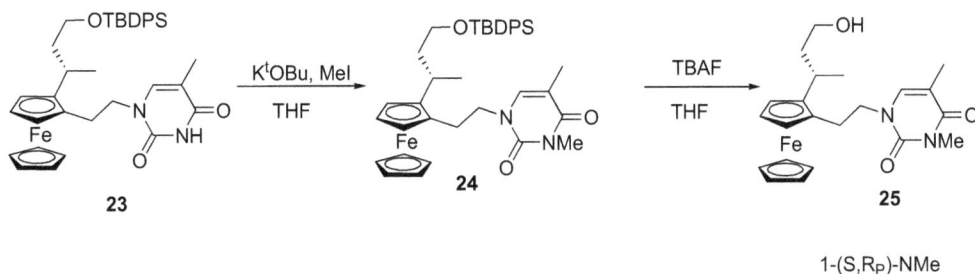

Scheme (5). Synthesis of bis-substituted ferrocene **19**, **22** and **25**.

The resulting ferrocene derivatives, **1-(S,Rp)-Me$_2$ 19** **1-(S,Rp)-OMe 22** and **1-(S,Rp)-NMe 25** and lead compound **1-(S,Rp)** showed IC$_{50}$ values 2.1, 2.7, 1.4 and 2.6 µM respectively on human osteosarcoma (HOS) cells. Further, Tucker *et al.* assessed the influence of nucleotide kinases in the cytotoxicity of these compounds –on 143B osteosarcoma cells, which are isogenic with HOS cells, except for the fact they are thymidine kinase (TK)-negative. A significant drop in cytotoxicity values of **1-(S,Rp)-OMe 22, 1-(S,Rp)-NMe 25** and **1-(S,Rp)-Me$_2$ 19** (IC$_{50s}$ = 5.8, 4.3 and 7.3 µM respectively) was observed. Further, these derivatives that required phosphorylation by thymidine kinase showed resistance to 143B cells, and 156-fold enhancement in resistance to the standard, gemcitabine (IC$_{50}$ = 0.1 nM), confirming the observed results.

Cyclometallated Pt (II) complexes having ferrocenyl moiety have been reported by Lopez *et al.* [18]. The resulting complexes were *in vitro* investigated for anticancer potential towards MCF-7, MDA-MB-231 (breast cancer) and HCT116 (colon) cancer cells. The synthetic procedure entails refluxing compound **26** with cis-[PtCl$_2$(dmso)$_2$] and sodium acetate in MeOH:toluene (1:5) for a time period of 3 days. It formed a deep-brown solution, carrying a mixture of **27a** and **28** – which upon careful chromatographic separation yielded **27a** and **28** as a minor and major product, respectively. The synthesis of new orange Pt(II) complex **29** was achieved by treating **28** with triphenylphosphine in chloroform. The complex **30** was obtained by reacting **28** with an excess of AgNO$_3$ in acetonitrile and then treating it with triphenylphosphine in DCM (Scheme **6**).

The resulting ferrocene-derivatized Pt(II) complexes demonstrated good anticancer profile. Interestingly, ferrocene derivative **26** was observed to be more cytotoxic in nature when compared to complex, platinacycle **28**, with an IC$_{50}$ value <6 µM against both the tested breast cancer (MCF-7 and MDA-MB-231) cell lines. Besides, ferrocene-derivatized Pt(II) metal complexes **29** and **30,** with only difference in the monoanionic ligand (Cl and NO$_2$ respectively), possessed good intrinsic cytotoxicities. However, **29** was approximately 3 and 6 fold more active than **30** against MCF-7 and MDA-MB-231 cells, respectively.

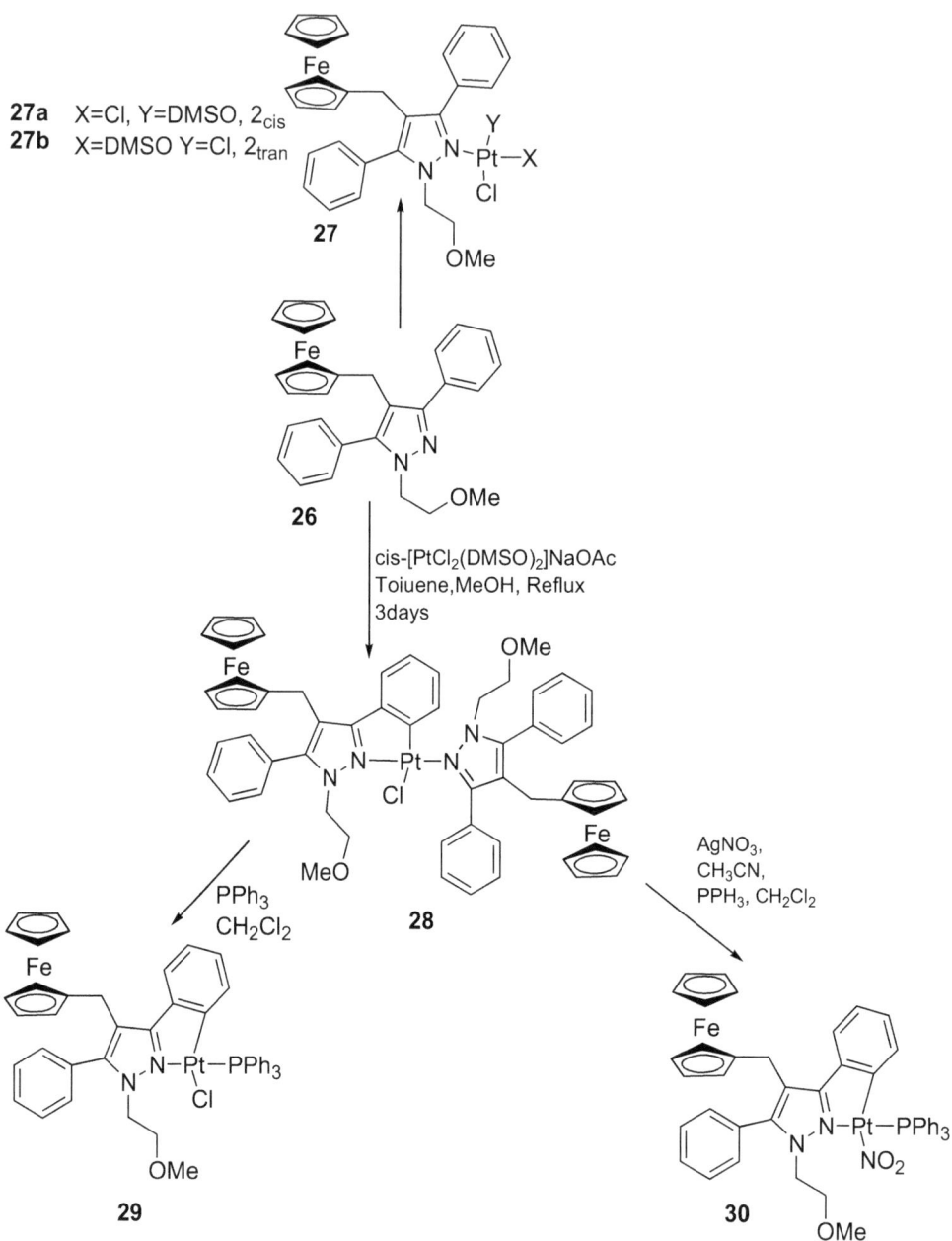

27a   X=Cl, Y=DMSO, 2$_{cis}$
27b   X=DMSO Y=Cl, 2$_{tran}$

Scheme (6). Synthesis of cyclometallated Pt (II) complexes **29** and **30**.

Jaouen *et al.* [19] disclosed the synthetic methodology and antiproliferative potential of ferrocenyl-Podophyllotoxin analogues on breast cancer cells. 6-bromopiperonal **31** was reacted with ferrocenyl lithium in THF, affording alcohol **32**, which upon treatment with ethyl acetoacetate in the catalytic amounts of *p*-toluenesulfonic acid provided keto-ester **33**. The allylation reaction of **33** in the presence of 15-C-5 crown ether provided **34**. The exocyclic alkene **35** was formed by an intramolecular Mizoroki-Heck reaction with a catalytic amount of Palladium(II) acetate, which was then treated with sodium ethoxide in EtOH/THF mixture, yielding the ferrocene-based ester **36**. The intermediate *γ*-oxo ester **38** was obtained by dihydroxylating **36** with OsCl$_3$, followed by a reaction with NaIO$_4$. Intermediate **38** was further reduced with NaBH$_4$, which led to the formation of the desired alcohol **39**, as shown in Scheme 7.

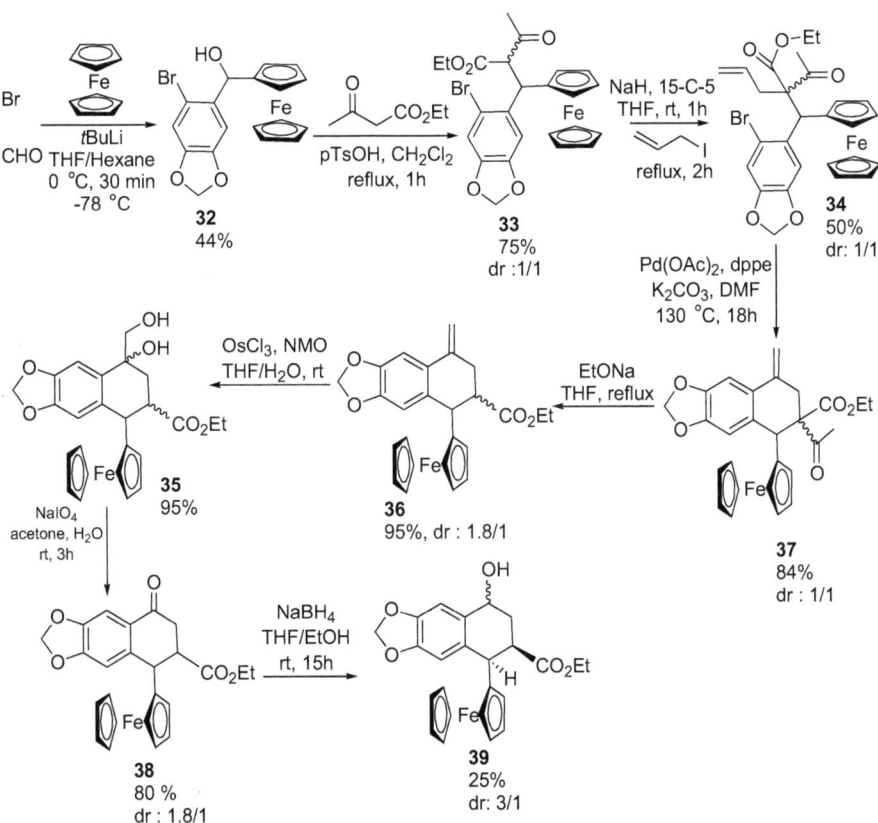

Scheme (7). Synthesis of ferrocenyl analogues of new Podophyllotoxin **39**.

Another Podophyllotoxin-ferrocene derivative **41** was obtained by reacting **40** with ferrocenyl chloride, in the presence of catalytic amounts of triethylamine and *N,N*-dimethyl propionamide (DMAP), as shown in Scheme **8**.

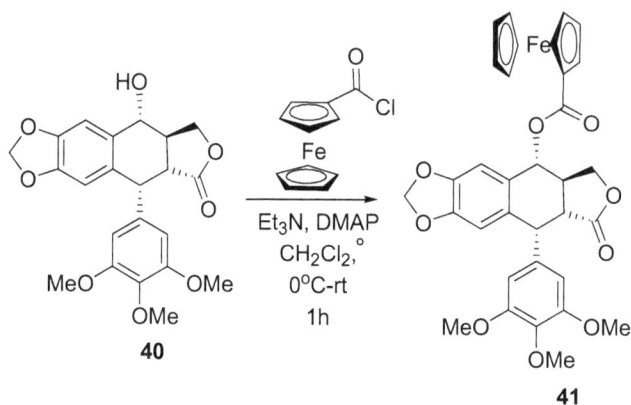

Scheme (8). Synthesis of ferrrocene-Podophyllotoxin derivative **41**.

The antiproliferative activities of the synthesized derivatives, *i.e., O*-ferroceny-
-podophyllotoxin (OFCP) **41** and ferrocene-Podophyllotoxin derivative **39** were
investigated against breast (MCF-7 and MDA-MB-231) cells, and results were
compared to standard podophyllotoxin. The addition of the ferrocene significantly
lowered the anti-breast cancer activity of Podophyllotoxin, with derivative **41**
showing IC_{50s} 0.93 μM and 0.43 μM, against MCF-7 and MDA-MB-231 cells,
respectively. On the other hand, the alcohol analogue **39** exhibited higher IC_{50}
values of 39.75 μM (MCF-7) and 27.6 μM (MDA-MB-231).

Plazuk and co-workers [20] synthesized the ferrocenyl-biotin conjugates **44**, **45**,
and **47** by reacting ferrocene with biotin analogues **42**, **43**, and **46**, respectively,
under Friedel Craft acylation conditions and assessed their antiproliferative
activities (Scheme **9**).

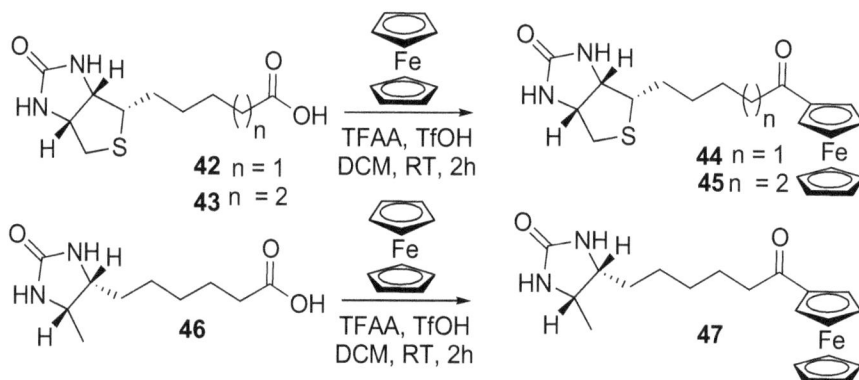

Scheme (9). Synthesis of ferrocenyl-biotin conjugates **44**, **45** and **47**.

The biotin-ferrocene conjugates **44**, **45** and **47** were treated with (*R*)-Me-Co-y-Bakshi-Shibata (CBS), leading to (*S*)-alcohol derivatives **48**, **49** and **58** and their isomers (*R*)-alcohols **50**, **51** and **61** by using (*S*)-Me-CBS. Further treatment of these diastereomeric alcohols with NaN$_3$ in AcOH afforded corresponding azides *viz.* (*S*)- azides **50**, **51** and **62** and (*R*)-azides **54**, **55** and **63**, which upon subsequent reduction with LiAH$_4$ as a reducing agent in dry THF afforded corresponding (*S*)-amines **54**, **55** and **64** and (*R*)-amines **56**, **57** and **65** (Scheme **10**).

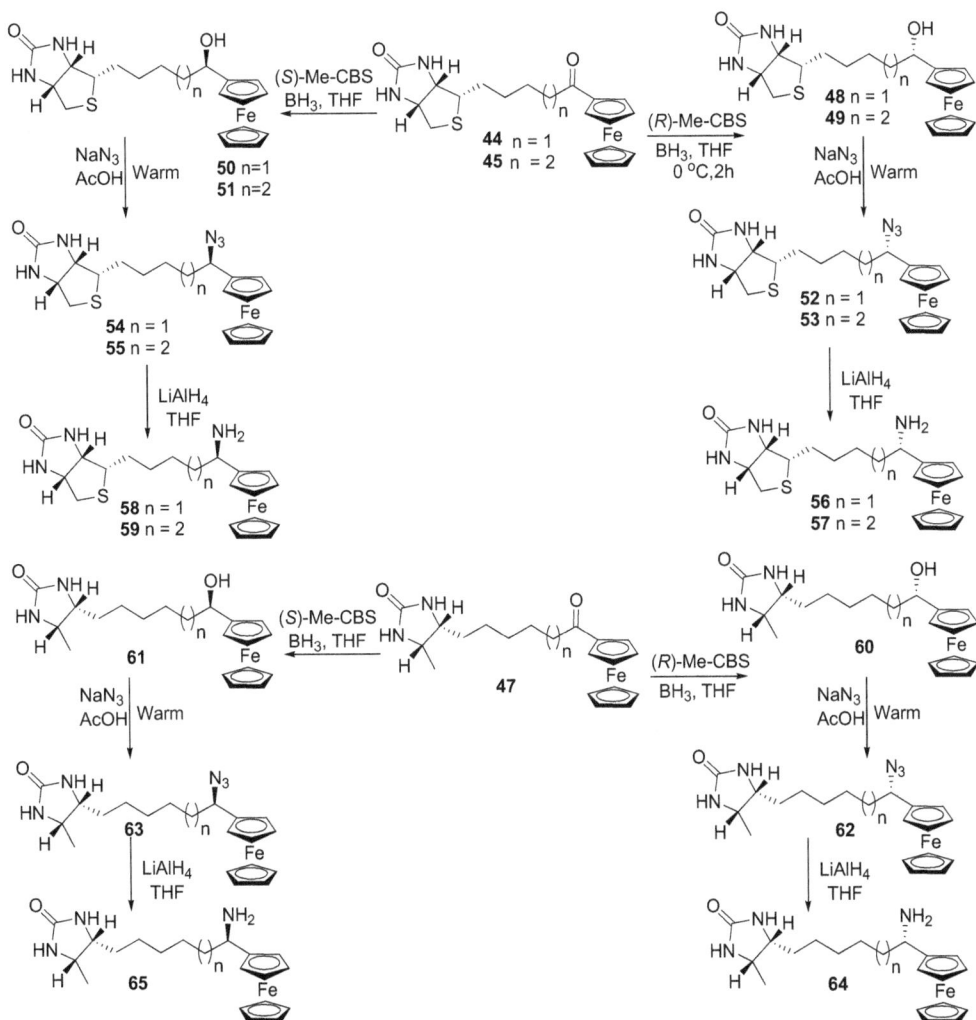

Scheme (10). Synthesis of ferrocenyl-biotin conjugates **56-59, 64** and **65**.

Scheme **11** shows the conversion of ferrocenyl ketone **44**, **45** and **47** to corresponding alkenes derivatives **66**, **67** and **68**.

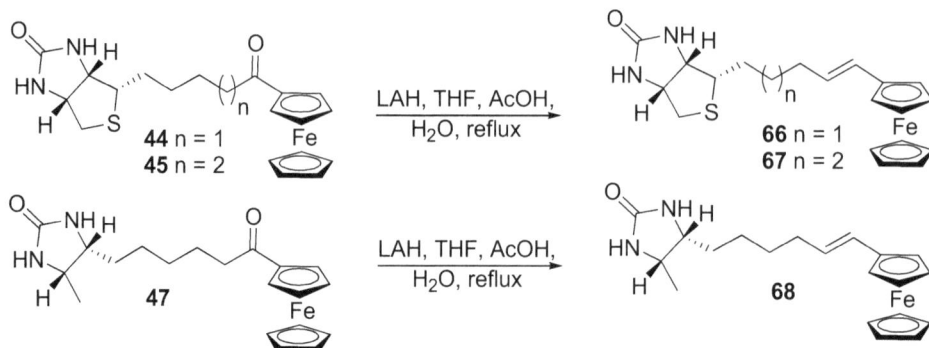

Scheme (11). Synthesis of ferrocenyl-biotin conjugates **66-68**.

The antitumor activity of the synthesized compounds was tested against three human colorectal cancer cell lines (Colo205, HCT116 and SW620). The compounds **54** (IC_{50} =54.0, 84.8 μM), **61** (IC_{50} =54.0, 84.8 μM) and **66** (IC_{50} =60.2, 74.0 μM) were shown to be toxic for both HCT116 and Colo205, whereas **51** was toxic for Colo205 (IC_{50} =3.60 μM) and SW620 (99.2 μM) cell line.

Pelinski and coworkers [21] synthesized ferrocene- indeno-isoquinolines hybrids **81-82** and tested them for anticancer activity towards breast cancer cells and DNA binding inhibition of topoisomerase I and II. Compounds **73-75** resulted from the condensation reaction of ferrocenyldiamines **70-72** with benzo[d] indeno[1,2-b] Pyrane-5,11 dione **69**. Indenoisoquinolines analogues **76a-d** carrying free primary amine were subjected to alkylation reaction with (ferrocenylmethyl) trimethyl ammonium iodide **77** with mild base potassium carbonate, resulting in ferrocenyl-indeno-isoquinolines **78 a-d** along with its disubstituted derivative **79a-d**. Subsequently, the reductive amination reaction of **76b-c** with 2-(*N,N*-dimethyl amino methyl) ferrocenecarboxyaldehyde **110** and sodium triacetoxyborohydride afforded desired ferrocenyl derivative **81a,b**. On the other hand, ferrocene derivative **82a-c** were synthesized by reacting **78b,c** with formaldehyde/ acetaldehyde under similar conditions as shown in Scheme **12**.

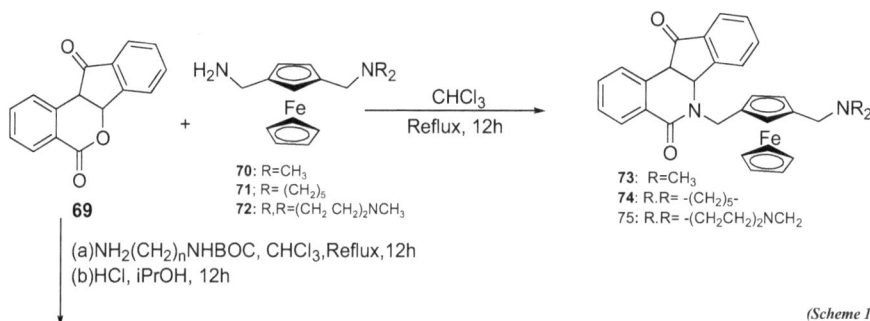

(Scheme 12) contd.....

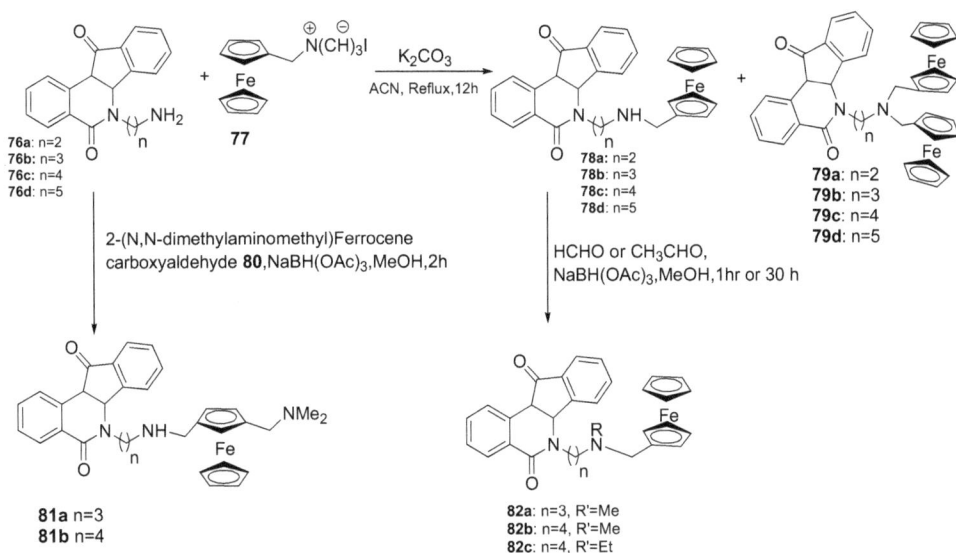

Scheme (12). Synthesis of indeno[1,2-C] isoquinolines-ferrocene conjugates **81-82**.

Ferrrocenyl amides **83a,b** were obtained by condensation of amines **76b,c** with ferrocene carboxylic acid in the presence of standard coupling agents, EDCI/HOBT (Scheme **13**).

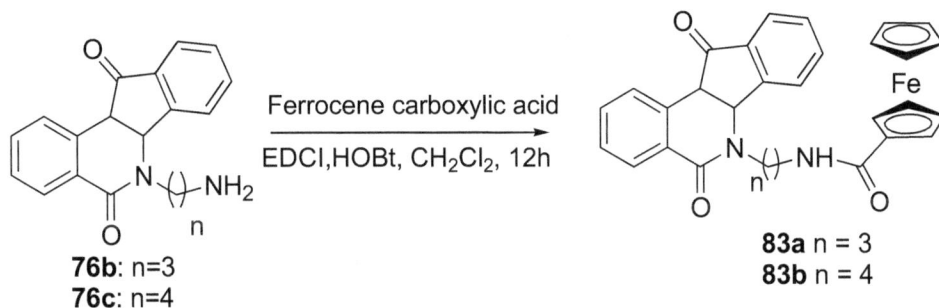

Scheme (13). Synthesis of indeno[1,2-C] isoquinolines-ferrocene conjugates **83**.

Using etoposide as a reference standard, all synthesized hybrids were screened for anti-breast cancer potential towards triple-negative MDA-MB-231 breast cancer cells. The data revealed that ferrocene derivatives **78b,c** and **81a,b** had a greater activity with an IC_{50} value <1.3 μM. However, hybrid-carrying two ferrocene units resulted in reduced activity, for instance, in **79b,c**. The existence of a second protonable site (**78b** *vs* **78c**, **81a** *vs* **81b**) and the *N6*-lactam spacer length (**78b** *vs.*

78c) showed no significant influence on cytotoxicity. The high toxicity of chemicals **78b,c** and **81a,b** is thought to be owing in part to their strong DNA contacts.

Rashinkar and colleagues [22] devised a method for making ferrocene tethered ionic liquids (ILs), which were examined for their antibreast cancer potential towards MCF-7 cells. The ferrocenylmethyl-benzimidazole analogues **84, 86** and ferrocenylmethyl-triazole analogue **88** were synthesized by reacting them with alkyl bromides in CH_3CN at room temperature. Following that, the addition of a significant amount of diethyl ether resulted in the creation of ferrocene tethered ionic liquids (ILs) **85, 87,** and **89**, which were purified by column chromatography, as illustrated in Scheme **14**.

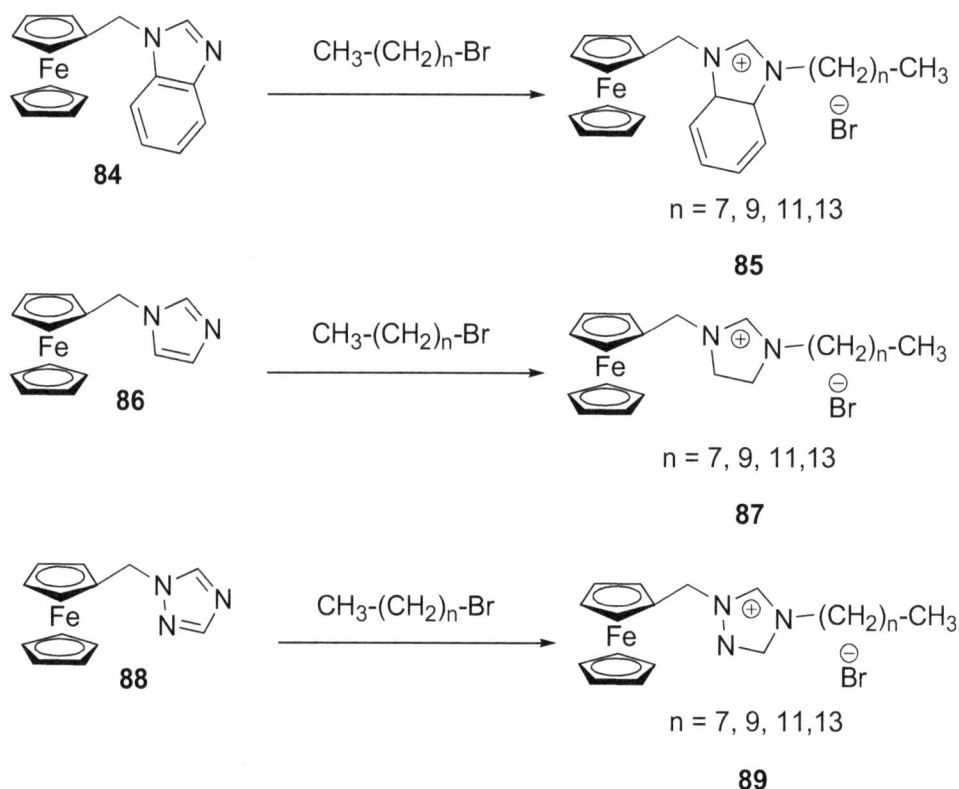

Scheme (14). Synthesis of ferrocene tethered ionic liquids(ILs) **85, 87,** and **89**.

Ferrocene tethered ILs **85, 87,** and **89** were investigated for antibreast cancer potential on MCF-7 cells by employing the SRB (sulforhodamine B) assay. All of the ferrocene-tailored ILs demonstrated appreciable activity (GI_{50s} = 0.016 to

0.174 µM), in comparison to the conventional drug doxorubicin. The ILs **85** (n = 7, 9, 11, 13) and **87** (n = 9) were the most promising anti-breast cancer agents (GI_{50s} = 0.019, 0.018, 0.040, 0.016, and 0.064 µM, respectively) equipotent to standard doxorubicin (GI_{50} 0.018 µM). Further exploration of the structure-activity relationship established that ferrocene-tailored ILs carrying long-chain alkyl spacers such as octyl and decyl possessed substantially better anticancer profile, compared to those bearing docecyl and tetradecyl spacers. This implies that alkyl chain length is a crucial parameter in developing ILs as anti-breast cancer agents. To assess selectivity, ferrocene tethered ILs were also tested against normal (Vero) cells utilising an *in vitro* SRB assay. With GI_{50} values ranging from 0.036-0.198 µM, the majority of the ferrocene-tailored ILs showed modest to strong selectivity towards MCF-7 cells. The most promising ILs, **85** (n = 7, 9, 11, 13) and **87** (n = 9) had GI_{50s} 0.036, 0.111, 0.068, 0.148, and 0.102 µM, respectively as shown in Fig. **3**.

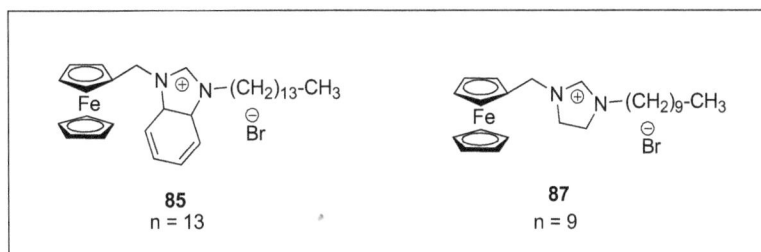

Fig. (3).   The most potent ILs **85** (n = 13) and **87** (n = 9).

Badshah and co-workers [24] prepared ferrocene derivative with a nitrophenyl moiety and tested it against drug-resistant and parental human ovarian tumour cell lines, such as A2780 and A2780 [cisR], and A2780[ZD0473R]. Diazonium salt of mono-/di-substituted aniline **90** was reacted with ferrocene in the presence of phase transfer catalyst, cetyltrimethyl ammonium bromide (CTAB) in ethanol:water (1:1) mixture, resulting in the desired compound, nitrophenyl ferrocene **91** (Scheme **15**).

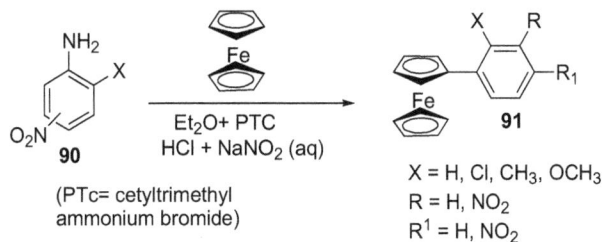

Scheme (15). Synthesis of ferrocene nitrophenyl conjugates **91**.

Ferrocenyl conjugates **91** were tested against ovarian cancer cells by employing cisplatin as a standard drug. Cisplatin was more active than these compounds against A2780[cisR] (cisplatin-resistant type), A2780[ZD0473R] (ZD0473-resistant type), and A2780 (parent) with IC_{50} values of 13.39, 14.08, and 1.30 μM, respectively. The decreased activity of the target conjugates **91** resulted from the fact that **91** interacts differently with DNA than cisplatin. However, it was interesting to observe that **91** showed substantially different IC_{50} values towards A2780 (parent) cells, which might be attributed to the differences in DNA binding ability and membrane contact.

Csampai *et al.* [24] synthesized ferrocene-cinchona hybrids using Cu-promoted cycloaddition reaction and the Sonogashira coupling, which was then *in vitro* analyzed on cancer (HepG-2 and HT-29) cells. The synthetic strategy involved the Cu-catalyzed azide-alkyne cycloaddition reaction of alkyne precursor **92** with azido-ferrocenylchalcone, resulting in a diastereomeric mixture of **93a-d** and **94a-d**. The Cu-catalyzed [3+2] cycloaddition was used to create a set of chalchone-free triazoles **97**, **98**, **101**, and **102**. Sonogashira coupling between **92b** and iodoferrocene, in the catalytic amount of copper iodide and $PdCl_2(PPh_3)_2$, afforded the hybrid **102b.** The complex CpRuCl(COD) was used as a catalyst in the cycloaddition reaction between **90c,d** and **91** to afford hybrid **103,** as shown in Schemes **16** and **17**.

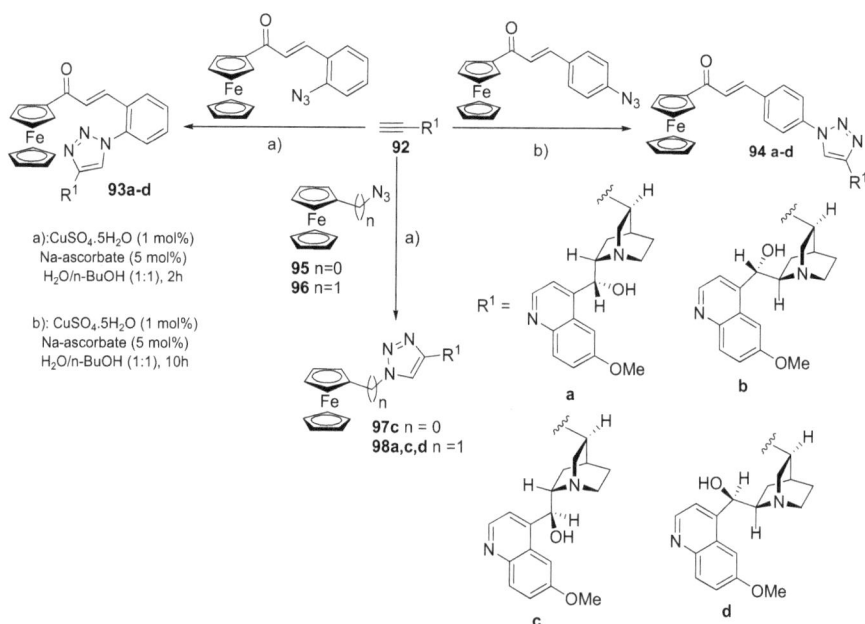

Scheme (16). Synthesis of ferrocene-cinchona hybrids **94**, **97** and **98**.

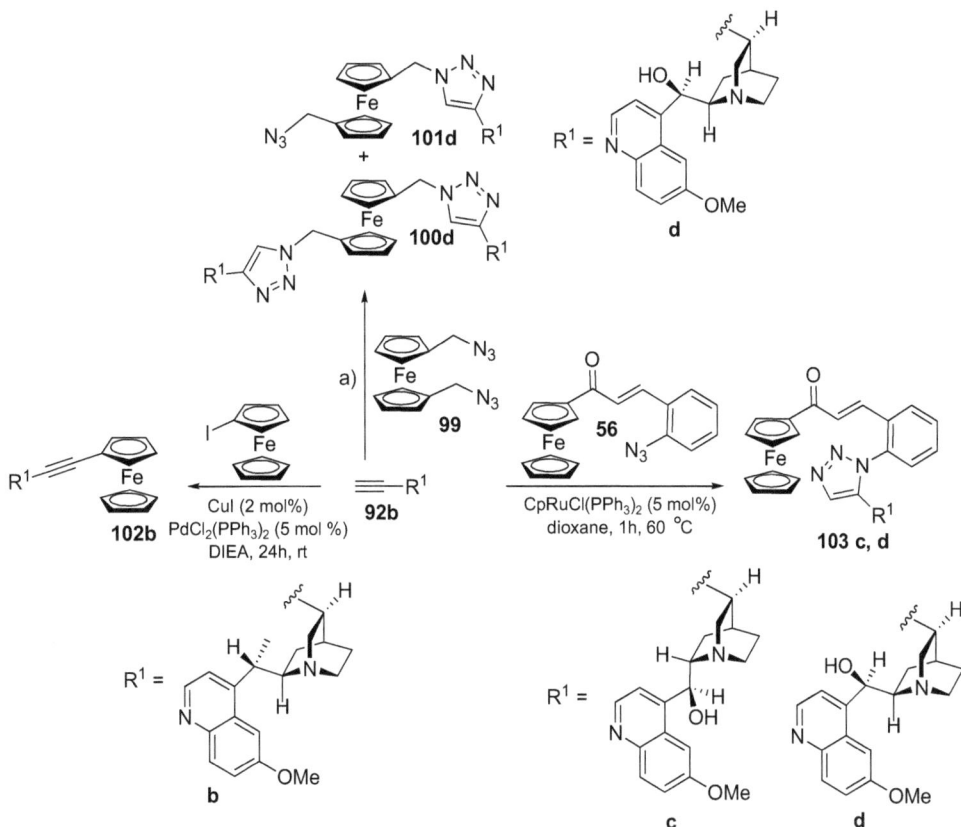

Scheme (17). Synthesis of ferrocene-cinchona hybrids.

The cytotoxic effect of these synthesized hybrids was then examined on hepatoma (HepG2) and colorectal (HT-29) adenocarcinoma cells. Hybrids **93a,c**, **94a,d**, and **103c,d** had sound cytotoxicities on both cancer cell lines, with IC_{50s} ranging between 0.7 to 1.5 μM, significantly better than the standard, Tamoxifen. Disubstituted benzene derivatives **93b** (IC_{50} = 3.9 μM) and **93d** (IC_{50} = 4.5 μM) showed moderate activities towards HT-29 cells, but were less promising towards HepG2 cells. Hybrids **103c** and **103d** were equipotent on both the cell cultures, with IC_{50s} (HT-29, HepG2) = 0.7 and 0.2 μM, as shown in Fig. **4**.

Kumar *et al.* [25] devised a method for synthesising $1H$-1,2,3-triazole-tailored uracil-ferrocenyl chalcone hybrids and examined their anticancer effectiveness in human leukaemia (CCRF-CEM) and human breast cancer (MDA-MB-468) cell cultures. The synthetic methodology involved an initial base (NaH) catalyzed alkylation reaction of 5-substituted uracil analogues with various dibromoalkanes, followed by subsequent treatment with NaN$_3$, yielding the *N*-alkylazido-5-substituted uracil derivatives **106**. Final set of hybrids *viz.* 1*H*-1,2,3 triazole-tailored uracil-ferrocenyl-chalcone hybrids **108** were then obtained *via* a Cu-catalyzed Click reaction between azido-uracils **106** and *O*-propargylated ferrocenyl-chalcones **107** as shown in Scheme **18**.

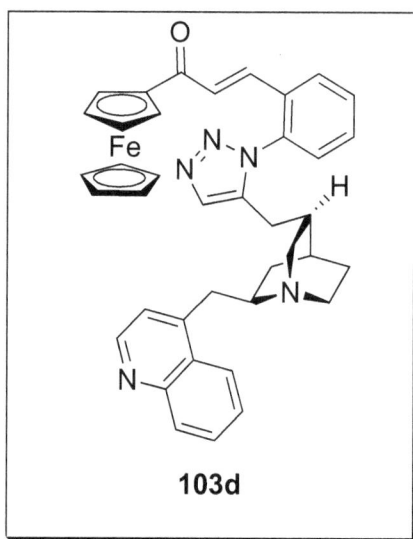

Fig. (4). The most promising 1*H*-1,2,3-triazole-linked cinchona-ferrocenylchalcone conjugate **103d**.

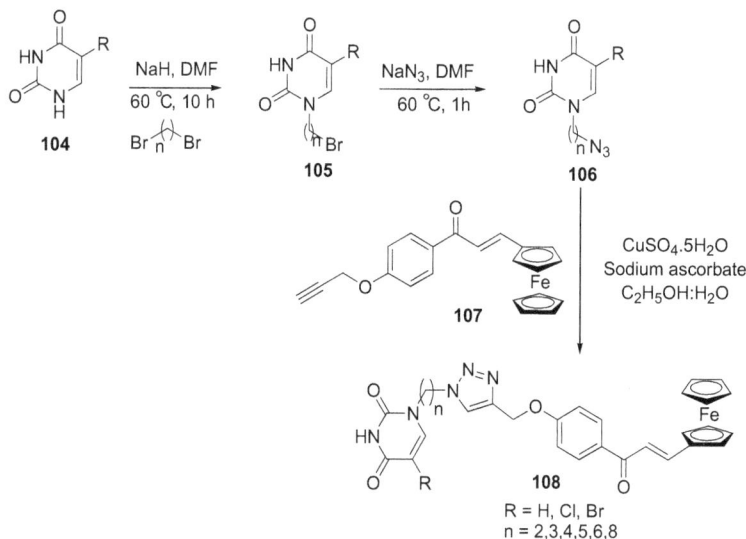

Scheme (18). Synthesis of 1*H*-1,2,3 triazole-tailored uracil-ferrocenyl chalcones **108**.

Using doxorubicin as a standard reference, the hybrids were tested for their cytotoxic characteristics towards human leukaemia (CCRF-CEM) and human breast cancer (MDA-MB-468) cell cultures. The alkyl chain length was found to affect cytotoxicity, with longer chains preferred over shorter ones, although the C-5 substitution at uracil did not influence activity profiles. After 3 days, hybrids carrying longer alkyl chains (n = 5, 6, and 8) reduced CCRF-CEM cell proliferation by approximately 70%. In a similar manner, cytotoxicity analysis on MDA-MB-468 revealed that the majority of the hybrids showed no activity even after 3 days, except for (n = 2, 5 R = H), which decreased cancer cell proliferation by 59 and 62%, respectively. The cytotoxic profiles of the most promising hybrids were evaluated and compared to doxorubicin against a normal kidney cell line (LLC-PK1 ATCC CL-101). After 24 hours of incubation at a concentration of 50 M, the hybrids showing promising anticancer potential against CCRF-CEM cell line exhibited no significant toxicity, as expected. Furthermore, in comparison to CCRF-CEM cells, these hybrids were substantially less cytotoxic towards LLC-PK1 (non-tumorigenic) cells, after 3 days.

Kumar and co-workers [26] used a Cu-catalyzed Click reaction to synthesize various isatin-ferrocene conjugates in order to explore their structure-activity relationship against estrogen-responsive (MCF-7) and triple-negative (MDA-M--231) breast cancer cells. In dry DMF, 5-substituted isatin **109** was first *N*-alkylated with MeI or PhCH$_2$Br, with NaH as a base, to obtain *N*-alkylated isatin **110**, which upon further reaction with trimethylsilyl iodide yielded spiroindoline-oxirane-2-ones **111**. The epoxide ring of **111** was opened using sodium azide in a 90:10 ethanol/water combination, yielding hybrid **112**. Furthermore, azide-alkyne cycloaddition reactions of **112** with ethynyl ferrocene, *O*-propargylated ferrocene methanol **114**, and *O*-propargylated ferrocenylchalcone **107**, with a catalytic amount of copper sulphate and sodium ascorbate, resulted in the formation of target isatin-ferrocene hybrids **113**, **115**, and **116**, as shown in Scheme **19**.

The antibreast cancer activities of the target hybrids **113-116** against estrogen-responsive (MCF-7) and triple-negative (MDA-MB-231) established that 16 isatin-ferrocene conjugates were highly efficacious towards MCF-7 than MDA-MB-231 cells. In particular, hybrid **113** (R^1 = CH$_3$, R^2 = H) was selective towards estrogen-responsive (MCF-7) cells over the unresponsive (MDA-MB-231) cells, however, the latter needed an IC$_{50}$ value of 97.11 μM to inhibit the cancer cell growth. On the contrary, hybrid **115** demonstrated excellent selectivity towards triple-negative (MDA-MB-231) cells over MCF-7 cell lines, as evidenced by lower IC$_{50}$ values towards MDA-MB-231 cells and high IC$_{50}$ values against estrogen-responsive cells. It was found that compound **113** (R^1 = CH$_2$Ph, R^2 = CH$_3$) was equally effective towards both the breast cancer cell lines with IC$_{50s}$ of 34.99 (MCF-7) and 33.73 μM (MDA-MB-231). Adding the chalcone moiety to hybrid **116** reduced their antiproliferative activities, whereas conjugates **116** bearing (R^1 = CH_2Ph, R^2 = CH_3) were ineffective against both cell lines.

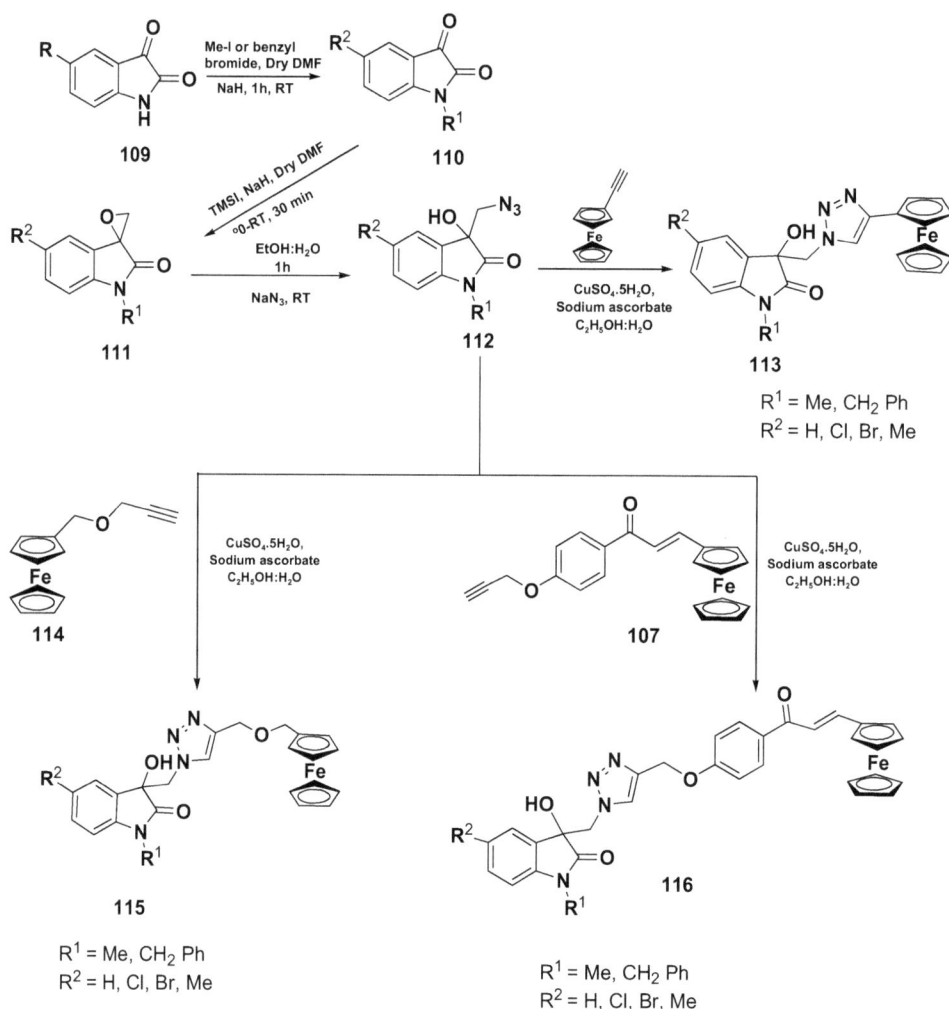

Scheme (19). Synthesis of Isatin-ferrocene conjugates **113**, **115** and **116**.

Hao *et al.* [27] synthesized *m*-ferrocenylbenzoylthiadiazoles **120** and tested them against four human cell lines for anticancer efficacy (EC-9706, SGC-7901, CCK-8 and Eca-109). The desired *m*-ferrocenylbenzoylthiadiazole derivatives **120** were obtained by reacting *m*-ferrocenyl benzoic acid **117** with oxalyl chloride to afford *m*-ferrocenyl benzoyl chloride **118**, which was then treated with **119** in the presence of Et$_3$N to yield *m*-ferrocenylbenzoylthiadiazole **120** as shown in Scheme **20**.

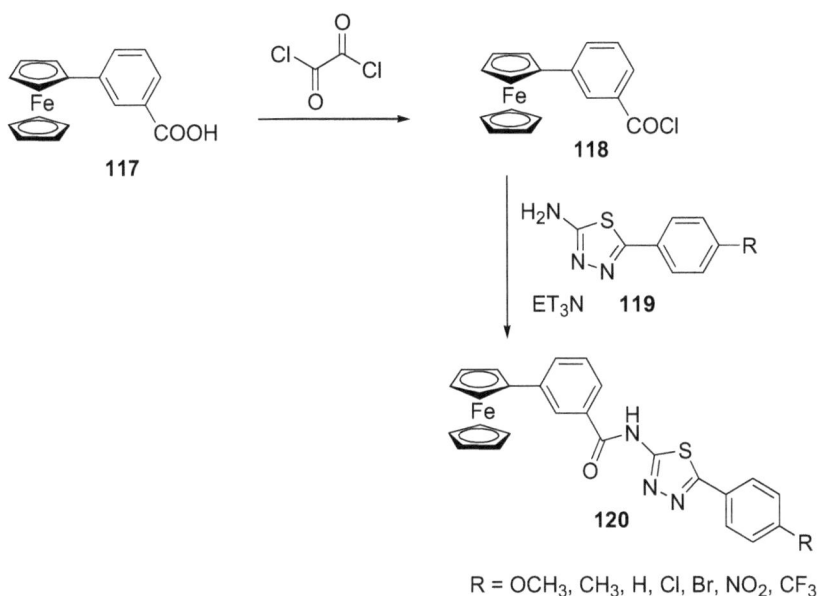

Scheme (20). Synthesis of *m*-ferrocenylbenzoylthiadiazole **120**.

The influence of various substitution patterns over the phenyl ring of **120** was studied against a panel of cancer cells (SGC7901, EC9706, Eca109, and CCK-8). Adding the H, Cl, Br, NO$_2$, or CF$_3$ increased the inhibitory action against the human esophageal cell lines EC9706 and Eca109, but the addition of OCH$_3$ or CH$_3$ had no effect on the derivative **120**. The NO$_2$ substituted compound (IC$_{50}$ 18.61 µmol L^{-1}) exhibited the best potency against EC-9706 cells. Further, studies revealed that the introduction of CF$_3$ afforded the highest Eca-109 inhibitory activity with an IC$_{50}$ 15.51 µmol L^{-1}. The compound with R = Br (IC$_{50}$ value of 28.17 µmol L^{-1}) displayed the highest efficiency against SGC-7901 (human gastric carcinoma) cell line. The most potent derivatives **120a** and **120b** as depicted in Fig **5**.

Melendez *et al.* [28] developed the route for synthesizing ferrocenyl esters and evaluated their anti-breast cancer potential in MCF-7 cells. 1,1'-ferrocenyl dichloride **122** upon reaction with pyrrol-phenol derivative **123** afforded Fc-(CO$_2$-Ph-4-Py)$_2$**124** along with two side-products Fc-CO$_2$-Ph-4-Py **125** and Fc-(CO$_2$-Ph-4-Py)CO$_2$H **126** (Scheme **21**).

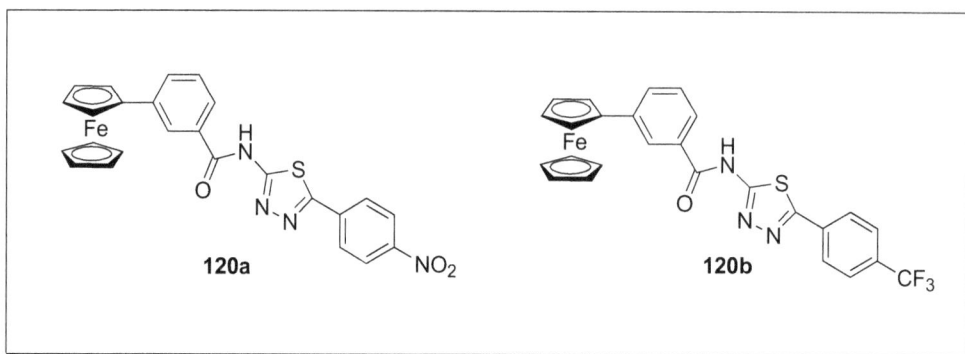

Fig. (5). The most potent *m*-ferrocenylbenzoylthiadiazole **120a** and **120b**.

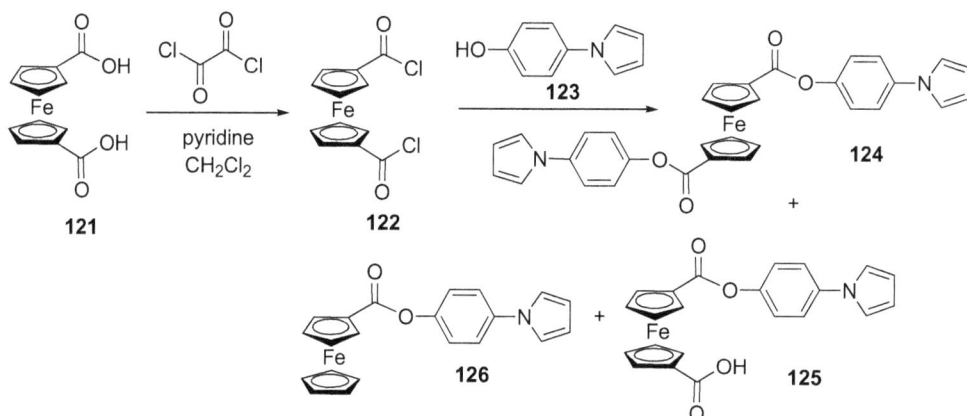

Scheme (21). Synthesis of ferrocenyl esters **124-126**.

Further, the reaction between ferrocene acetyl chloride **77** and pyrrol-phenol derivative **123** afforded the Fc-CH$_2$-CO$_2$-Ph-4-Py **129** (Scheme **22**).

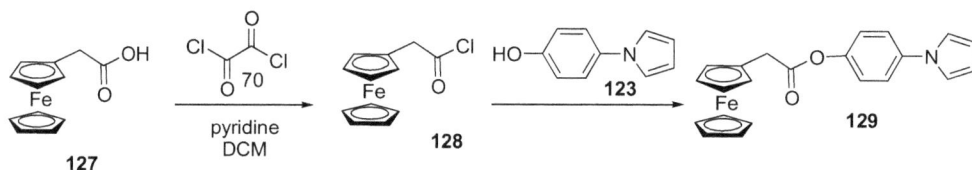

Scheme (22). Synthesis of ferrocenyl esters **129**.

The synthesized ferrocene derivative and its precursors were evaluated on the breast (MCF-7), colon (HT-29) and normal (MCF-10A) cell lines. The ferrocenyl ester displayed a dose-dependent anticancer impact on the breast cancer (MCF-7) cells and was more effective than its precursors. Compounds **124** and **125** displayed modest antibreast cancer profile and low activity towards non-tumorigenic MCF-10A cells. The ester Fc-CH$_2$-CO$_2$-Ph-4-Py **129** was moderately effective in all the three tested cancer cell lines *viz.* MCF-7, HT-29, and MCF-10A.

Selenoureas are bioactive chemicals existing as enzyme inhibitors, free radical scavengers, and DNA binding agents. Badshah *et al.* [29] described the synthetic methodology and anticancer effectiveness of ferrocene tethered selenoureas, shown in Scheme **23**. Ferrocenyl phenyl aniline **133** upon reaction with isoselenocyanate **134** produced desired FIS (ferrocene integrated selenoureas) **135**.

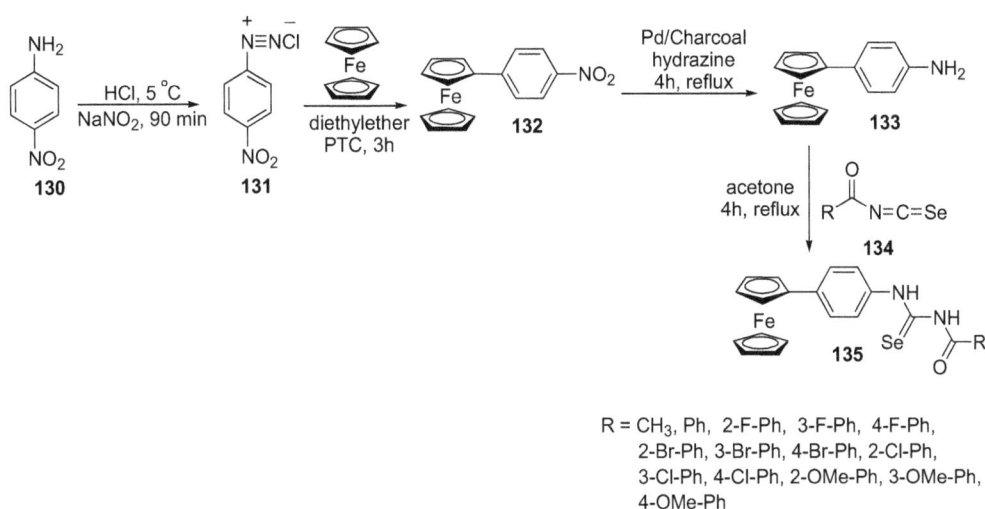

R = CH$_3$, Ph, 2-F-Ph, 3-F-Ph, 4-F-Ph,
2-Br-Ph, 3-Br-Ph, 4-Br-Ph, 2-Cl-Ph,
3-Cl-Ph, 4-Cl-Ph, 2-OMe-Ph, 3-OMe-Ph,
4-OMe-Ph

Scheme (23). Synthesized the ferrocene incorporated selenoureas **135**.

Screening of these hybrids on a panel of cancer cell lines explicitly established that the hybrids **135a**, **135b**, **135c**, and **135d** were the most potent towards neuroblastoma (MYCN-2) cells, as shown in Fig. **6**. The cytotoxic findings against Hepa IcIc7 and liver cells were dismal, with only 27-37% of cells dying. The compound **135a** inhibited cell proliferation by 73% and was shown to be particularly effective towards the MCF-7 cells.

Kowalski and colleagues [30] reported the synthesis of ferrocenyl derived nucleobases with cyclic thiopyrane rings and investigated their anti-tumor effects.

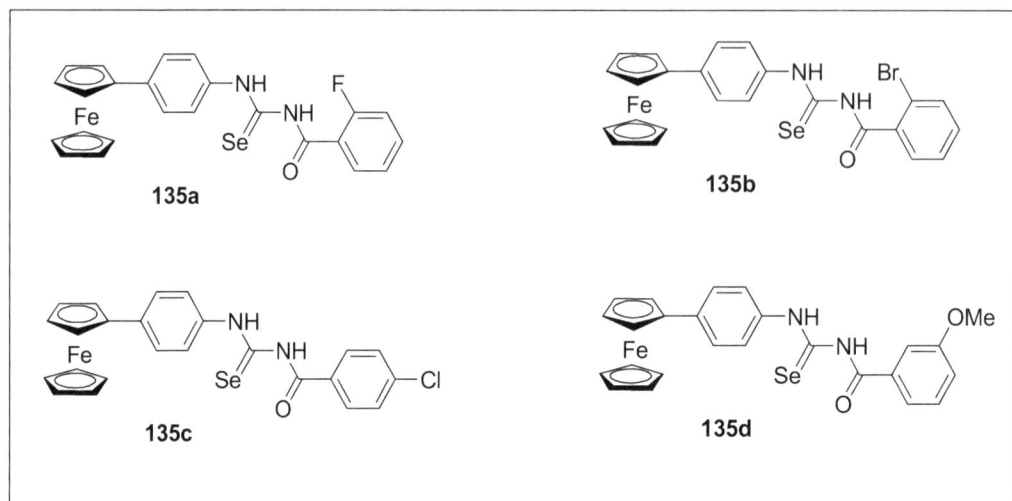

Fig. (6). Active ferrocene-selenoureas against MYCN-2 cell line **135a-d**.

Ketone **136** was treated with Lawesson's reagent **140** to produce thioketone intermediate **141**, which was then subjected to a Diels-Alder cycloaddition with butadinene analogue **136** to produce the target hybrid **142**. Treatment of ferrocenyl ketone **136** with phosphorous pentasulfide afforded **137**, which upon subsequent Diels-Alder cycloaddition with 2,3-dimethyl-1,3-butadiene **138** afforded target hybrid **139** (Scheme **24**).

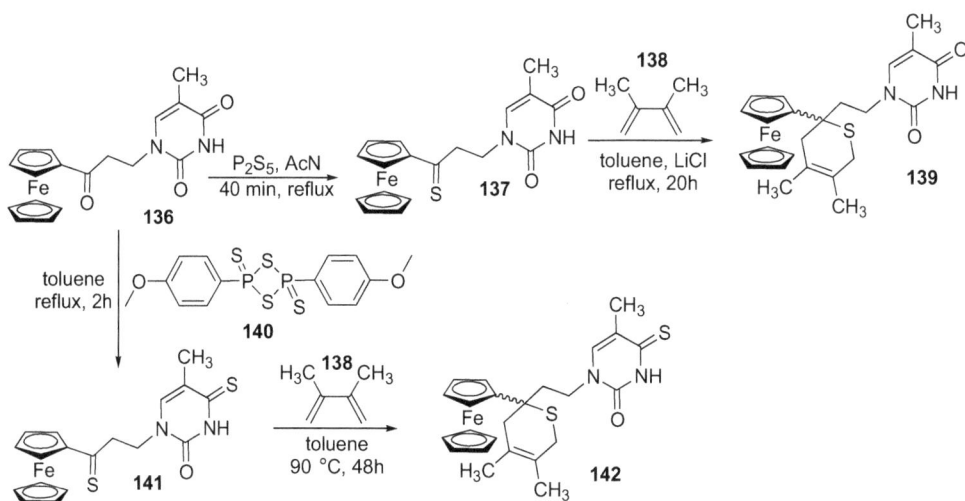

Scheme (24). Synthesis of ferrocenyl nucleobase derivatives **139** and **142**.

In vitro anticancer investigation of hybrids **139** and **142** was performed on a panel of human cancer cells: HT-29 (colon), MCF-7 and MDA-MB-231 (breast), HL-60 (promyelocytic leukemia), MonoMac6 (human monocytic) cell lines. Results established that hybrid **142** showed moderately good anticancer effects in monocytic and breast cancer cell lines, with IC_{50} values 28.2 and 11.1 μM, respectively, whereas the cytostatic effects were quite weak in MDA-MB-231 (IC_{50} >100 μM) and HT-29 (IC_{50} >83.3 μM) cells. Compound **139**, on the other hand, exhibited a strong cytostatic effect on all investigated cell lines, including MCF-7, HL-60, and MDA-MB-231 cells with IC_{50} < 10 μM.

Peres *et al.* [31] created ferrocene-embedded chalcones, aurones, and flavones in order to test their collateral selectivity against resistant tumour cells. In terms of GSH efflux and selectivity ratio > 9.1, the most effective combination appeared to be ferrocene-linked flavone, which produced CS-analogues that were nearly as good as verapamil. At room temperature, aldol condensation of ortho-hydroxyacetophenone **143** and ferrocene carboxaldehyde 1 resulted in the synthesis of ferrocenylchalcones **144**. Condensation of benzo-furanones **145** with ferrocene carboxaldehyde 1 was performed under different reaction conditions: aurone **146a** was obtained using aluminium oxide in DCM at room temperature, and aurones **146b-d** were obtained using potassium hydroxide in a methanol/water mixture under reflux.The chalcones **146b,c** were rearranged in pyridine at 80°C using mercury(II)acetate to yield the desired aurones **146e,f**. Furthermore, the hydroxylated ferrocenylaurone **146g** was obtained by demethylating **146e** in a reflux of DCM with aluminium chloride. Finally, as shown in Scheme **25**, the corresponding flavones **147a-c** were synthesized by isomerizing ferrocenylaurones **146a,b,f** with potassium cyanide in refluxing ethanol.

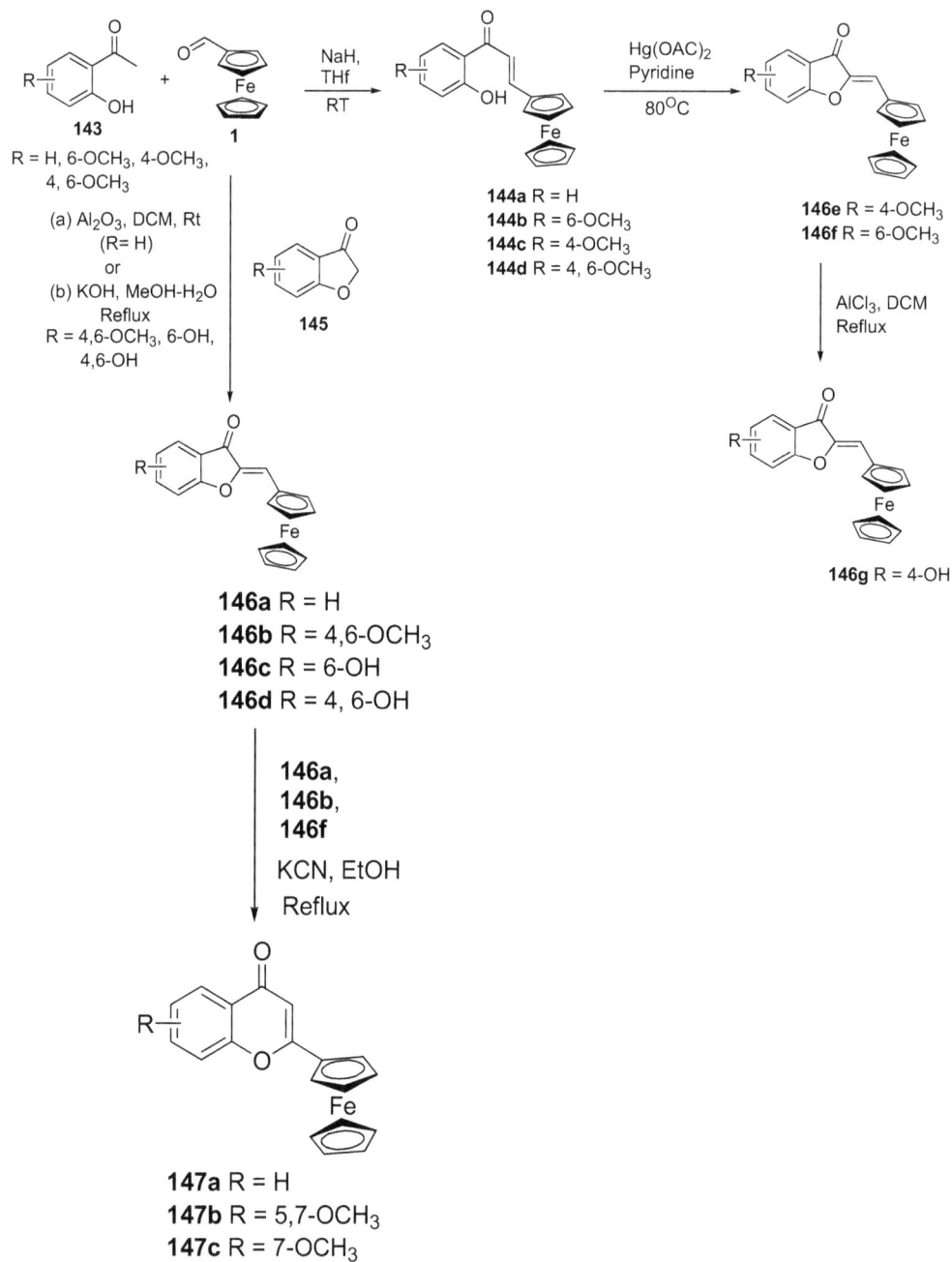

Scheme (25). Synthesis of ferrocene-embedded chalcones **144**, aurones **146**, and flavones **147**.

BHK-21wt (wild-type) and BHK-21 cells transfected with the MRP1 gene were used as a control to test the compounds (BHK-21-MRP1). The most promising compounds of the series were analyzed for toxicity on BHK-21 and H69AR cell lines using verapamil as a standard reference, to test the selective behaviour of resistant cells that show overexpression of MRP1. When compared to verapamil, the majority of the ferrocenyl compounds generated had improved GSH efflux results. The most active conjugate **144a** produced 76 percent GSH efflux at concentration of 20 µM. For the 4-methoxylated aurone 146e, GSH efflux was also good, with 73 percent at 20 µM and 52 percent at 5 µM. On the same note, the unsubstituted ferrocenyl analogue **147a** displayed a good efflux percentage (64%) at 20 µM. Di-methoxylation of phenyl moiety, as in **144d**, **146b**, and **147c**, aurone generated the best GSH efflux of 60% at 20 µM. At 5 µM, compound **147b** exhibited the highest GSH outflow.

The positive GSH efflux results in chalcones **144a,c** aurones **146a, b** and **146e, g** and flavones **147b,c** at 20 µM and 5 µM encouraged evaluation of their cytotoxicity in both cell lines: BHK21 and BHK21-MRP1 and NCI-H69 and H69AR. In terms of collateral sensitivity, the chalcones **144a,c** outperformed verapamil with a selective ratio (SR) of 3.5 (BHK-21) and >2.7 (BHK-21-MRP1). However, these were ineffective against cancer cell lines NCI-H69/H69AR. For BHK-21-MRP1, the aurones **146a,b,e**, and **g** showed no appreciable collateral sensitivity. The selective ratio (SR) for H69AR was found to be > 8.2 and > 4.2 for **146e** and **146g,** respectively. Furthermore, the most promising derivatives were found to be ferrocenyl flavones **147c** and **147b** with selective ratios (SR) of 7 (BHK-21-MRP1) and > 9.1 (H69AR) and **147b** with selective ratios (SR) of 5.3 (BHK-21-MRP1) and > 4 (H69AR), respectively, with **147c** being the best, with efflux maximal from 5 µM range. Fig. (**7**) shows the structure of the most potent derivatives **147c**.

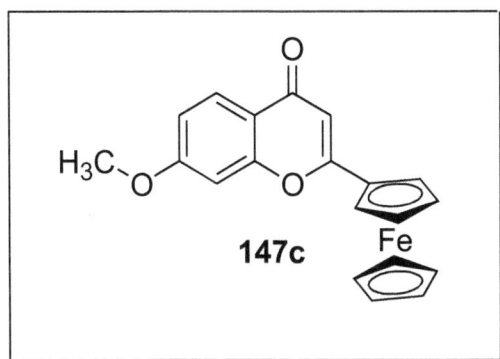

Fig. (7). The most potent derivatives **147c**.

Clotrimazole ferrocenyl analogues were recently tested for anticancer efficacy against the HT 29 (colorectal) and the MCF-7 (breast cancer) cell lines by Patti *et al.* [32]. The reaction of ferrocene with 2-/4-chlorobenzophenone in the presence of strong base nBuLi at 0 °C yielded the required alcohols **149a-b**. The subsequent treatment of **149a-b** with acetic acid in MeOH at 80°C for a duration of 3 hours generated the **150a-b**. Next step involved the introduction of heteroaromatic ring on **150a-b** by reacting it with imidazole, resulting in the formation of desired hybrid **151a-b,** as mentioned in Scheme **26**.

Hybrids **151a** and **151b** were subjected to anticancer profiling against the breast (MCF-7) and colon (HT29) cancer cell lines. These hybrids (GI$_{50}$ (**151a, 151b**): 23.84, 20.04 µM) showed antibreast cancer activity, comparable to standard clotrimazole (GI$_{50}$ = 21.44 µM). However, on HT-29 cells, **151a** (GI$_{50}$ = 27.51 µM) and **151b** (GI$_{50}$ = 28.13 µM) demonstrated better results than clotrimazole (GI$_{50}$ = 64.19 µM).

Jaouen *et al.* [33] recently investigated the reactivity of ferrocenequinonemethides **153** to cellular nucleophiles as well as their antiproliferative activities. The nucleophilic attack of thiols on ferrocenequinonemethides **153** was used to create the corresponding ferrocenequinonemethide-thioladducts **154,** as depicted in Scheme **27**.

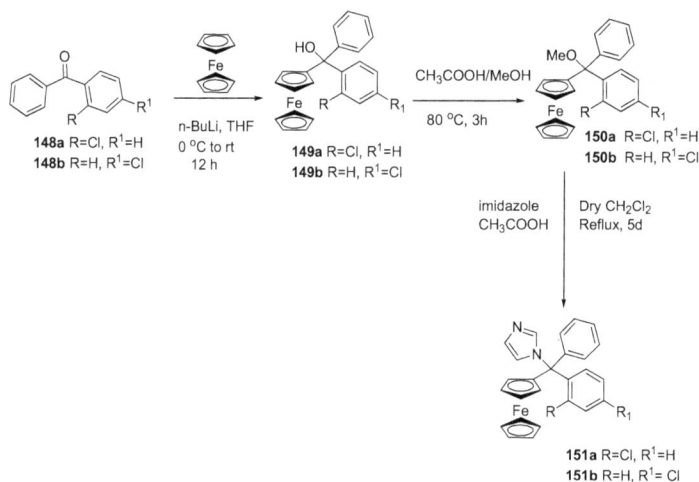

Scheme (26). Synthesis of ferrocenyl analogues of clotrimazole **151**.

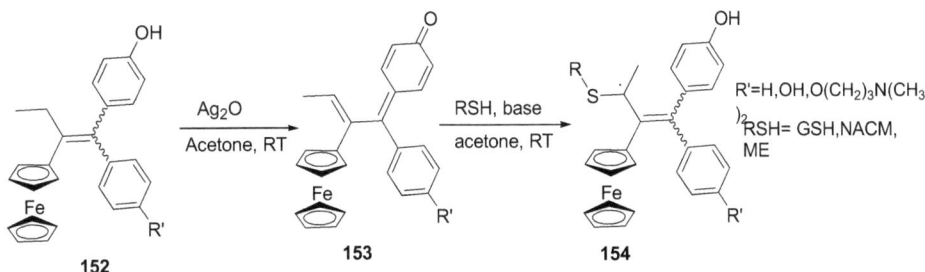

Scheme (27). Synthesis of ferrocenequinonemethide-thiol adducts **154**.

A library of ferrocene-thiols was prepared by an enzymatic process, where ferrociphenols were incubated with various liver microsomes (LM) in NADPH and a variety of thiols (Scheme **28**). This also led to the formation of ferrocenyl indenes and allylic alcohols as bye-products.

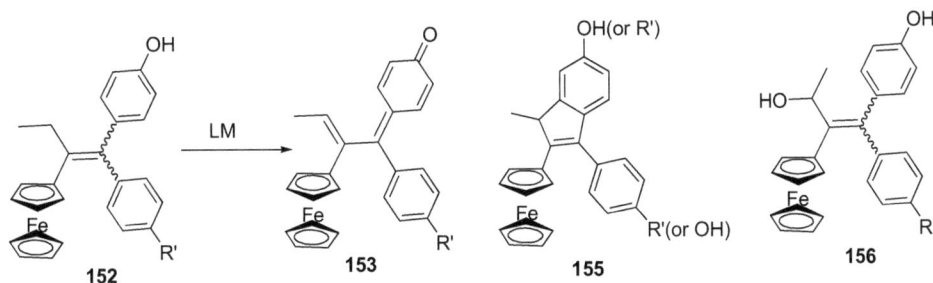

Scheme (28). Liver Microsome-promoted incubation of ferrociphenols **156**.

Some of the ferrocene-thiol adducts displayed anticancer effects comparable to ferrociphenols against hormone-resistant cancer cells. Quinone-methide metabolites can be generated by the oxidation of these adducts, which play a crucial role in apoptosis and cell death.

Jaouen and colleagues [34] described the synthesis of hydroxypropyl-ferrociphenols and investigated their antiproliferative properties. The desired compounds **159** and **165** were obtained by McMurry cross-coupling of dihydroxybenzophenone **158** with either substituted ferrocenyl butane-1-one **157** or ferrocenyl butanoate **164**. The compound **160**, which contains a terminal thio unit, was created using an isothiuronium intermediate, while **161** was created by nucleophilic attack of 2-hydroxyisoindoline-1,3-dione on **159**. Similarly, as shown in Schemes **29** and **30**, compounds **162** and **163** were synthesized from **161** *via* alcoholysis or hydrazine hydrate treatment, respectively.

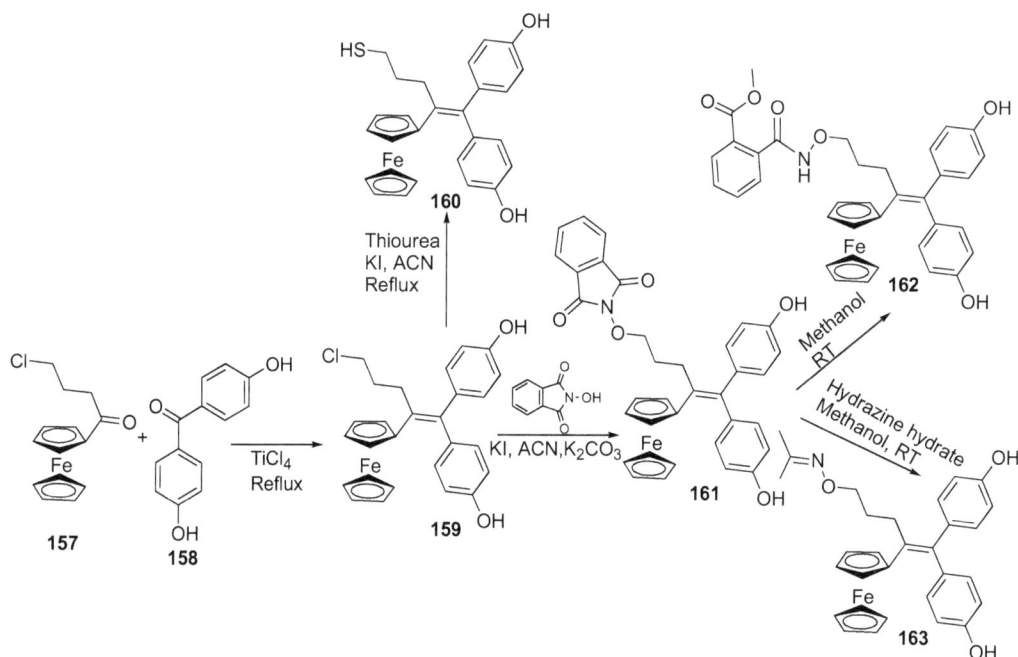

Scheme (29). Synthesis of hydroxypropyl-ferrociphenols.

Ferrociphenol **166** was obtained, when **165** reduced with LAH in anhydrous THF, which upon further reaction with acetic anhydride/pyridine afforded the corresponding derivative **167** (Scheme **30**).

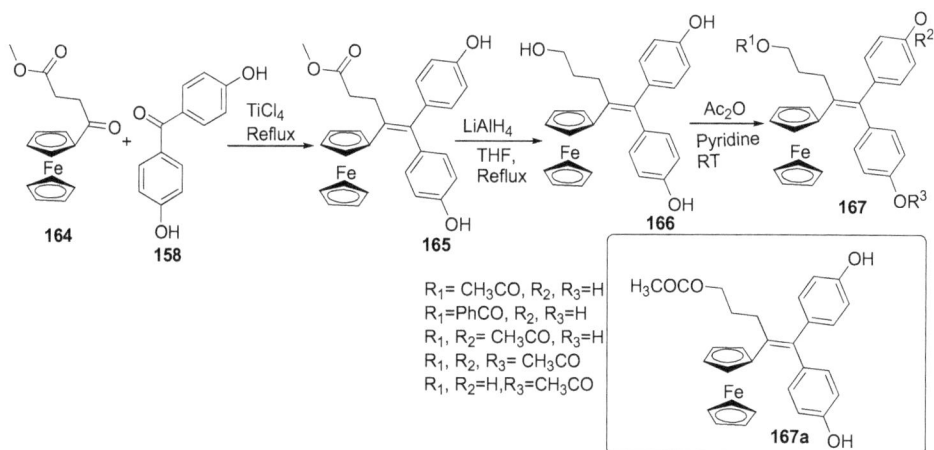

R_1 = CH_3CO, R_2, R_3=H
R_1=PhCO, R_2, R_3=H
R_1, R_2= CH_3CO, R_3=H
R_1, R_2, R_3= CH_3CO
R_1, R_2=H, R_3=CH_3CO

Scheme (30). Synthesis of aceylated-hydroxypropyl-ferrociphenols.

All of the conjugates were tested for demonstrated good anti-breast cancer activity against triple-negative MDA-MB-231 cells with IC_{50s} in the low micromolar range. Due to the ease of hydrolysis of acetyl group, compound **167a** displayed the best results (IC_{50} (MDA-MB-231) = 0.28 µM). As the acetyl group is thought to be useful for the protection of aliphatic alcohols, its impact on the hydroxypropyl-ferrociphenol's structure was investigated. A library of compounds, with one or more acetylated phenol/alcohol group, was synthesized and tested against MDA-MB-231 cancer cells. The synthesized derivatives inhibited the growth of MDA-MB-231 cells with an excellent inhibitory activity of 0.26-0.38 µM.

The methodology described above was expanded to include the synthesis and anticancer investigation of succinimido-ferrociphenols [35]. The synthetic protocol proceeded with McMurry cross-coupling reaction between **168** and **158** to produce **169**, which upon subsequent treatment with various imides, produced the desired hybrid **170**. Compounds **170a** (IC_{50} = 49 nM) and **170b** (IC_{50} = 74 nM) emerged as the most promising against A2780-Cis (cisplatin resistant) cells (Scheme **31**).

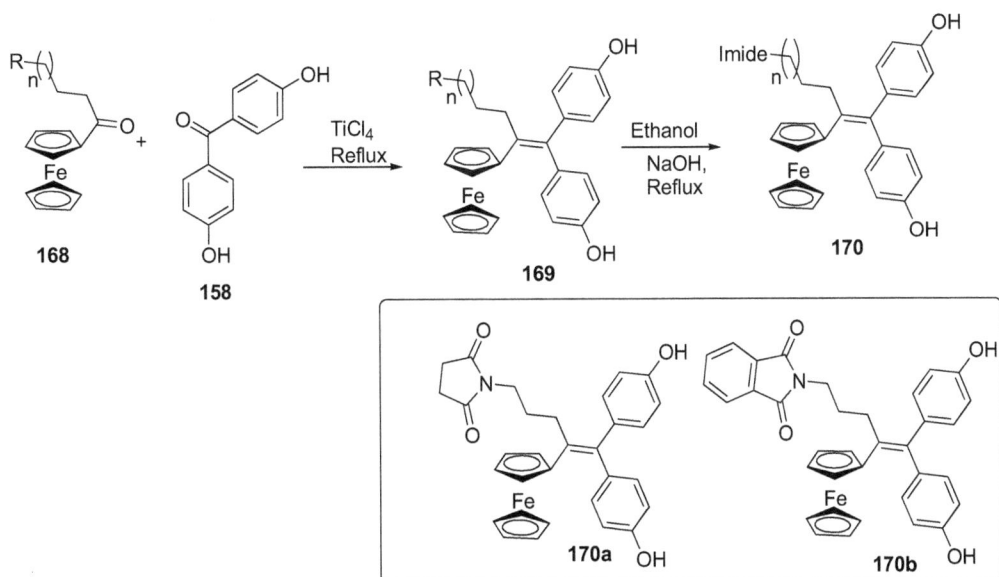

Scheme (31). Synthesis of succinimido-ferrociphenols.

Zhang *et al.* [36] created ferrocene-coumarin conjugates and screened them for anticancer potential against human breast cancer (BIU-87, MCF, EC-9706, SGC-

7901, Eca-109, and Jurkat) cells. The key step in the synthesis is stirring ferrocene derivatives **171**, **176** and coumarin intermediates **172**, **174** in DCM at room temperature until the complete dissolution of the solid is achieved, then adding a dropwise mixture of DAMP and DCC in DCM and stirring the reaction mixture for 16 hours. The desired ferrocene-coumarin conjugates **173**, **175**, and **177** were obtained by purification of the crude mixture on silica gel-based column chromatography (Scheme **32**).

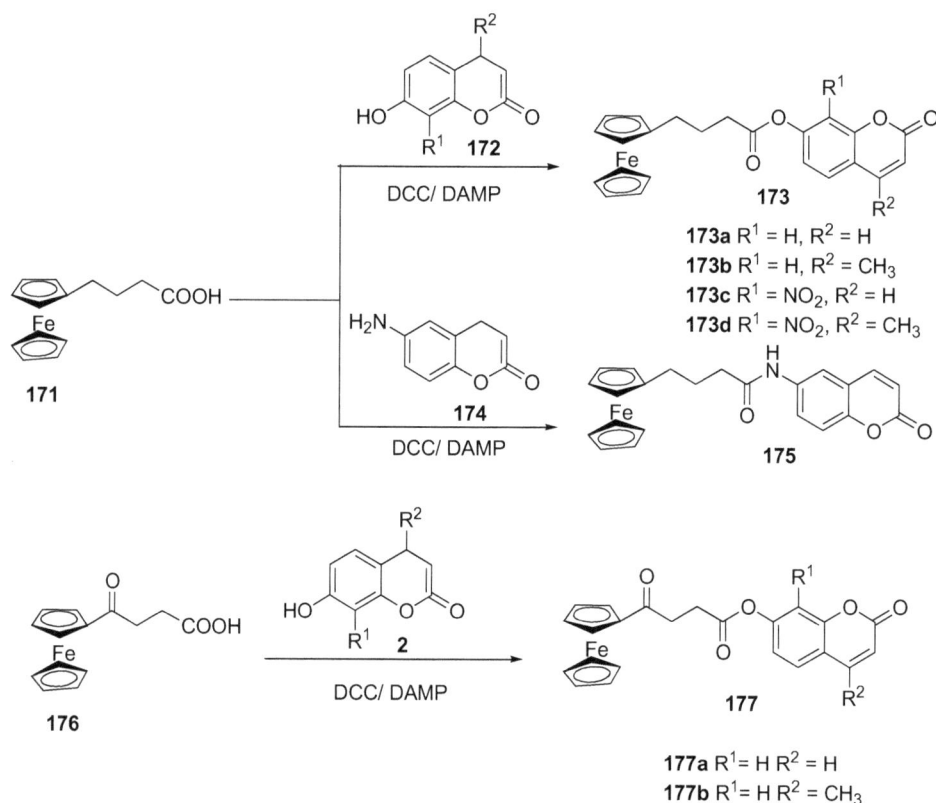

Scheme (32). Synthesis of ferrocene-coumarinconjugates **173**, **175** and **177**.

The anticancer potential of ferrocene-coumarin conjugates **173** and **175** was determined towards breast cancer MCF-7 cells, and the results (IC_{50} values) were obtained in the range of 12.10-28.10 umol/L, which were significantly higher than that of **177a** and **177b** (IC_{50}> 50 umol/L). The anticancer activity of ferrocene-coumarin conjugates **173** and **175** was also higher than that of **177a** and **177b** against BIU-87, SGC-7901, and Jurkat cancer cell lines. Compounds **173a**, **173b**, and **173d** displayed IC_{50} values of 5.24, 4.48, and 1.09 umol/L, respectively, against BIU-87, which was higher than the Adriamycin (IC$_{50}$ = 6.09 mol/L). All of the compounds displayed excellent SGC-7901 inhibitory potential, with **175**, showing the most promising results (IC$_{50}$ value of 3.56 μmol/L). Furthermore, SAR established that the C-4 methyl substitution at coumarin positively affected the anticancer activity of the ferrocene-coumarin hybrids. Compounds **173b**, **173d**, and **177b**, for example, had lower IC$_{50}$ values against EC-9706, BIU-87, MCF-7, and Eca-109 than hybrids **173a**, **173c**, and **177a,** respectively. Compounds **173d** and **177b** for SGC-7901 also exhibited superior antiproliferative activity when compared to compounds **173c** and **177a,** respectively.

Delogu and colleagues [37] created a series of antiproliferative compounds by conjugating ferrocene and curcumin-based biomolecules, namely zingerone, dehydrozingerone, and their biphenyl dimers, and tested their antiproliferative activity against the PC 12 cell line. The synthetic methodology involved Claisen-Schmidt reaction of the appropriate aromatic aldehydes **179** and **183** with acetylferrocene1 yielded the desired ferrocenyl chalcones **180** and **184** under different reaction conditions. By employing Pd/charcoal as a catalyst, simultaneously reducing the ethylene fragment and deprotecting the *O*-benzyl ether (in **180a** and **184b**) resulted in a release of zingerone-like derivative **181** and its corresponding dimer **185.** As demonstrated in Scheme **33**, selectively deprotecting the *O*-benzyloxygroups in **180b** and **184b** in the presence of Me3SiI at room temperature generated **182** and **186** with preserved ethylenic linkages.

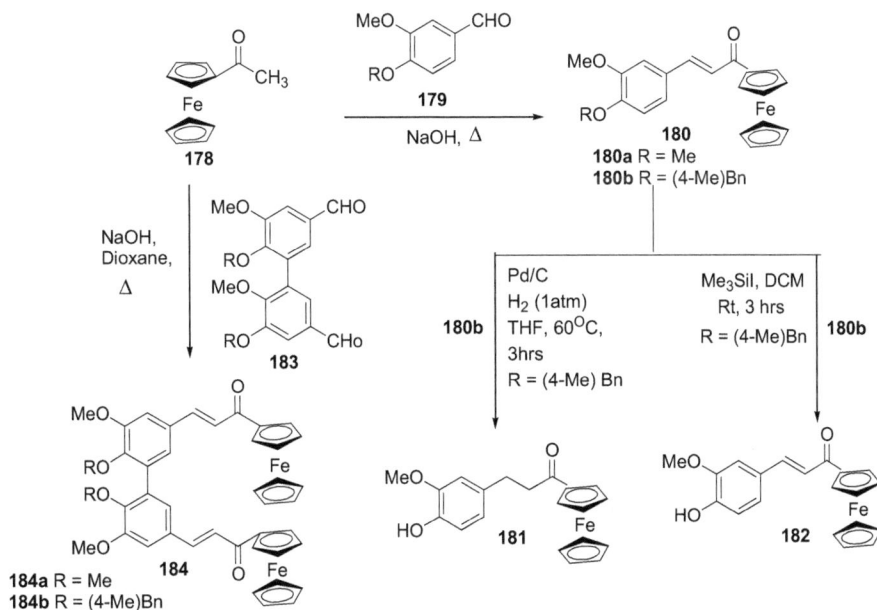

(Scheme 33) contd.....

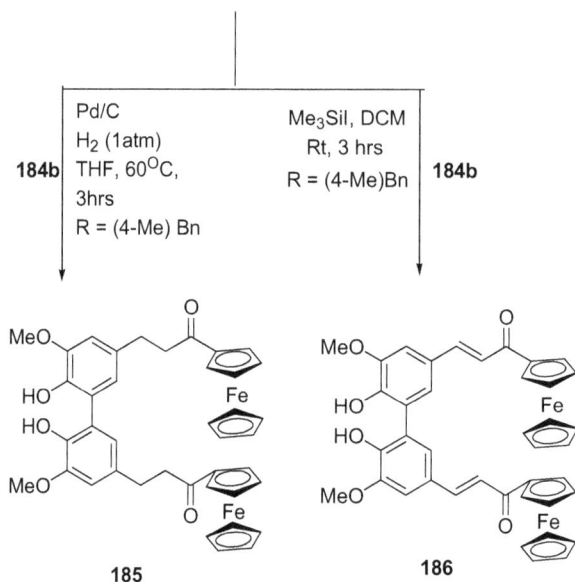

Scheme (33). Synthesis of ferrocenyl chalcones.

The influence of ferrocene inclusion on cytotoxic potency, in case of PC12 cells, was determined. All the tested compounds showed increased cell viability when compared to ferrocene alone. It was observed that ferrocenyl chalcones **184b**, **181**, **182**, **185**, and **186** enhanced the oxidative stress caused by H_2O_2 – a molecule commonly found in cancer cells and has recently been investigated as a potential prodrug.

CONCLUDING REMARKS

Despite the fact that millions of people die every year from cancer, the disease's morbidity and mortality will continue to climb for a long time. Anticancer drugs are an important and effective treatment strategy for cancer, but the growing problem of drug resistance, as well as the low specificity of currently available anticancer drugs, necessitates the development of novel anticancer drugs with high specificity and potency against drug-resistant cancers. Organometallic chemistry has gained popularity in recent years, thanks to the discovery of ferrocene as a bio-active core. Ferrocifen had a more diverse mechanism of action than platinum-based anticancer drugs, indicating that ferrocene hybrids could be utilised to treat a range of cancers, including multidrug-resistant tumours. Many ferrocene hybrids have been developed in the last 10 years, and some of them have demonstrated activity *in vitro* and *in vivo* against both drug-sensitive and drug-resistant cancers, including multidrug-resistant tumours. Finally, the present chapter discusses the many methods for inserting ferrocene core into drug/drug-like compounds that have recently surfaced (2015-20), as well as the evaluation studies that have been conducted.

REFERENCES

[1]   Ren, J.; Wang, S.; Ni, H.; Yao, R.; Liao, C.; Ruan, B. Synthesis, Characterization and Antitumor Activity of Novel Ferrocene-Based Amides Bearing Pyrazolyl Moiety. *J. Inorg. Organomet. Polym.,* **2015**, *25*, 41-426.
        [http://dx.doi.org/10.1007/s10904-014-0056-6]

[2]   van Staveren, D.R.; Metzler-Nolte, N. Bioorganometallic chemistry of ferrocene. *Chem. Rev.,* **2004**, *104*(12), 5931-5986.
        [http://dx.doi.org/10.1021/cr0101510] [PMID: 15584693]

[3]   Heinze, K.; Lang, H. Ferrocene-Beauty and Function. *Organometallics,* **2013**, *32*(20), 5623-5625.
        [http://dx.doi.org/10.1021/om400962w]

[4]   Fouda, M.F.R.; Abd-Elzaher, M.M.; Abdelsamaia, R.A.; Labib, A.A. On the medicinal chemistry of ferrocene. *Appl. Organomet. Chem.,* **2007**, *21*(8), 613-625.
        [http://dx.doi.org/10.1002/aoc.1202]

[5]   Singh, A.; Lumb, I.; Mehra, V.; Kumar, V. Ferrocene-appended pharmacophores: an exciting approach for modulating the biological potential of organic scaffolds. *Dalton Trans.,* **2019**, *48*(9), 2840-2860.rsc.li/Dalton
        [http://dx.doi.org/10.1039/C8DT03440K] [PMID: 30663743]

[6]   Wang, R.; Chen, H.; Yan, W.; Zheng, M.; Zhang, T.; Zhang, Y. Ferrocene-containing hybrids as potential anticancer agents: Current developments, mechanisms of action and structure-activity relationships. *Eur. J. Med. Chem.,* **2020**, *190*, 112109-112129.
        [http://dx.doi.org/10.1016/j.ejmech.2020.112109] [PMID: 32032851]

[7]   Fuertes, M.A.; Alonso, C.; Pérez, J.M. Biochemical modulation of Cisplatin mechanisms of action: enhancement of antitumor activity and circumvention of drug resistance. *Chem. Rev.,* **2003**, *103*(3), 645-662.
        [http://dx.doi.org/10.1021/cr020010d] [PMID: 12630848]

[8]   Larik, F.A.; Saeed, A.; Fattah, T.A.; Muqadar, U.; Channar, P.A. Recent advances in the synthesis, biological activities and various applications of ferrocene derivatives. *Appl. Organomet. Chem.,* **2017**, *31*(8), e3664.
        [http://dx.doi.org/10.1002/aoc.3664]

[9]   Lee, H.Z.S.; Buriez, O.; Labbé, E.; Top, S.; Pigeon, P.; Jaouen, G.; Amatore, C.; Leong, W.K. Oxidative Sequence of a Ruthenocene-Based Anticancer Drug Candidate in a Basic Environment. *Organometallics,* **2014**, *33*(18), 4940-4946.
        [http://dx.doi.org/10.1021/om500225k]

[10]   Görmen, M.; Pigeon, P.; Hillard, E.A.; Vessières, A.; Huché, M.; Richard, M.A.; McGlinchey, M.J.; Top, S.; Jaouen, G. Synthesis and Antiproliferative Effects of [3]Ferrocenophane Transposition Products and Pinacols Obtained from McMurry Cross-Coupling Reactions. *Organometallics,* **2012**, *31*(16), 5856-5866.
        [http://dx.doi.org/10.1021/om300382h]

[11]   Sharma, B.; Kumar, V. Has Ferrocene Really Delivered Its Role in Accentuating the Bioactivity of Organic Scaffolds? *J. Med. Chem.,* **2021**, *64*(23), 16865-16921.
        [http://dx.doi.org/10.1021/acs.jmedchem.1c00390] [PMID: 34792350]

[12]   Jaouen, G.; Vessières, A.; Top, S. Ferrocifen type anti cancer drugs. *Chem. Soc. Rev.,* **2015**, *44*(24), 8802-8817.
        [http://dx.doi.org/10.1039/C5CS00486A] [PMID: 26486993]

[13]   Muenzner, J.K.; Biersack, B.; Albrecht, A.; Rehm, T.; Lacher, U.; Milius, W.; Casini, A.; Zhang, J.J.; Ott, I.; Brabec, V.; Stuchlikova, O.; Andronache, I.C.; Kaps, L.; Schuppan, D.; Schobert, R. Ferrocenyl-Coupled *N*-Heterocyclic Carbene Complexes of Gold(I): A Successful Approach to Multinuclear Anticancer Drugs. *Chemistry,* **2016**, *22*(52), 18953-18962.
        [http://dx.doi.org/10.1002/chem.201604246] [PMID: 27761940]

[14]   Liu, Y.W.; Cheng, H.J.; Ruan, B.F.; Hu, Q. Synthesis, characterization and antitumor activity of (*E*)--methyl-3-ferrocenyl-*N*-acrylamide derivatives. *J. Organomet. Chem.,* **2019**, *887*, 71-79.
        [http://dx.doi.org/10.1016/j.jorganchem.2019.02.021]

[15]   Mahmoud, W.H.; Deghadi, R.G.; Mohamed, G.G. Metal complexes of ferrocenyl-substituted Schiff base: Preparation, characterization, molecular structure, molecular docking studies, and biological investigation. *J. Organomet. Chem.,* **2020**, *917*, 121113-121127.
        [http://dx.doi.org/10.1016/j.jorganchem.2020.121113]

[16]   Ismail, M. K.; Khan, Z.; Rana, M.; Horswell, S. L.; Male, L.; Nguyen, H. V.; Ismail, M.K.; Khan, Z.; Rana, M.; Horswell, S.L.; Male, L.; Nguyen, H.V.; Perotti, A.; Romero-Canelón, I.; Wilkinson, E.A.; Hodges, N.J.; Tucker, J.H.R. Effect of regiochemistry and methylation on the anticancer activity of a ferrocene-containing organometallic nucleoside analogue. *ChemBioChem,* **2020**, *21*(17), 2487-2494.
        [http://dx.doi.org/10.1002/cbic.202000124] [PMID: 32255248]

[17]   Hall, D.W.; Richards, J.H. The Acetylation of Some Substituted Ferrocenes. *J. Org. Chem.,* **1963**, *28*(6), 1549-1554.
        [http://dx.doi.org/10.1021/jo01041a026]

[18]   Guillén, E.; González, A.; Basu, P.K.; Ghosh, A.; Font-Bardia, M.; Calvet, T.; Calvis, C.; Messeguer, R.; López, C. The influence of ancillary ligands on the antitumoral activity of new cyclometallated Pt(II) complexes derived from an ferrocene-pyrazole hybrid. *J. Organomet. Chem.,* **2017**, *828*, 122-132.
        [http://dx.doi.org/10.1016/j.jorganchem.2016.11.031]

[19]   Beauperin, M.; Polat, D.; Roudesly, F.; Top, S.; Vessieres, A.; Oble, J.; Jaouen, G.; Poli, G. Metallocene-uracil conjugates: Approach to ferrocenyl-podophyllotoxin analogs and their evaluation as anti-tumor agents. *J. Organomet. Chem.,* **2017**, *839*, 83-90.
        [http://dx.doi.org/10.1016/j.jorganchem.2017.02.005]

[20]   Błauż, A.; Rychlik, B.; Makal, A.; Szulc, K.; Strzelczyk, P.; Bujacz, G.; Zakrzewski, J.; Woźniak, K.; Plażuk, D. Ferrocene-Biotin Conjugates: Synthesis, Structure, Cytotoxic Activity and Interaction with Avidin. *ChemPlusChem,* **2016**, *81*(11), 1191-1201.
        [http://dx.doi.org/10.1002/cplu.201600320] [PMID: 31964109]

[21]   Wambang, N.; Schifano-Faux, N.; Aillerie, A.; Baldeyrou, B.; Jacquet, C.; Bal-Mahieu, C.; Bousquet, T.; Pellegrini, S.; Ndifon, P.T.; Meignan, S.; Goossens, J.F.; Lansiaux, A.; Pélinski, L. Synthesis and biological activity of ferrocenyl indeno[1,2-c]isoquinolines as topoisomerase II inhibitors. *Bioorg. Med. Chem.,* **2016**, *24*(4), 651-660.
        [http://dx.doi.org/10.1016/j.bmc.2015.12.033]

[22]   Bansode, P.; Patil, P.; Choudhari, P.; Bhatia, M.; Birajdar, A.; Bansode, P.; Patil, P.; Choudhari, P.; Bhatia, M.; Birajdar, A.; Somasundaram, I.; Rashinkar, G. Anticancer activity and molecular docking studies of ferrocene tethered ionic liquids. *J. Mol. Liq.,* **2019**, *290*, 111182-111192.
        [http://dx.doi.org/10.1016/j.molliq.2019.111182]

[23]   Altaf, A.A.; Lal, B.; Badshah, A.; Usman, M.; Chatterjee, P.B.; Huq, F.; Ullah, S.; Crans, D.C. Synthesis, structural characterization, modal membrane interaction and anti-tumor cell line studies of nitrophenyl ferrocenes. *J. Mol. Struct.,* **2016**, *1113*, 162-170.
        [http://dx.doi.org/10.1016/j.molstruc.2016.02.045]

[24]   Kocsis, L.; Szabó, I.; Bősze, S.; Jernei, T.; Hudecz, F.; Csámpai, A. Synthesis, structure and *in vitro* cytostatic activity of ferrocene—Cinchona hybrids. *Bioorg. Med. Chem. Lett.,* **2016**, *26*(3), 946-949.
        [http://dx.doi.org/10.1016/j.bmcl.2015.12.059] [PMID: 26739780]

[25] Singh, A.; Mehra, V.; Sadeghiani, N.; Mozaffari, S.; Parang, K.; Kumar, V. Ferrocenylchalcone–uracil conjugates: synthesis and cytotoxic evaluation. *Med. Chem. Res.,* **2018**, *27,* 1260-1268.
[http://dx.doi.org/10.1007/s00044-018-2145-5]

[26] Singh, A.; Saha, S.T.; Perumal, S.; Kaur, M.; Kumar, V. Azide–Alkyne Cycloaddition En Route to 1 *H*-1,2,3-Triazole-Tethered Isatin–Ferrocene, Ferrocenylmethoxy–Isatin, and Isatin–Ferrocenylchalcone Conjugates: Synthesis and Antiproliferative Evaluation. *ACS Omega,* **2018**, *3*(1), 1263-1268.
[http://dx.doi.org/10.1021/acsomega.7b01755] [PMID: 30023800]

[27] Li, B.; Zhu, X.; Guo, Y.; Ren, Y-P.; Jia, Z-D.; Wei, J-N.; Fu, D-X.; Xu, Y.; Hao, X-Q. Synthesis and properties of m-ferrocenylbenzoylthiadiazole derivatives. *Appl. Organometal. Chem.,* **2018**, e4265-4272.
[http://dx.doi.org/10.1002/aoc.4265]

[28] Pérez, W.I.; Soto, Y.; Ortíz, C.; Matta, J.; Meléndez, E. Ferrocenes as potential chemotherapeutic drugs: Synthesis, cytotoxic activity, reactive oxygen species production and micronucleus assay. *Bioorg. Med. Chem.,* **2015**, *23*(3), 471-479.
[http://dx.doi.org/10.1016/j.bmc.2014.12.023] [PMID: 25555734]

[29] Hussain, R.A.; Badshah, A.; Pezzuto, J.M.; Ahmed, N.; Kondratyuk, T.P.; Park, E.J. Ferrocene incorporated selenoureas as anticancer agents. *J. Photochem. Photobiol. B,* **2015**, *148*, 197-208.
[http://dx.doi.org/10.1016/j.jphotobiol.2015.04.024] [PMID: 25966308]

[30] Skiba, J.; Karpowicz, R.; Szabó, I.; Therrien, B.; Kowalski, K. Synthesis and anticancer activity studies of ferrocenyl-thymine-3,6-dihydro-2*H*-thiopyranes – A new class of metallocene-nucleobase derivatives. *J. Organomet. Chem.,* **2015**, *794*, 216-222.
[http://dx.doi.org/10.1016/j.jorganchem.2015.07.012]

[31] Peres, B.; Nasr, R.; Zarioh, M.; Lecerf-Schmidt, F.; Pietro, A.D.; Pérès, B.; Nasr, R.; Zarioh, M.; Lecerf-Schmidt, F.; Di Pietro, A.; Baubichon-Cortay, H.; Boumendjel, A. Ferrocene-embedded flavonoids targeting the *Achilles heel* of multidrug-resistant cancer cells through collateral sensitivity. *Eur. J. Med. Chem.,* **2017**, *130*, 346-353.
[http://dx.doi.org/10.1016/j.ejmech.2017.02.064] [PMID: 28273561]

[32] Pedotti, S.; Ussia, M.; Patti, A.; Musso, N.; Barresi, V.; Condorelli, D.F. Synthesis of the ferrocenyl analogue of clotrimazole drug. *J. Organomet. Chem.,* **2017**, *830*, 56-61.
[http://dx.doi.org/10.1016/j.jorganchem.2016.12.009]

[33] Wang, Y.; Richard, M.A.; Top, S.; Dansette, P.M.; Pigeon, P.; Vessières, A.; Mansuy, D.; Jaouen, G. Ferrocenyl Quinone Methide-Thiol Adducts as New Antiproliferative Agents: Synthesis, Metabolic Formation from Ferrociphenols, and Oxidative Transformation. *Angew. Chem. Int. Ed.,* **2016**, *55*(35), 10431-10434.
[http://dx.doi.org/10.1002/anie.201603931] [PMID: 27276169]

[34] Wang, Y.; Pigeon, P.; McGlinchey, M.J.; Top, S.; Jaouen, G. Synthesis and antiproliferative evaluation of novel hydroxypropyl-ferrociphenol derivatives, resulting from the modification of hydroxyl groups. *J. Organomet. Chem.,* **2017**, *829*, 108-115.
[http://dx.doi.org/10.1016/j.jorganchem.2016.09.005]

[35] Pigeon, P.; Wang, Y.; Top, S.; Najlaoui, F.; Garcia Alvarez, M.C.; Bignon, J.; McGlinchey, M.J.; Jaouen, G. A New Series of Succinimido-ferrociphenols and Related Heterocyclic Species Induce Strong Antiproliferative Effects, Especially against Ovarian Cancer Cells Resistant to Cisplatin. *J. Med. Chem.,* **2017**, *60*(20), 8358-8368.
[http://dx.doi.org/10.1021/acs.jmedchem.7b00743] [PMID: 28895732]

[36]   Wei, Z-D.; Jia, J-Na.; Zhou, Y-Q.; Chen, P-H.; Li, B.; Zhang, N. Hao, Y. X-Qi.; Zhang, X. B. Synthesis, characterization and a ferrocene-coumarin conjugates. *J. Organomet. Chem.,* **2019**, *902*, 12096-120102.
[http://dx.doi.org/10.1016/j.jorganchem.2019.120968]

[37]   Pedotti, S.; Patti, A.; Dedola, S.; Barberis, A.; Fabbri, D. Pedotti, S.; Patti, A.; Dedola, S.; Barberis, A.; Fabbri, D.; Dettori, M.A.; Serra, P.A.; Delogu, G. Synthesis of new ferrocenyl dehydrozingerone derivatives and their effects on viability of PC12 cells. *Polyhedron,* **2016**, *117*, 80-89.
[http://dx.doi.org/10.1016/j.poly.2016.05.039]

CHAPTER 2

Synthesis of Fused Nitrogenated Heterocycles: Intramolecular Povarov Reaction

Carme Masdeu[1], Jesús M. de los Santos[1], Francisco Palacios[1] and Concepcion Alonso[1,*]

[1] *Departamento de Química Orgánica I, Facultad de Farmacia and Centro de Investigación Lascaray (Lascaray Research Center). Universidad del País Vasco/Euskal Herriko Unibertsitatea (UPV/EHU). Paseo de la Universidad 7, 01006 Vitoria-Gasteiz, Spain*

Abstract: Nitrogenated heterocycles take part in the structure of many natural products and agents with important biological activity, such as antiviral, antibiotic and antitumor drugs. For this reason, heterocyclic compounds are one of the most desired synthetic targets nowadays. In this review work, the literature related to the preparation of polyheterocyclic compounds by using the intramolecular Povarov reaction will be collected. The Povarov reaction is a process in which aromatic amines, carbonyl compounds and olefins or acetylenes participate to give rise to the formation of the nitrogenated compounds. Then, intramolecular Povarov reactions to carry out these syntheses are described according to the key processes involved; catalytic reactions with transition metals will be included discussing the reaction mechanisms and examining the effect of catalysts and solvents in the preparation of the products, thus reflecting the synthetic potential of this strategy. Moreover, applications of prepared compounds will also be considered.

Keywords: Aldimines, Aza-Diels-Alder Reaction, Brönsted and Lewis Acid Catalysis, Cycloaddition, Intramolecular Povarov Reaction, Nitrogen-containing Fuse Heterocycles.

1. INTRODUCTION

Heterocyclic systems play an important biological role and are essential in various aspects of life and material science. The vast majority of naturally occurring compounds, as well as a significant number of compounds synthesized in the laboratory, have a nitrogenated heterocyclic structure, which has led to various applications in different fields [1 - 4].

---
* **Corresponding author Concepcion Alonso:** Departamento de Química Orgánica I, Facultad de Farmacia and Centro de Investigación Lascaray (Lascaray Research Center). Universidad del País Vasco/Euskal Herriko Unibertsitatea (UPV/EHU). Paseo de la Universidad 7, 01006 Vitoria-Gasteiz, Spain; Email:concepcion.alonso@ehu.eus

Shazia Anjum (Ed.)

In other words, many of the pharmacologically significant compounds and agrochemicals are nitrogenated heterocyclic compounds.

Therefore, the development of new methodologies for the construction of nitrogenated heterocyclic compounds has been a central issue in the synthesis of organic and natural products, pushed especially by the increasing demand for these compounds in the last decades. In particular, among the different strategies developed for the synthesis of heterocyclic compounds, the Povarov reaction has been described as a very versatile tool for the construction of nitrogenated heterocycles [5 - 9].

In the middle of the 20th century, Povarov and co-workers [10] described the preparation of tetrahydroquinolines by the reaction between aromatic Schiff bases (aromatic aldimines) and electron-rich alkenes. The early works of Povarov described reactions of ethyl vinyl ether **2** or ethyl vinyl sulfide **3** and *N*-aryl aldimine **1** under acid catalysis ($BF_3 \cdot OEt_2$) to give 2,4-substituted tetrahydro-quinolines **4**, **5**, which were converted after oxidation to the corresponding quinoline **6** Scheme (**1**).

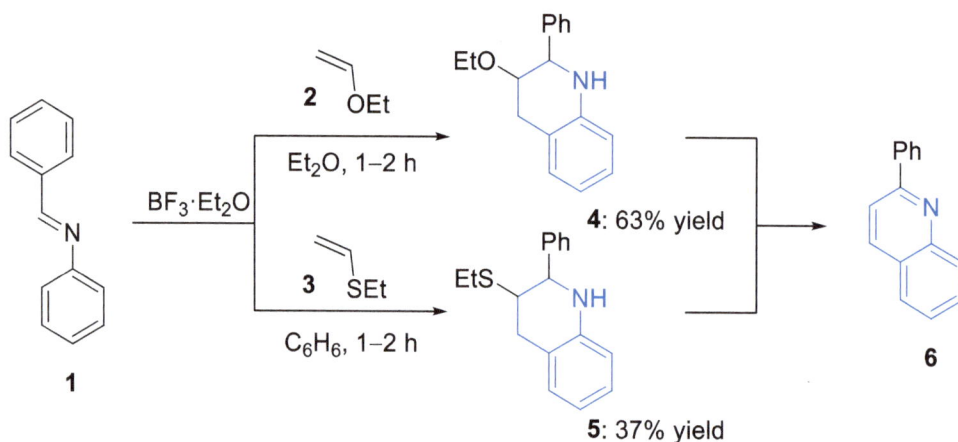

Scheme (1). Reaction described by Povarov for the first time.

In general terms, the formation of tetrahydroquinoline derivatives **VI** by Povarov's reaction could therefore be explained by a formal [4+2] cycloaddition between aldimines **III**, obtained by condensation between aromatic aldehydes **I** and aromatic amines **II**, and olefins **IV** to give adducts **V** whose subsequent tautomerization would give the tetrahydroquinoline **VI**. Aromatization of these tetrahydroquinolines **VI** would generate the corresponding quinolines **VII** Scheme (**2**). This strategy represents a very powerful tool for the preparation of quinoline derivatives by generating three stereocenters in a single step. In general, the reaction shows a high regio- and diastereoselectivity with the isolation in most cases of a major isomer. In those cases where mixtures of diastereoisomers are observed, the ratio of the *endo/exo* diastereisomers formed is modulated and determined by the catalyst and solvent used in the process.

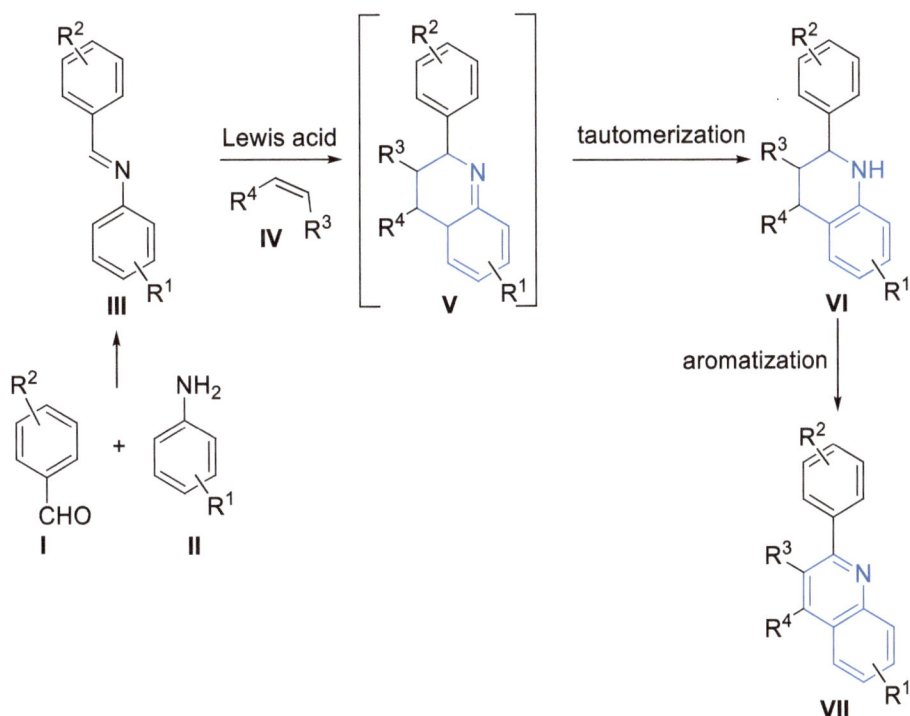

Scheme (2). General scheme of the Povarov reaction.

Regarding the mechanistic aspects of this reaction, Kobayashi *et al.* suggested a stepwise reaction mechanism for the imino-Diels-Alder reaction of aniline-derived imines with alkenes [11]. On the other hand, experimental results pointed toward a concerted asynchronous cycloaddition process [12]. Moreover, a combined theoretical and experimental study of Povarov-type cycloaddition reactions between *N*-(3-pyridyl) imines and ethylene, or substituted ethylenes (styrene, cyclopentadiene or indene) catalyzed by $BF_3 \cdot Et_2O$ was carried out in our research group [13]. These computational studies gave light to understand the mechanism of the Povarov reaction between aldimines **VIII** Scheme (**3**) derived from 3-aminopyridine and benzaldehyde with different olefins. In this case, we studied the most favorable orientation of the pyridine ring in the cycloaddition reaction, the influence of the $BF_3 \cdot Et_2O$ and if this Lewis acid preferentially coordinates to the pyridinic nitrogen or to the iminic nitrogen. In addition, we extended the calculations to study the regio- and diastereoselectivity of the reaction towards simple olefins such as styrene, cylopentadiene and indene. According to these studies, the use of 2 equivalents of $BF_3 \cdot Et_2O$ activates the azadienic system, making this process faster than when 1 equivalent of $BF_3 \cdot Et_2O$ was used, due to double coordination of the two nitrogen atoms. Moreover, the lower activation barriers correspond to the *endo* transition-state structures that lead to the formation of the *endo* adducts **X** in a regio- and stereoselective way with the control of two or three stereocenters. The mechanism may be explained through an exothermic, concerted and asynchronous process yielding the corresponding tetrahydro-1,5-naphthyridines **XI** (Scheme **3**).

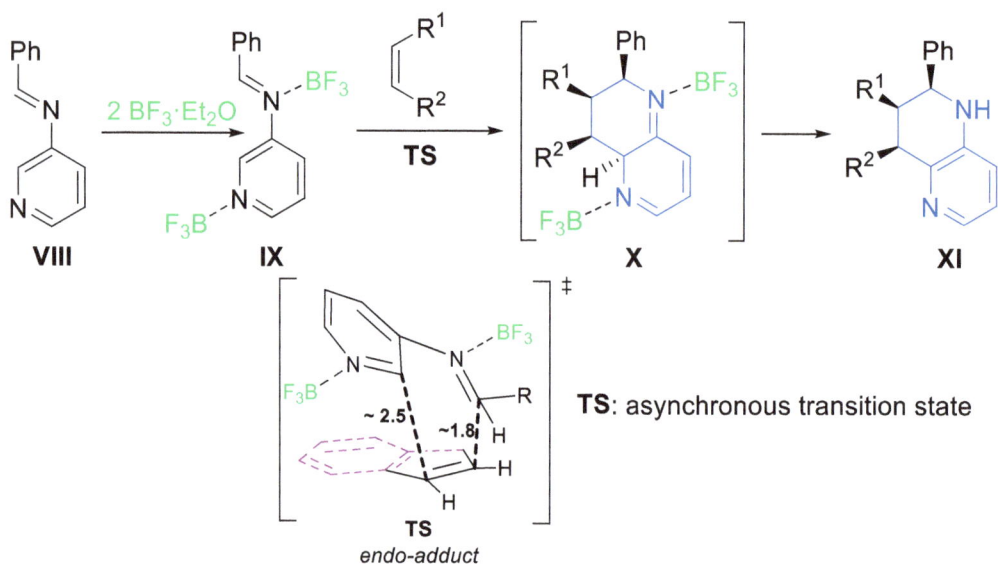

Scheme (3). Cycloaddition reactions in the presence of $BF_3 \cdot Et_2O$ through an asynchronous transition state.

As collected in their review [14], Povarov and co-workers discussed the extent of the reaction and some mechanistic aspects of this cycloaddition, such as the important role of the Lewis acid catalysts. Then, taking into account that tetrahydroquinolines were obtained in relatively low yields influenced by the amount of Lewis acid, new catalysts have been used and developed. For example, Kobayashi and co-workers demonstrated that lanthanide triflates are excellent catalysts [11]. Currently, in order to improve the selectivity and yield of these aza-Diels-Alder reactions, although $BF_3 \cdot OEt_2$ is still successfully used, in addition to lanthanide metal chlorides [15], several other Lewis acids are known to catalyze these reactions, including $LiClO_4$ [16], $AlCl_3$ [17], $BiCl_3$ [18], $GaCl_3$ [19], $InCl_3$ [20] and CAN [21] among others. Not only Lewis acids can catalyze the Povarov reaction, but also the preparation of some quinoline derivatives has been described by using a variety of Brønsted acid catalysts, such as TFA [22, 23], phosphoric acids [24, 25], CF_3SO_3H [26, 27] and *p*-TsOH [28].

With the aim of improving yields and obtaining greater structural diversity, the acid-catalyzed multicomponent protocols of this reaction were developed. Multicomponent reactions (MCRs) represent a versatile alternative since lower costs, shorter reaction time and less energy are required [29 - 32]. In the Povarov multicomponent reaction, the three components, amine, aldehydes and dienophiles, are reacted in the presence of a Lewis or Brønsted acid and the corresponding adducts are obtained regio- and diastereoselectively. In other words, the *in situ* formation of *N*-arylaldimine and its 'domino' reaction with a dienophile, present in the reaction medium, enables the Povarov reaction to be performed in a single step. In recent years, numerous papers [33 - 36], including some in our research group [37 - 41], have been published on the Povarov reaction, in which the preparation of heterocyclic compounds has been comparatively analyzed using both, stepwise and multicomponent protocols.

Although Povarov initially described this reaction with aldimines using electron-rich olefins, if acetylenes are used as dienophiles when reacting with aldimines, the corresponding aromatized quinoline derivatives can be obtained directly instead of the tetrahydroquinoline derivatives [42 - 46].

When the aldehyde carries functionalized unsaturated bonds in its structure and is reacted with aromatic and/or heteroaromatic amines, a particular type of multicomponent Povarov reaction is involved. In other words, the Diels-Alder reaction takes place intramolecularly when a molecule contains both diene and dienophile functionalities connected by a chain. As a result of these interactions, besides forming the bicyclic ring obtained by the Povarov reaction as described so far, other fused carbonated or heterocyclic cycles are generated where the size of the second ring depends on the chain broadening (Scheme **4**).

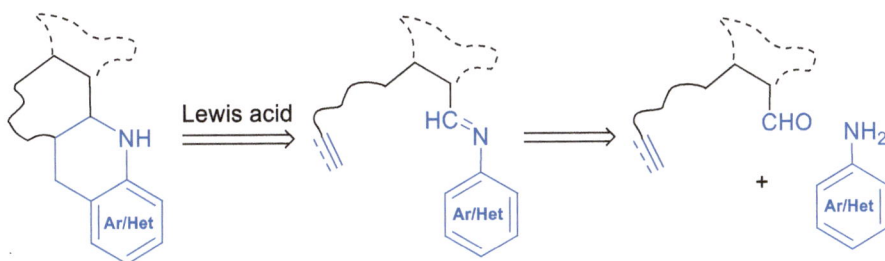

Fused nitrogenated heterocycles
Ar/Het: aromatic or heteroaromatic

Scheme (4). Retrosynthetic scheme for the intramolecular Povarov reaction.

In this particular case, we are talking about the intramolecular Povarov reaction, which is the focus of this chapter. This review therefore focuses on new efficient approaches based on the intramolecular Povarov reaction and covers the literature up to May 2021. The main objective is to show the utility of this process for the preparation of fused nitrogenated heterocyclic compounds. For this purpose, reactions have been classified according to the number of fused cycles in the compound obtained by this methodology, taking as a reference the bicyclic system formed directly by the Povarov reaction to which other cyclic structures are added.

2. SYNTHESIS OF TRICYCLIC FUSED HETEROCYCLES

One of the applications of the intramolecular Povarov reaction found in the literature has been the preparation of fused nitrogen heterocycles such as acridine, as well as other tricyclic quinoline derivatives.

2.1. Formation of the Acridine Skeleton

Acridine derivatives form an important class of heterocycles due to their wide range of pharmaceutical properties [47]. Among these biological properties, acridine derivatives are known anticancer drugs and cytotoxic agents, representing a very interesting synthetic goal [48]. An intramolecular Povarov reaction of *N*-arylimines-derived from aliphatic aldehydes tethered to non-activated olefins would afford 1,2,3,4,4a,9a,10-octahydroacridine derivatives (OHA). Laschat's group carried out the preparation of acridine derivatives using a wide range of Lewis or Brønsted acids, which would catalyze the formation of octahydroacridines **10** Scheme (**5**) [49]. In this case, the cycloaddition of *N*-arylimines **9** derived from aliphatic aldehydes functionalized with unactivated olefins **7** generated the OHA derivatives in high yields and in the authors' opinion, by a cycloaddition mechanism through an endocyclic transition state. The selectivity was found to be more dependent on the substrate structure than on the type of catalyst used. The multicomponent version, by successive addition of Lewis acid and aldehyde **7** to a precooled (-78 °C) solution of amine **8**, was also investigated, with yields and *cis/trans* ratios quite similar to cyclization when isolated imines **9** were used.

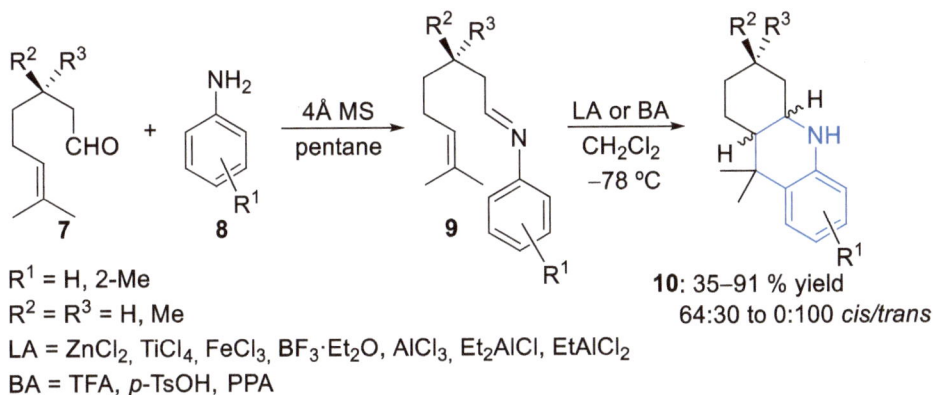

R^1 = H, 2-Me
R^2 = R^3 = H, Me
LA = ZnCl$_2$, TiCl$_4$, FeCl$_3$, BF$_3$·Et$_2$O, AlCl$_3$, Et$_2$AlCl, EtAlCl$_2$
BA = TFA, *p*-TsOH, PPA

10: 35–91 % yield
64:30 to 0:100 *cis/trans*

Scheme (5). Synthesis of octahydroacridines **10** by using Lewis or Brønsted acids.

The same group studied the Povarov reaction between 3-methylcytronellal **11** and anilines **8** in the presence of molecular sieves and observed that the reaction proceeded differently depending largely on the type of molecular sieve used Scheme (**6**) [50]. When using powdered molecular sieves, anilines **8** gave very pure imines **12** in almost quantitative yield after 15 min. However, while 4-nitroaniline **8** (R^1 = 4-NO$_2$) could not be converted to imine **12** with powdered molecular sieves, when molecular sieve beads were used, the formation of a mixture of the *trans*-acridines **13** together with functionalized aniline **14** was observed.

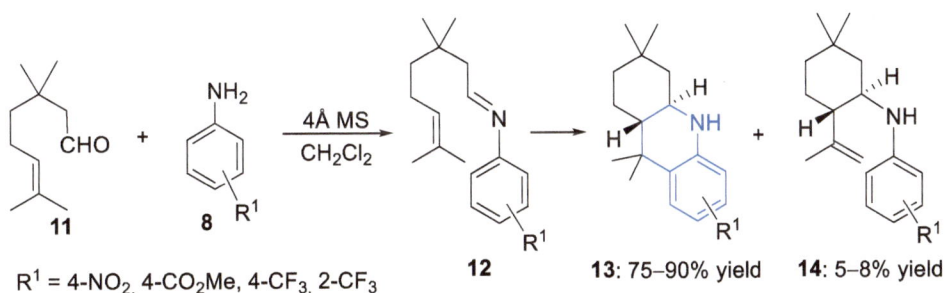

R^1 = 4-NO$_2$, 4-CO$_2$Me, 4-CF$_3$, 2-CF$_3$

12   13: 75–90% yield   14: 5–8% yield

Scheme (6). Synthesis of octahydroacridines **13** and/or amines **14** with molecular sieves.

Not only the intramolecular Povarov reaction of anilines but also diamines were studied for the synthesis of acridine derivatives [50]. The bis-cyclization reaction occurred when diamines with "separated" aromatic systems were used and activated with molecular sieve beads to obtain the corresponding acridine heterocycles. But when powdered molecular sieve beads were used, the corresponding di-imines were formed, but not the bis-cyclization products.

However, when the diamines presented two amine groups in the same aromatic system, the bis-cyclization reaction was not possible neither with enhanced molecular sieves, nor beads. Probably because the presence of a second imine function in the same aromatic ring decreases the reactivity of the first imine, and a stronger activation with $MeAlCl_2$ Lewis acid was needed for cyclization to obtain the bis-cycloadducts.

The authors extended this process to the use of chromium tricarbonyl substituted amine, and the corresponding chromium OHA complexes were obtained by the same mechanism in a highly *trans*-selective manner, which seems to be in agreement with a concerted hetero-Diels-Alder type mechanism [51]. These (octahydroacridine)chromium complexes were also prepared by a direct complexation of the octahydroacridines with chromium tricarbonyl.

When bismuth(III) chloride was used as Lewis acid, this reaction proceeds in a highly stereoselective manner giving the *trans*-products in a diastereoselective manner. Sabitha *et al.* studied the intramolecular hetero-Diels-Alder reaction of citronellal **7** (R^2 = Me, R^3 = H) and anilines **8** using bismuth(III) chloride Scheme (7) [52]. When the reaction is carried out at 0 °C in acetonitrile, *trans*-adducts are obtained stereoselectively. These *trans*-adducts **15** were obtained exclusively when using unsubstituted amines **8** (R^1 = H).

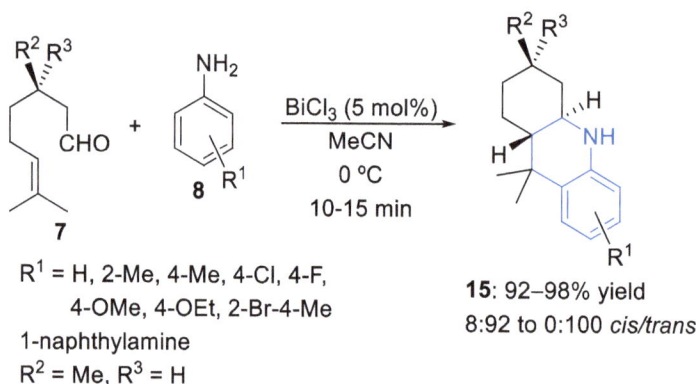

R^1 = H, 2-Me, 4-Me, 4-Cl, 4-F,
4-OMe, 4-OEt, 2-Br-4-Me
1-naphthylamine
R^2 = Me, R^3 = H

15: 92–98% yield
8:92 to 0:100 *cis/trans*

Scheme (7). Synthesis of octahydroacridines with bismuth (III) chloride.

The synthesis of octahydroacridines has also been reported using a solid-supported catalyst ($SiO_2/ZnCl_2$), under MW irradiation and without any solvent Scheme (7) [53]. When the reaction was performed between (+)-citronellal **7** (R^2 = Me, R^3 = H) and a variety of anilines **8** (R^1 = H, 2-Me, 4-Me, 2-CO_2H, 4-Cl) or 1-naphthylamine in the presence of 10% $SiO_2/ZnCl_2$ catalyst, the corresponding OHA derivatives were obtained in good yields (75-92% yield) and different diastereoselectivities (3:1 to 1:2 *cis*/*trans*). If thio-functionalized anilines (R^1 = 4-$SCH_2(p$-$ClC_6H_4)$, 4-$SC_{12}H_{25}$) are used in the Povarov intramolecular reaction octahydroacridines are obtained in moderate yields (65-72%) and as stoichiometric mixtures of *cis*- and *trans*-diastereoisomers [54]. However, if using 3-(phenylthio)citronellal (R^2 = Me, R^3 = SC_6H_5, S-(4-$MeOC_6H_4$)), the formation of a mixture of *trans*- and *cis*-adducts was observed (45-92% yield), with good selectivity to the *trans*-fused thio-octahydroacridines in most of the cases (up to 4:96 *cis*/*trans*) [54].

Several other catalytic materials were studied as well in the intramolecular Povarov reaction with (*R*)-(+)-citronellal [55]. With the unreduced CuO/SiO_2 (CuO/Si) catalyst at room temperature in the presence of air, silica-alumina cracking catalysts with a 13% content of Al_2O_3 (SiAl 13), 0.6% alumina on silica (SiAl 0.6), Montmorillonite K10 and KSF the corresponding octahydroacridines were obtained on varied yields (59-97%) and selectivities (73:27 to 37:63 *cis*/*trans*).

Some other groups developed the synthesis of acridine derivatives by using different synthetic methodologies. In this sense, taking advantage of the benefits offered by the solid-phase synthesis methodology, octahydroacridine **19** has been prepared in excellent yield (88%) as a single diastereoisomer by intramolecular Povarov reaction of (*R*)-(+)-citronellal **16** with aniline **17** in the presence of Yb(OTf)$_3$ (Scheme **8**) followed by acid hydrolysis (TFA) [56].

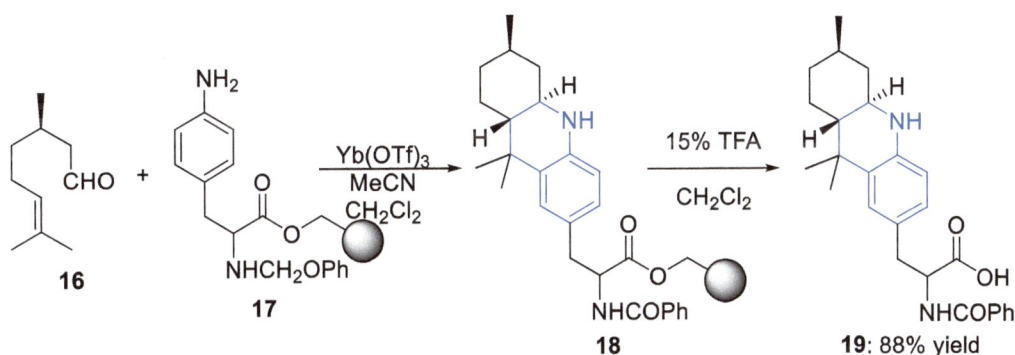

Scheme (8). Solid-phase synthesis of octahydroacridine derivative.

The preparation of octahydroacridines was also accomplished by fluorous phase synthesis [57]. Arylamines and citronellal were reacted at room temperature in the presence of trifluoroethanol (TFE), without any additional catalyst affording the corresponding octahydroacridines in higher yields (82-95%), than when the reaction was performed in the presence of 10 mol% TiCl$_3$ (46-72%). However, worse stereoselectivities were observed when applying fluorous technologies.

As an alternative to the preparation of octahydroacridines, ionic liquids have also been used. Ionic liquids have the advantage that they act both as solvent and catalyst of the reaction, being at the same time easily recoverable after the reaction. In this sense, the hetero-Diels-Alder reaction between (*R*)-citronellal **16** and anilines **8** (R^1 = H, 4-Me) in the presence of ionic liquids based on selenium and tellurium as solvents and/or catalysts generated the corresponding OHA derivatives in good yield (74-76% yield, Scheme **9**) [58]. Under microwave irradiation, the reaction time diminished and the expected products **21** were obtained in better yields (73-80%).

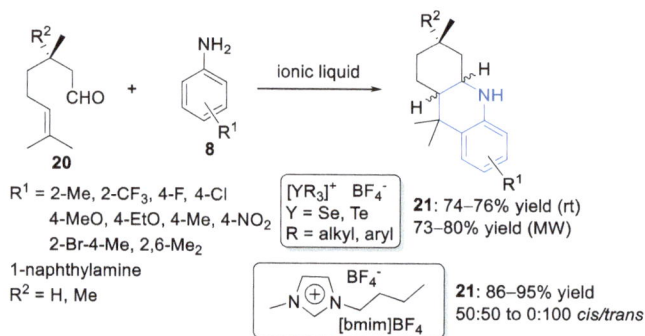

Scheme (9). Synthesis of octahydroacridines **21** by using selenium- and tellurium-based ionic liquids.

Likewise, other ionic liquids, such as 1-butyl-3-methylimidazolium tetrafluoroborate, [bmim] BF$_4$, 1-hexyl-3-methylimidazolium tetrafluoroborate, [hmim]BF$_4$, and 1-octyl-3-methylimidazolium tetrafluoroborate, [octmim] BF$_4$, have been used in the preparation of octahydroacridine derivatives **21** Scheme (**9**) [59].

On the other hand, the asymmetric version for the preparation of acridine derivatives could be attributed to Jørgensen's group [60]. In this case, a one-pot domino Michael addition/intramolecular Povarov reaction led to the synthesis of optically active octahydroacridines with four stereocenters. The reaction of malononitrile **22** with α,β-unsaturated aldehydes **23**, in the presence of diarylprolinol **24**, and subsequent acid-catalyzed (TFA) amine condensation/ intramolecular Povarov reaction with *p*-substituted anilines **8**, produced OHA **26** with high yields and excellent enantio- and diastereomeric control (Scheme **10**).

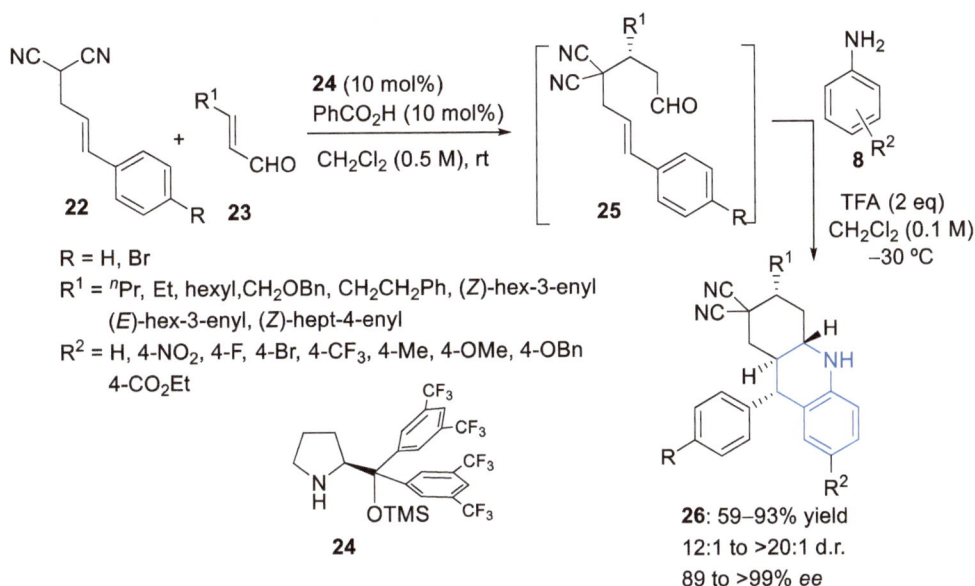

Scheme (10). Enantioselective organocatalytic one-pot domino Michael/intramolecular Povarov reaction.

2.2. Formation of Quinoline Derivatives Fused with Carbo- and Heterocycles

By means of the intramolecular Povarov reaction, tricyclic quinoline derivatives fused with five- and six-membered carbocycles or heterocycles have been prepared. In this sense, cyclopenta[*b*]quinoline derivatives **29**, part of the isoschizozygane alkaloid core, were prepared in highly diastereoselectivity and yields by intramolecular hetero-Diels-Alder reaction starting from imine **28** obtained from unsaturated aldehyde **27** and aniline **8** (Scheme **11**) [61]. In this case, a fused 1*H*-cyclopenta[*b*]quinoline derivative is formed with four contiguous stereocenters.

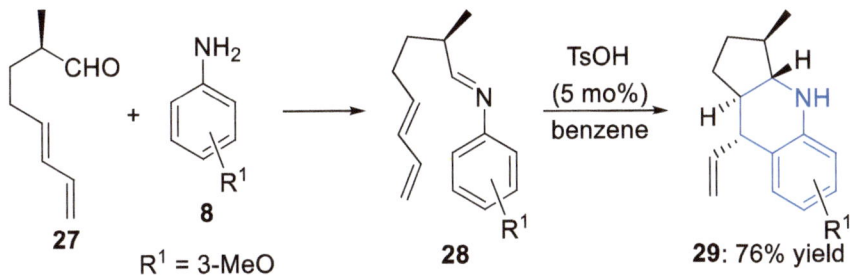

Scheme (11). Synthesis of quinoline derivative fused with a five-membered carbocycle.

When the *O*-allylic or propargylic ester derivative of glyoxylic acid was used in the intramolecular Povarov reaction, quinolines fused with lactone heterocycles were obtained [62]. For example, the intramolecular cycloaddition of aldimines **31**, obtained by condensation between aldehydes **30** and anilines **8** in the presence of 1 equivalent of $BF_3 \cdot OEt_2$ in CH_2Cl_2 at room temperature, generated the fused lactone-quinoline derivatives **32** in low yields (Scheme **12**). In addition, equimolecular formation of the amine derived from the reduction of the carbon-nitrogen double bond of aldimine **31** was observed. This result suggests that the initial imine **31**, present in the reaction medium, acts as an oxidant of the tetrahydroquinoline to give the corresponding quinoline **32**. In order to convert all the imine to quinoline, the reaction was carried out in the presence of 2 equivalent of 2,3-dichloro-5,6-dicyano-*p*-benzoquinone (DDQ) oxidant and the corresponding fused quinoline-lactones **32** were obtained in good yields. On the other hand, when the reaction was carried out in the presence of a Brønsted acid, such as TFA, the isolation of the fused furo[3,4-*b*]quinolin-3(1*H*)-one derivatives **32** was feasible [63]. More recently, Muthukrishnan's group reported an intramolecular dehydrogenation promoted by oxone followed by imino-Diel-Alder reaction (Povarov cyclization) of alkyne-tethered *N*-aryl glycine esters and amides for the preparation of quinoline fused lactones and lactams [64].

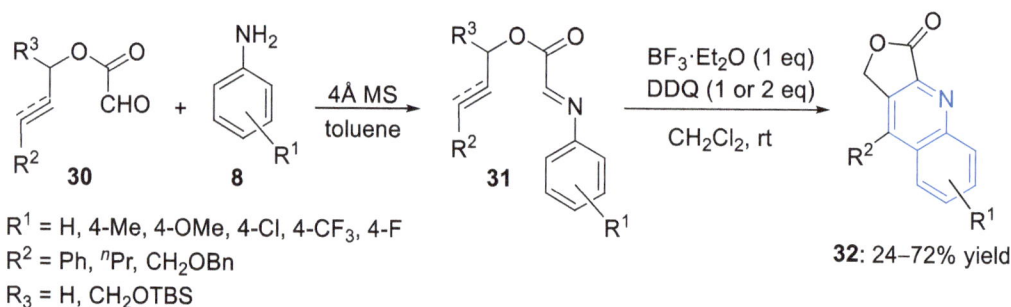

R^1 = H, 4-Me, 4-OMe, 4-Cl, 4-CF$_3$, 4-F
R^2 = Ph, ^{n}Pr, CH$_2$OBn
R_3 = H, CH$_2$OTBS

32: 24–72% yield

Scheme (12). Quinoline fused-lactones **32** obtained by intramolecular Povarov reaction.

In addition, the preparation of quinolines fused to pyrrol has also been described starting from aldehydes derived from amino acids. Thus, Raghunathan *et al.* prepared pyrroloquinolines **35** by reaction between *N*-prenylated aliphatic aldehydes **33** and anilines **8** catalyzed by Lewis acid [65, 66]. In this report, an efficient synthesis of pyrroloquinolines in excellent chemical yields and *trans*-selectivity (40:60 to 23:77 *cis/trans*) using 20 mol% InCl$_3$ in MeCN (Scheme **13**) is described. These authors also reported the synthesis of enantiopure *trans*-pyrroloquinolines **35** in 86-97% yield when using alkene-linked aldehydes derived from (*S*)-phenylalanine [67].

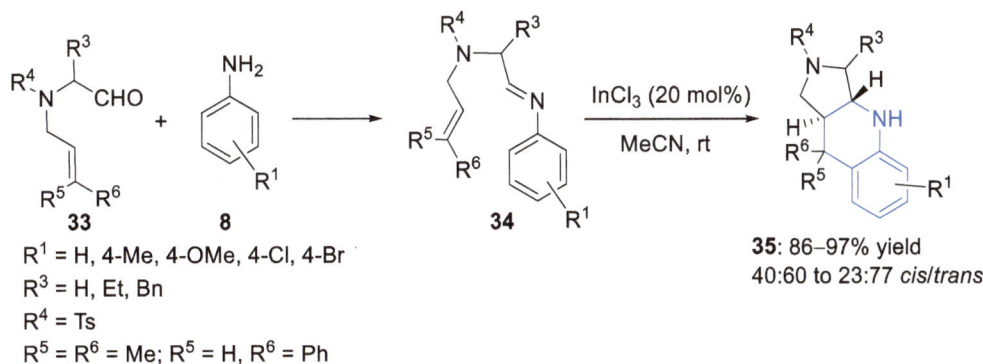

R[1] = H, 4-Me, 4-OMe, 4-Cl, 4-Br
R[3] = H, Et, Bn
R[4] = Ts
R[5] = R[6] = Me; R[5] = H, R[6] = Ph

35: 86–97% yield
40:60 to 23:77 *cis/trans*

Scheme (13).   InCl$_3$ catalyzed intramolecular Povarov reaction for the preparation of pyrrolo [4,3-*b*] quinolines.

When a strategically positioned aldehyde moiety tethered to an alkynyl group is used in the intramolecular Povarov reaction, more aromatized pyrrolo [3,4-*b*]quinolines can be prepared. Likewise, the reaction between *N*-propargyl aldehyde derived from α-amino acids and anilines carried out in the presence of BF$_3$·Et$_2$O in dry CH$_2$Cl$_2$ afforded a series of pyrrolo[3,4-*b*]quinolines in excellent yields [68].

On the other hand, the preparation of quinolines fused with lactam heterocycles has also been described. For this purpose, for example, cinnamoylaminoaldehyde derived from amino acids or *N*-benzylglyoxamides were used. First, the reaction of aldehyde **36** with aniline **8** using the soft Lewis acid, ytterbium triflate, yielded the thermodynamically more stable anti-*trans*-anti products **37** (Scheme **14**) [63]. When the reaction was performed with the *N*-benzyl-*N*-cinnamyl-2-oxoacetamide **39**, the corresponding 3*H*-pyrrolo[3,4-*b*]quinolin-3-one **38** was obtained with the *trans*-anti configuration as the major isomer (Scheme **14**) [63].

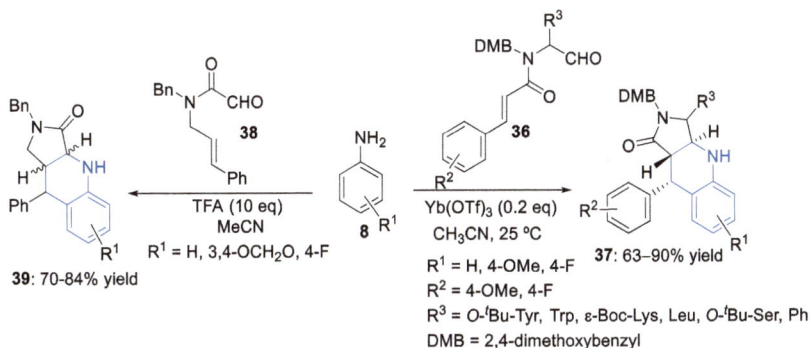

39: 70-84% yield

38

TFA (10 eq)
MeCN
R[1] = H, 3,4-OCH$_2$O, 4-F

36

8

Yb(OTf)$_3$ (0.2 eq)
CH$_3$CN, 25 °C
R[1] = H, 4-OMe, 4-F
R[2] = 4-OMe, 4-F
R[3] = O-[t]Bu-Tyr, Trp, ε-Boc-Lys, Leu, O-[t]Bu-Ser, Ph
DMB = 2,4-dimethoxybenzyl

37: 63-90% yield

Scheme (14).   Synthesis of quinolines fused with lactam heterocycles.

Likewise, fused derivatives between quinolines and pyridines were obtained by intramolecular Povarov reaction between aldehydes **40** and anilines **8** (Scheme **15**) [69]. For this transformation, various Lewis acids were used, *e.g.*, Yb(OTf)$_3$, InCl$_3$, Sc(OTf)$_3$, or BiCl$_3$, or even Brønsted acids such as TFA. However, in the presence of BiCl$_3$, the reaction proceeded efficiently to give the azaheterocycle benzo[*b*] [1, 6]naphthyridines **42** in good yields as a diastereomeric mixture of *cis*- and *trans*-adducts.

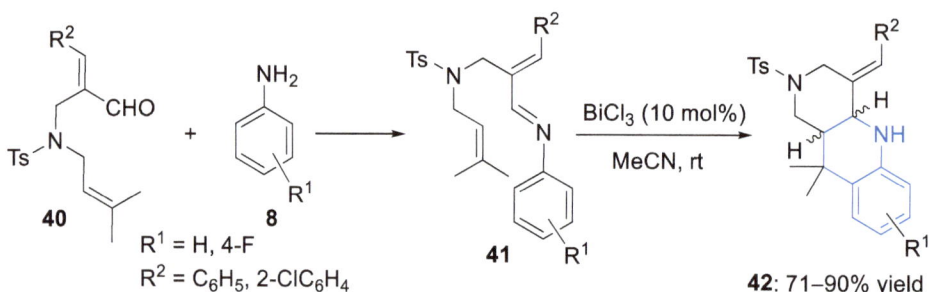

Scheme (15). BiCl$_3$-promoted intramolecular Povarov reaction.

Finally, one example reported the formation of tricyclic fused nitrogen pyrroloquinoline heterocycles from aminopyrrol **44**, a heterocyclic amine derived from pyrrol instead of the use of anilines. Vilches-Herrera *et al.* [70] prepared the tricyclic derivative **46** *via* cycloaddition of aldimine **45** obtained by condensation reaction between 2-amino pyrrole **44** and a non-aromatic aldehyde such as citronellal **43** (Scheme **16**). Water reaction is carried out under microwave irradiation at 200 °C and without catalyst, which meets all the requirements of sustainable chemistry.

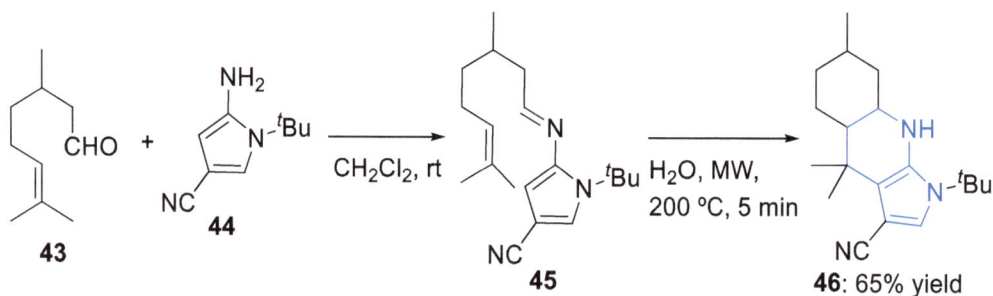

Scheme (16). Intramolecular Povarov reaction using 2-amino pyrrole under microwave conditions.

3. SYNTHESIS OF TETRACYCLIC FUSED HETEROCYCLES

3.1. Formation of Quinolines Fused with Carbocycles

Quinolines fused with carbocycles **50** have been obtained through a tandem allylation/intramolecular Povarov reaction [71]. The imine group in compound **49** acts as a directing group for the introduction of a pendant alkene, thus enabling a Lewis acid-catalyzed intramolecular Povarov reaction (Scheme 17). The *in situ* Povarov cyclization was catalyzed by silver (I), since other Lewis acid, *e.g.*, $BiCl_3$, $Sc(OTf)_2$ and $Zn(OTf)_2$, led to significant decomposition of the ketimine **49**. The reaction proceeds with a broad substrate scope, high regio- and *cis*-stereoselectivity, and 100% atom economy, affording polycyclic indeno[1,2-*b*]quinoline **50** as a single diastereoisomer.

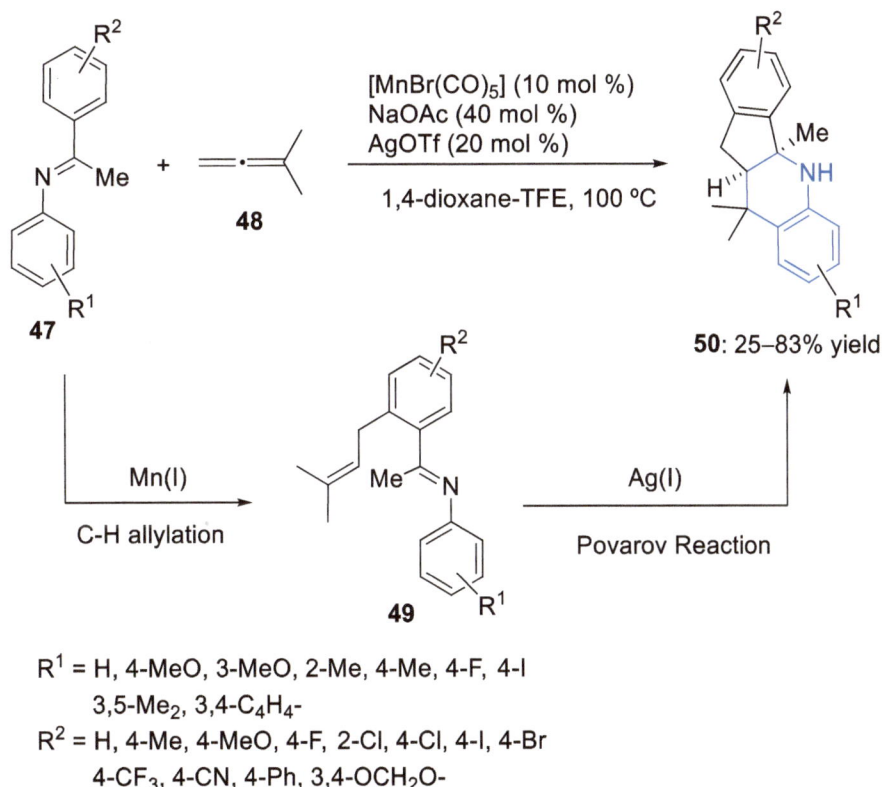

R^1 = H, 4-MeO, 3-MeO, 2-Me, 4-Me, 4-F, 4-I
         3,5-Me$_2$, 3,4-C$_4$H$_4$-
R^2 = H, 4-Me, 4-MeO, 4-F, 2-Cl, 4-Cl, 4-I, 4-Br
         4-CF$_3$, 4-CN, 4-Ph, 3,4-OCH$_2$O-

Scheme (17). Synthesis of polycyclic compounds through previous allylation followed by intramolecular Povarov reaction.

The ecofriendly synthesis of indeno[1,2-*b*]quinolines **55** *via* an intramolecular Povarov reaction of aldimines **52**, formed *in situ* by condensation of aromatic amines **8** with *o*-propargylbenzaldehydes **51**, was developed by Liu *et al.* (Scheme **18**) [72]. The intramolecular Povarov reaction between azadiene moiety and the alkyne group of **52** should afford intermediate **53**. After elimination of OR^2 group and subsequent double bond isomerization in **54** indenoquinolines **55** were attained (Scheme **18**).

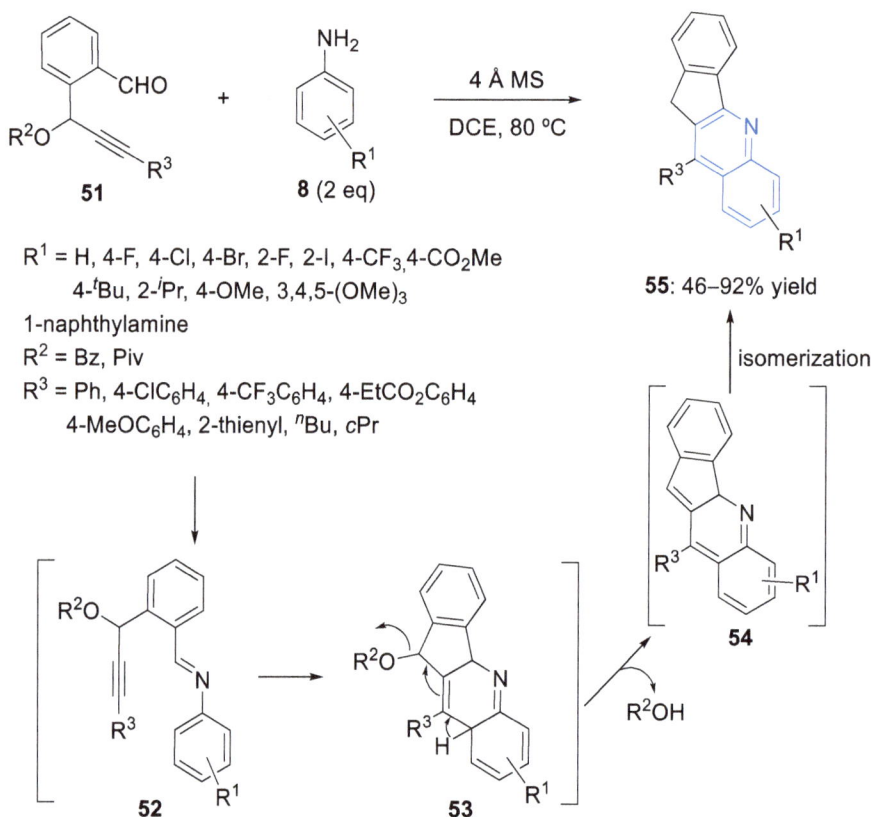

R^1 = H, 4-F, 4-Cl, 4-Br, 2-F, 2-I, 4-CF$_3$,4-CO$_2$Me
    4-^{t}Bu, 2-^{i}Pr, 4-OMe, 3,4,5-(OMe)$_3$
1-naphthylamine
R^2 = Bz, Piv
R^3 = Ph, 4-ClC$_6$H$_4$, 4-CF$_3$C$_6$H$_4$, 4-EtCO$_2$C$_6$H$_4$
    4-MeOC$_6$H$_4$, 2-thienyl, ^{n}Bu, cPr

55: 46–92% yield

Scheme (18). Synthesis of indeno[1,2-*b*]quinolines through the reaction of aromatic amines and *o*-propargylbenzaldehydes.

Intramolecular Lewis acid-catalyzed hetero-Diels-Alder reaction of functionalized aldimine **57**, generated *in situ* from the condensation of cyclopentanecarboxaldehyde derivative **56** with *o*-toluidine **8** (R^1 = 2-Me), gave to the formation of cyclopenta[*c*]acridine derivative **58** in a diastereoselective fashion (Scheme **19**) [73]. The formation of the *cis*- or *trans*-isomers was controlled depending on the Lewis acid used. However, diastereoselectivity is favored in the formation of the *trans*-isomers.

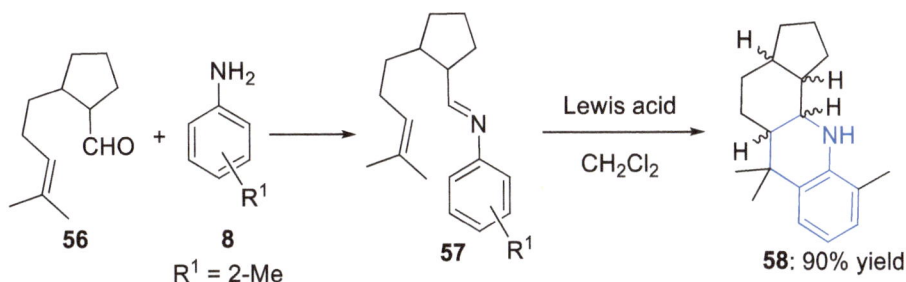

Scheme (19). Synthesis of cyclopenta[*c*]acridine derivatives.

Other quinolines fused with carbocycles **60** can be prepared by reaction of anilines **8** and trisubstituted cyclohexadienal **59** catalyzed by InCl₃ Lewis acid (Scheme **20**) [74]. The scope of substituted anilines **8** with a varied array of functional groups was studied, showing a strong reaction efficiency effect by electronic and steric factors. These octahydrobenzo[*c*]acridines **60** were obtained with high *trans*-selectivity as supported by the H-H coupling value of $J = 10$ Hz comparable to those described for *trans*-octahydroacridines.

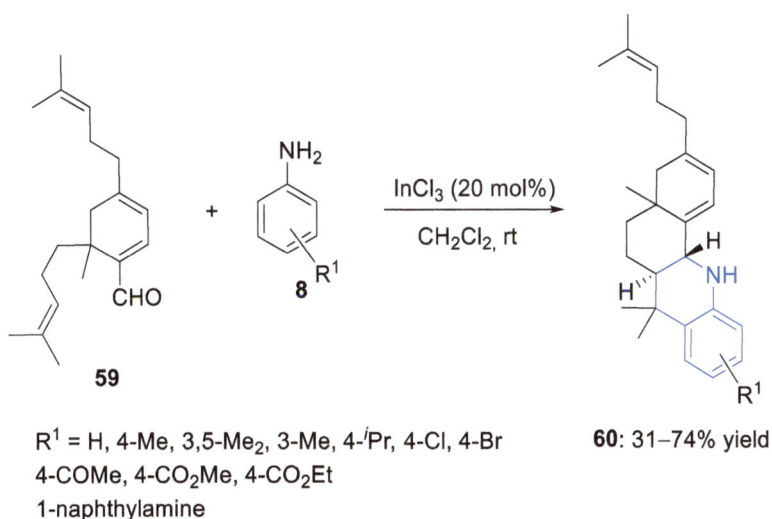

R¹ = H, 4-Me, 3,5-Me₂, 3-Me, 4-ⁱPr, 4-Cl, 4-Br
4-COMe, 4-CO₂Me, 4-CO₂Et
1-naphthylamine

60: 31–74% yield

Scheme (20). Preparation of octahydrobenzo[*c*]acridines.

cis-Annulated hexahydrobenzo[*c*]acridines **63** were achieved with selectivities up to 97:3 by bismuth(III) chloride-promoted intramolecular Povarov reaction. The corresponding aldimines **62** were obtained by condensation of aromatic amines **8** and 2-prenylated benzaldehyde **61** (Scheme **21**) [75]. This approach observed no effect on the reaction time or chemical yield when using amines bearing electron-donating or electron-withdrawing groups.

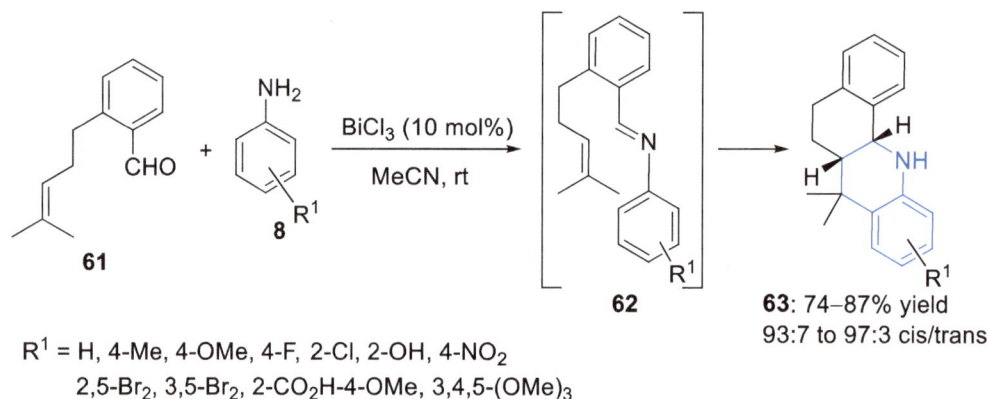

R^1 = H, 4-Me, 4-OMe, 4-F, 2-Cl, 2-OH, 4-NO$_2$
2,5-Br$_2$, 3,5-Br$_2$, 2-CO$_2$H-4-OMe, 3,4,5-(OMe)$_3$

Scheme (21). BiCl$_3$-promoted intramolecular Povarov reaction.

3.2. Formation of Quinolines Fused with Nitrogen-Containing Heterocycles

In the intramolecular Povarov reaction, other quinolines fused with nitrogen-containing heterocycles can be designed. For instance, *N*-cinnamyl pyrrole--carbaldehyde **64** reacted with aromatic amines **8** for the preparation of intermediate imines **65**, which after Lewis acid-catalyzed intramolecular cyclization afforded pyrrol annulated quinoline derivatives **66** in good yields (Scheme **22**) [76].

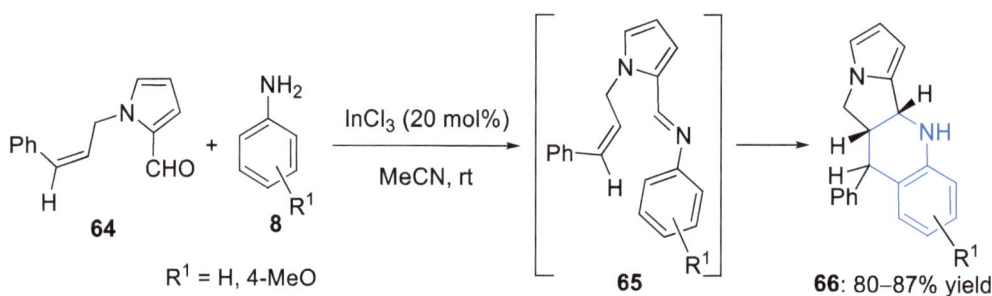

R^1 = H, 4-MeO

Scheme (22). InCl$_3$ promoted-intramolecular Povarov reaction for the preparation of pyrrol annulated quinolines.

The one-pot construction of substituted indolizino[1,2-*b*]quinolin-9(11*H*)-ones by the combination of visible-light-photoredox and Brønsted acid catalysis has been recently reported by Zhang's group [77]. The Brønsted acid promoted intramolecular Povarov cycloaddition reaction of *in situ* generated aldimines from pyridine derived 2-carbaldehydes **67** and anilines **8** affords tetrahydroquinolines **68** as intermediates (Scheme **23**). Subsequent visible-light-promoted dehydrogenation of **68**, using Ru(bpy)$_3$Cl$_2$·6H$_2$O or Ru(bpy)$_3$(PF$_6$)$_2$ as photocatalysts, gave to the formation of indolizino[1,2-*b*]quinolin-9(11*H*)-ones **69** in 49-96% yield. Electron-withdrawing groups (Cl, CN) at the *para*-position of the aniline ring showed a negative effect on the reaction yield. Conversely, both weakly and strongly electron-donating groups (Me, OMe, OBn) at this position undergo excellent yields of compound **69**.

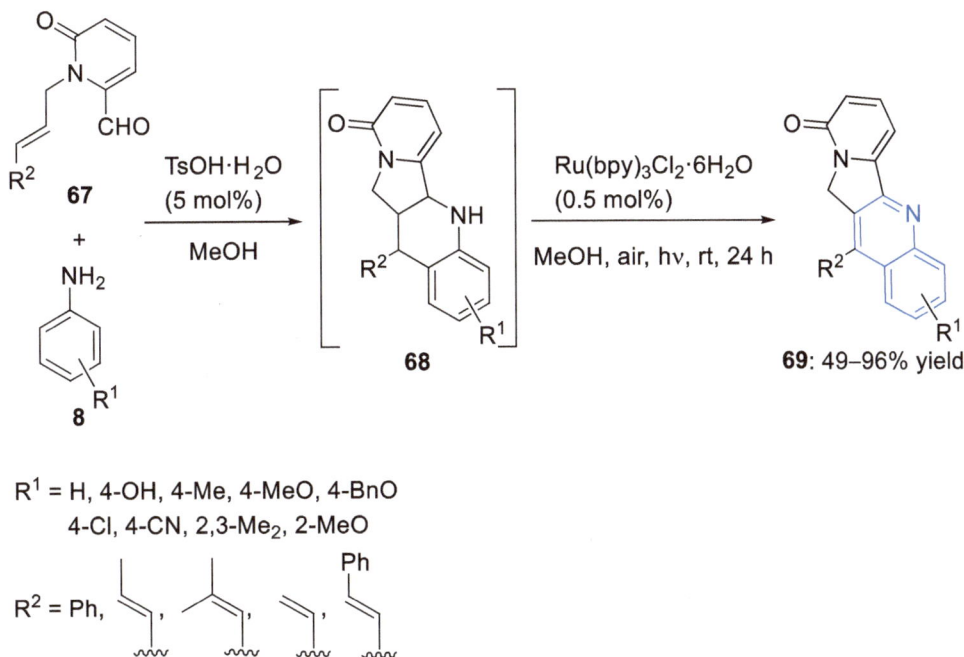

R^1 = H, 4-OH, 4-Me, 4-MeO, 4-BnO
4-Cl, 4-CN, 2,3-Me$_2$, 2-MeO

R^2 = Ph,

Scheme (23). Indolizino[1,2-*b*]quinolin-9(11*H*)-ones synthesized by the combination of visible-ligh--photoredox and Brønsted acid-catalyzed intramolecular Povarov reaction.

The intramolecular Povarov reaction has been applied for the preparation of alkaloids with fused heterocycles. For instance, Batey's group [78] reported the reaction using *N*-propargylic substituted aldehyde **70** for the synthesis of the pyrrolo[3,4-*b*]quinoline nucleus of camptothecin (Scheme **24**). Anilines **8** reacted with *N*-propargylpyridine derived aldehyde **70** in the presence of 10 mol% of Dy(OTf)$_3$ at room temperature to afford imine **71**. Conversely, quinoline **72** was directly obtained when the reaction was carried out at 50 °C. The formation of quinoline derivative **72** (R^1 = H) constitutes a formal synthesis of camptothecin, whereas compound **72** (R^1 = OMe) can be used as a precursor, which allows for the preparation of topotecan.

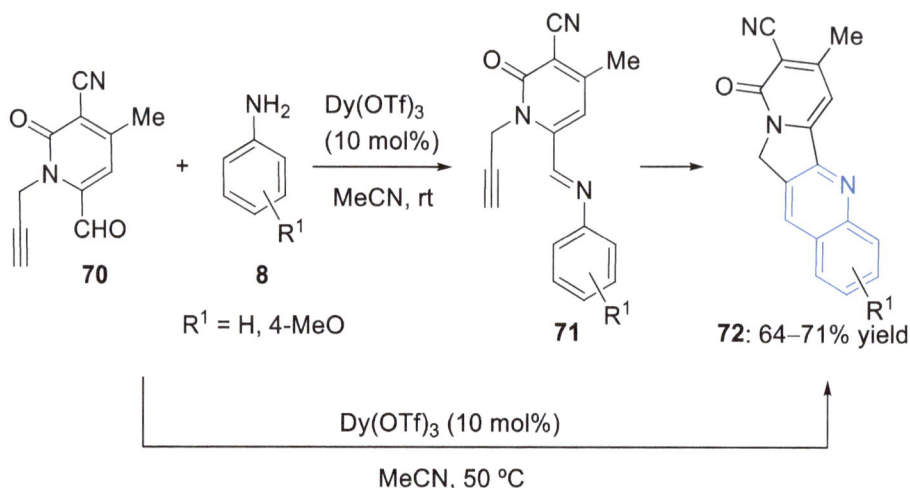

Scheme (24). Formal synthesis of camptothecin *via* intramolecular Povarov reaction.

Other quinolines fused with nitrogen-containing heterocycles, to obtain tetracyclic 1,7-naphthyridines, have been synthesized by intramolecular hetero-Diels-Alder reaction catalyzed by a Lewis or Brønsted acid by Laschat *et al*. [73, 79 - 81]. Hence, the condensation reaction of the L-proline-derived aldehydes **73** with *o*-toluidine **8** (R^1 = 2-Me) afforded aldimines **74**, whose subsequent intramolecular cyclization produced diastereoselectively the benzo[*b*]pyrrolo[1,2-*h*] [1, 7]naphthyridines **75** in good yields (Scheme **25**). This process exhibited a notable Lewis acid-dependent reversal of the diastereoselectivity. For instance, when using $BF_3 \cdot Et_2O$, $SnCl_4$, $FeCl_3$, *p*-TsOH, TFA, $AlCl_3$ and Et_2AlCl, the diastereoselectivity is favored in the formation of the *trans*-isomers, whereas when using $EtAlCl_2$, $MeAlCl_2$ and Me_2AlCl_2 the *cis*-stereoselectivity was observed.

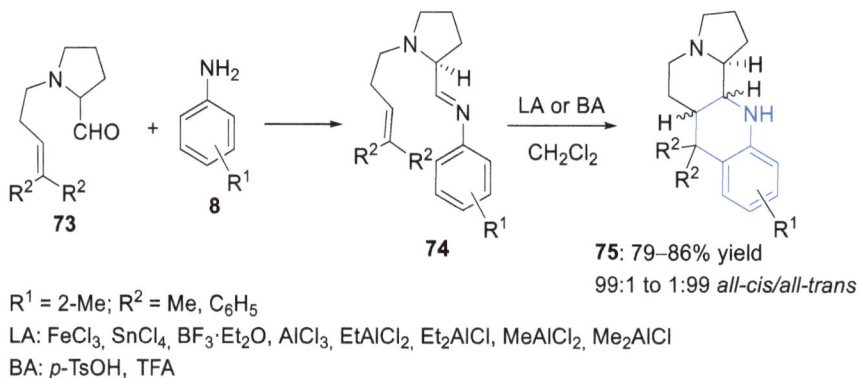

R^1 = 2-Me; R^2 = Me, C_6H_5
LA: $FeCl_3$, $SnCl_4$, $BF_3 \cdot Et_2O$, $AlCl_3$, $EtAlCl_2$, Et_2AlCl, $MeAlCl_2$, Me_2AlCl
BA: *p*-TsOH, TFA

Scheme (25). Benzo[*b*]pyrrolo[1,2-*h*] [1, 7]naphthyridine derivatives through intramolecular Povarov reaction.

Similar proline-derived aldehydes react with aromatic diamines, and after condensation, the *in situ* generated imines undergo intramolecular Povarov reactions affording bis(benzo[*b*]pyrrolo[1,2-*h*] [1, 7]naphthyridine)methane **76** or bis-1,7-naphthyridines **77** (Fig. **1**) [82].

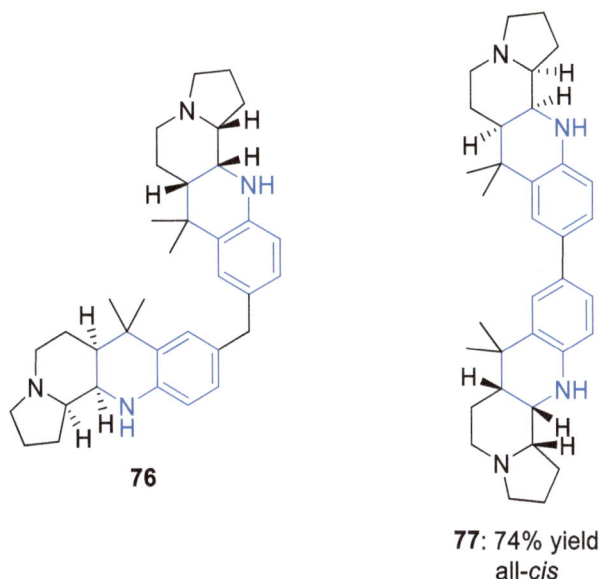

76

77: 74% yield
all-*cis*

Fig. (1). Fused bis(benzo[*b*]pyrrolo[1,2-*h*] [1, 7]naphthyridines).

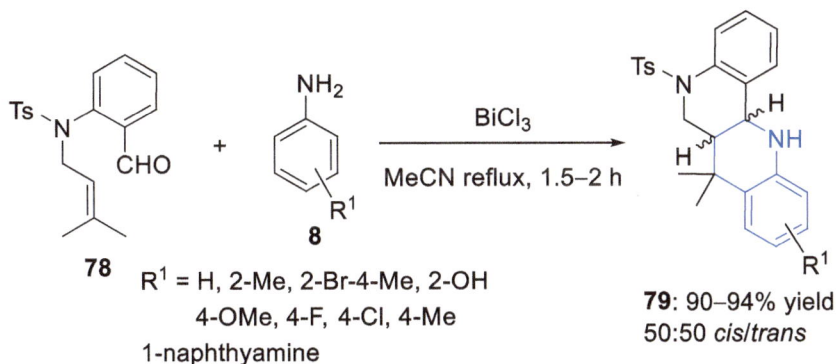

78 R¹ = H, 2-Me, 2-Br-4-Me, 2-OH
4-OMe, 4-F, 4-Cl, 4-Me
1-naphthyamine

79: 90–94% yield
50:50 *cis/trans*

Scheme (26). BiCl₃ Promoted intramolecular Povarov reaction for the synthesis of hexahydrodibenzo[*b,h*] [1, 6]naphthyridines.

The intramolecular [4+2] cycloaddition reaction might also be extended to the preparation of 1,6-naphthyridines. When aldimines derived from *o*-aminobenzaldehyde derivative **78** and aromatic amines **8** were used, the preparation of hexahydrodibenzo[*b,h*] [1, 6]naphthyridines **79** was reported by

Sabitha *et al.* [83] (Scheme **26**). The reaction was catalyzed by 10 mol% of BiCl$_3$ as Lewis acid yielding a mixture of *cis*- and *trans*-diastereoisomers of hexahydronaphthyridines **79** in a 1:1 ratio in excellent yields.

Likewise, dibenzo[*b,h*] [1, 6]naphthyridines may be obtained by a highly efficient synthesis between 2-(*N*-propargylamino)benzaldehydes **80** and arylamines **8** in the presence of CuBr$_2$ as catalyst [84] (Scheme **27**). The *in situ* generated electron-deficient aldimines **81** underwent an intramolecular inverse electron-demand hetero-Diels–Alder reaction, whose subsequent spontaneous dehydrogenation gave to the formation of 5,6-dihydrodibenzo[*b,h*] [1, 6]naphthyridines **82** in good yields. CuBr$_2$ was the most efficient catalyst tested in the reaction in terms of yield, since other copper halides, *e.g.*, CuCl, CuBr and CuBr$_2$, gave low yields of naphthyridines **82**. This protocol tolerated several substituents at the aromatic rings of anilines **8** and aldehydes **80**, as well as at the alkyne tethered to aldehyde, affording compounds **82** under mild reaction conditions.

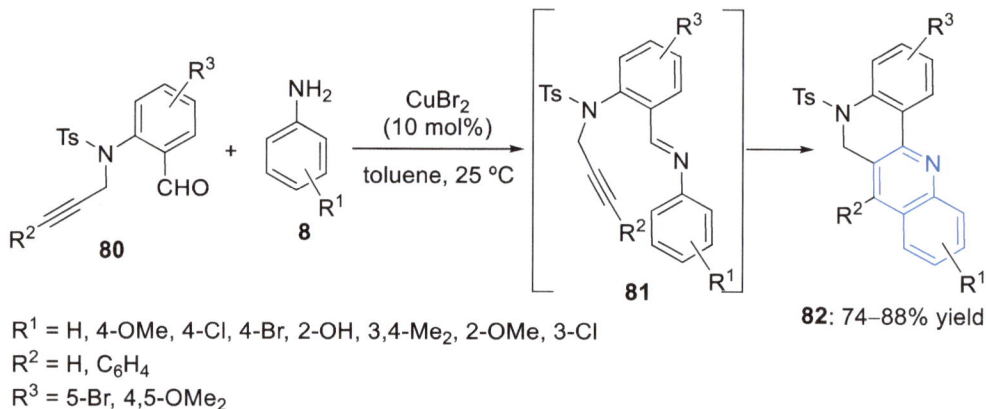

R^1 = H, 4-OMe, 4-Cl, 4-Br, 2-OH, 3,4-Me$_2$, 2-OMe, 3-Cl
R^2 = H, C$_6$H$_4$
R^3 = 5-Br, 4,5-OMe$_2$

Scheme (27). Copper catalyzed intramolecular Povarov reaction on the preparation of 5,6-dihydrodibenzo[*b,h*] [1, 6]naphthyridine derivatives.

Similar quinolines fused with nitrogenated heterocycles have been attained under mild reaction conditions involving the intramolecular aza-Diels-Alder reaction/aromatization cascade protocol [85]. This strategy catalyzed by 10 mol% of BF$_3$·Et$_2$O as Lewis acid affords hexahydronaphthyridine derivatives whose oxidative aromatization using Ru(bpy)$_3$(PF$_6$)$_2$ as photosensitizer in acetonitrile under an aerobic condition with the irradiation of visible light yields 5,6-dihydrodibenzo[*b,h*] [1, 6]naphthyridine derivatives in good yields.

In 2017 Masson's group developed for the first time an efficient asymmetric organocatalytic intramolecular Povarov reaction for an elegant preparation of optically active compounds fused with nitrogen-containing heterocyclic derivatives such as 1,6-naphthyridines [86]. (*R*)-3,3'-bis(2,4,6-triiso-propylphenyl)-1,1'-binaphthyl-2,2'-diyl hydrogen phosphate (**85**, TRIP), as a chiral phosphoric acid, catalyzes the enantioselective intramolecular reaction of functionalized aldehydes **83** and 2-hydroxy anilines **84** (Scheme **28**). Excellent yields and high diastereo-, and enantioselectivities ranging from 88 to 98% *ee* of tetrahydrodibenzo [1, 6]naphthyridin-6-ones **86** were obtained.

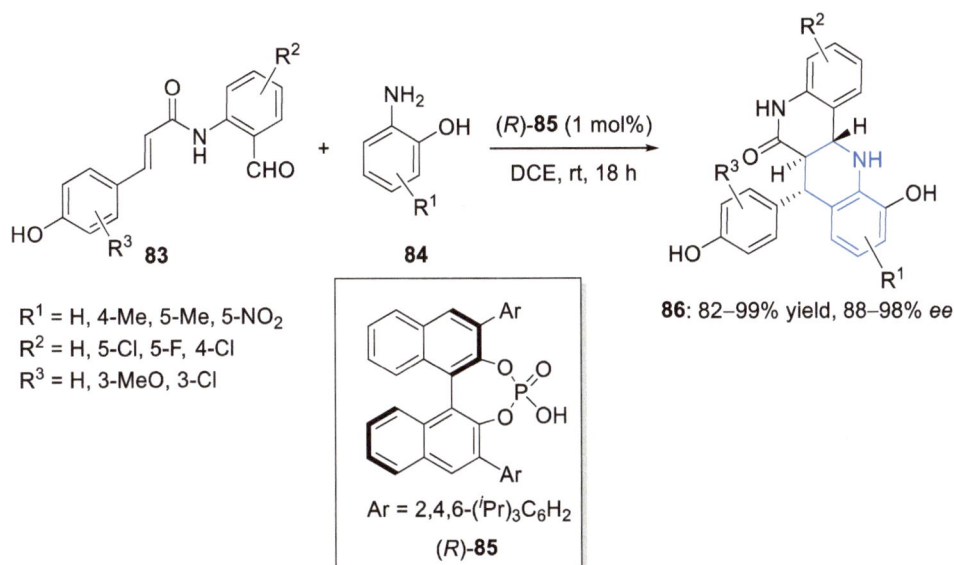

Scheme (28). Asymmetric intramolecular Povarov reaction for the preparation of tetrahydrodibenzo [1, 6]naphthyridin-6-ones.

These authors demonstrate recently the success and scope of the present methodology, applying the method to aldehydes **87** with other amino groups attached as a linker to the styrene moiety [87]. Precursors with an amine group as a linker were smoothly converted into the corresponding hexahydrodibenzo[*b,h*] [1, 6]naphthyridines **88-90**, in excellent yields, enantioselectivities ranging from 87 to 98% *ee*, and *trans,trans*-diastereoselectivities (Scheme **29**). Similar yields but slightly lower enantioselectivity have been attained for these naphthyridine derivatives varying the protecting group at the nitrogen atom of aldehydes **87**.

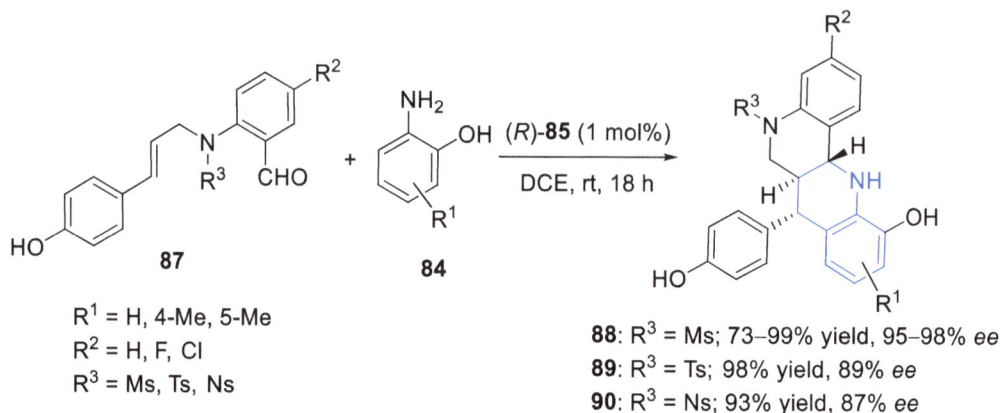

Scheme (29). Enantiomerically enrich hexahydrodibenzo[*b*,*h*] [1, 6]naphthyridines through asymmetric intramolecular Povarov reaction.

Bai's group [88] reported a series of libraries of tetracyclic 1,6-naphthyridines fused heterocycles. Imine intermediates formed *in situ* from the reaction of aromatic amines **8** and allylaminopyrimidine-5-carbaldehydes **91** undergo an intramolecular Povarov reaction (Scheme **30**). This synthetic strategy catalyzed by trifluoroacetic acid as Brønsted acid catalyst affords exclusively *cis*-benzopyrimido[4,5-*h*] [1, 6]naphthyridines **92** in good to excellent yields. Further transformations on the pyrimidine ring in **92** were demonstrated by oxidation of the phenylthio moiety to sulfoxide and subsequent nucleophilic substitution sequence.

3.3. Formation of Quinolines Fused with Heterocycles Containing Oxygen or Sulfur Atoms

In this section, we disclose the synthesis of quinoline ring fused with oxygenated heterocyclic compounds. For instance, polysubstituted tetrahydro-chromeno[4,3-*b*]quinolines **94** can be obtained by intramolecular [4+2] cycloaddition reaction of *in situ* generated imines derived from aniline derivatives **8** and *O*-allyl derived salicylaldehydes **93** (Scheme **31**). This process has been catalyzed in the presence of different Brønsted acids, such as trifluoroacetic acid (TFA) [89] and sulfamic acid [90] or in the presence of Lewis acids such as InCl$_3$ [91], BiCl$_3$ [92], Yb(OTf)$_3$ [89], lithium perchlorate in diethyl ether (LPDE) [93], triphenylphosphonium perchlorate (TPP) [94], and even in the presence of a recyclable ionic liquid as a reaction medium, [bmim]BF$_4$ [95]. Good to excellent yields of tetrahydrochromeno[4,3-*b*]quinolines **94** as a mixture of *cis*- and *trans*-diastereoisomers in a ratio of 50:50 were obtained in all cases.

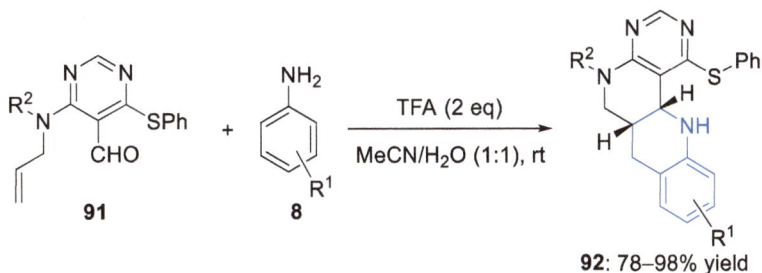

R¹ = H, 3-NO₂, 4-NO₂, 4-CO₂H, 4-F, 4-CO₂H, 2,4-F₂, 3-Cl,
     2-Me, 3-Me, 4-Me, 2,4-Me₂, 4-OH, 4-MeO, 3,4-(–OCH₂O–)
1-naphtylamine
R² = Me, Bn, CH₂CO₂Et

Scheme (30). Tetracyclic quinoline-fused heterocyclic libraries through intramolecular Povarov reaction.

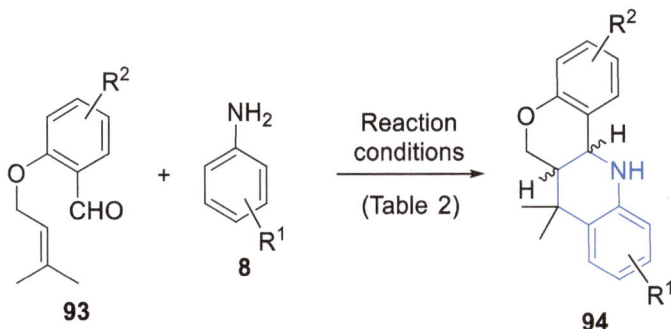

R¹ = H, 4-CO₂H, 2-CO₂H, 4-OMe, 4-F, 4-Cl, 4-Me,
     2-Me, 2-Cl, 2-Br, 4-NO₂, 4-CN, 3-OMe, 5-Cl
1- naphthylamine
R² = H, 5-OMe, 5-Br, 6-OMe, 3-Br, 3-OBn, 3-OMe, 5-Cl, 4-NO₂, 4-Br

Scheme (31). Synthesis of polysubstituted tetrahydrochromeno[4,3-*b*]quinolines using different catalysis conditions.

Other functionally and configurationally varied tetrahydrochromeno [4,3-*b*]quinolines have been prepared through a formal intramolecular aza-Diel--Alder cyclization. Starting materials include cinnamyl salicylaldehyde ethers as well as an electron-rich cinnamate ester and substituted anilines [63]. Modest to good yields of tetrahydroquinoline products have been obtained as single isomers (*trans*-configuration) by using trifluoroacetic acid (10 eq) in acetonitrile at 55 °C.

The selective preparation of *trans*-fused tetrahydrochromeno[4,3-*b*]quinolines has been also achieved through intramolecular Povarov reaction between nitrobenzenes and 2-(cinnamyloxy)benzaldehydes. The presence of iron as nitrobenzene reducing agent for the *in situ* generation of aniline and montmorillonite K10 as catalyst in aqueous citric acid at 80 °C afforded *trans*-tetrahydrochromeno[4,3-*b*]quinolines in good yields [96]. This domino process tolerates a large range of nitrobenzenes as well as 2-(cinnamyloxy)benzaldehydes.

An efficient asymmetric organocatalytic intramolecular Povarov reaction for the preparation of optically active chromeno fused quinoline derivatives was developed by Masson's group. The enantioselective intramolecular Povarov-type reaction of alkene-tethered aldehydes **95** and 2-hydroxy anilines **84** was catalyzed by a chiral phosphoric acid TRIP **85**, affording tetrahydrochromeno[4,3-*b*] quinolin-6-ones **96** in good to excellent chemical yields, high diastereo- (complete diastereoselectivity in favor of the *trans,trans*-tetrahydrochromeno[4,3-*b*] quinolin-6-one derivative **96** has been observed), and enantioselectivities ranging from 81 to 98% *ee* (Scheme **32**) [86]. This method was applied to a precursor with ether group as the linker between the aromatic aldehyde ring and the styrene group [87]. In this case, the reaction proceeded smoothly into the corresponding tetrahydro-6*H*-chromeno[4,3-*b*]quinolines in excellent yields (84-97%), enantioselectivities up to 99% *ee*, and *trans,trans*-diastereoselectivities.

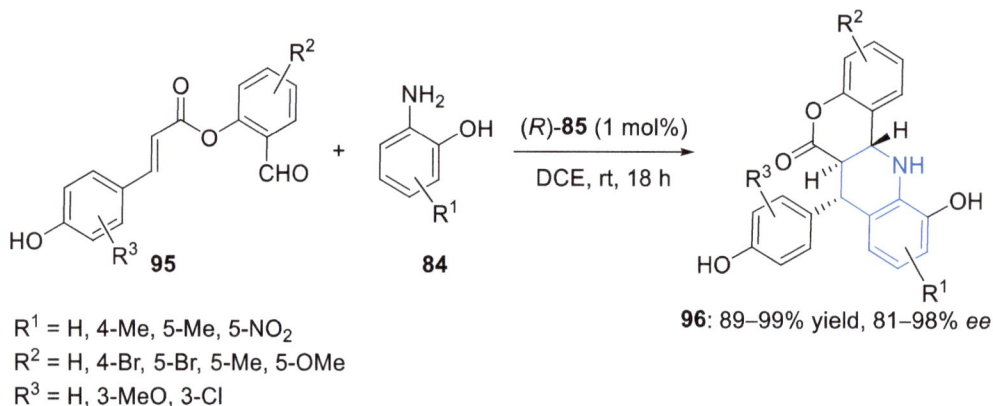

R^1 = H, 4-Me, 5-Me, 5-NO$_2$
R^2 = H, 4-Br, 5-Br, 5-Me, 5-OMe
R^3 = H, 3-MeO, 3-Cl

Scheme (32). Enantioselective organocatalytic intramolecular Povarov reaction for the synthesis of tetrahydrochromeno[4,3-*b*]quinolin-6-one derivatives.

The synthetic methodology for the preparation of tetrahydrochromeno[4,3-*b*]quinolines has been extended to the solid-phase synthesis using the AMEBA (acid-sensitive methoxy benzaldehyde polystyrene) resin [97] (Scheme **33**). Immobilized anilines **97** and salicylaldehydes **93** containing an electron-rich olefin substituent react in the presence of both 1% TFA and Yb(OTf)$_3$ catalysts to yield tetrahydroquinoline derivatives **98** as 50:50 mixtures of diastereoisomers.

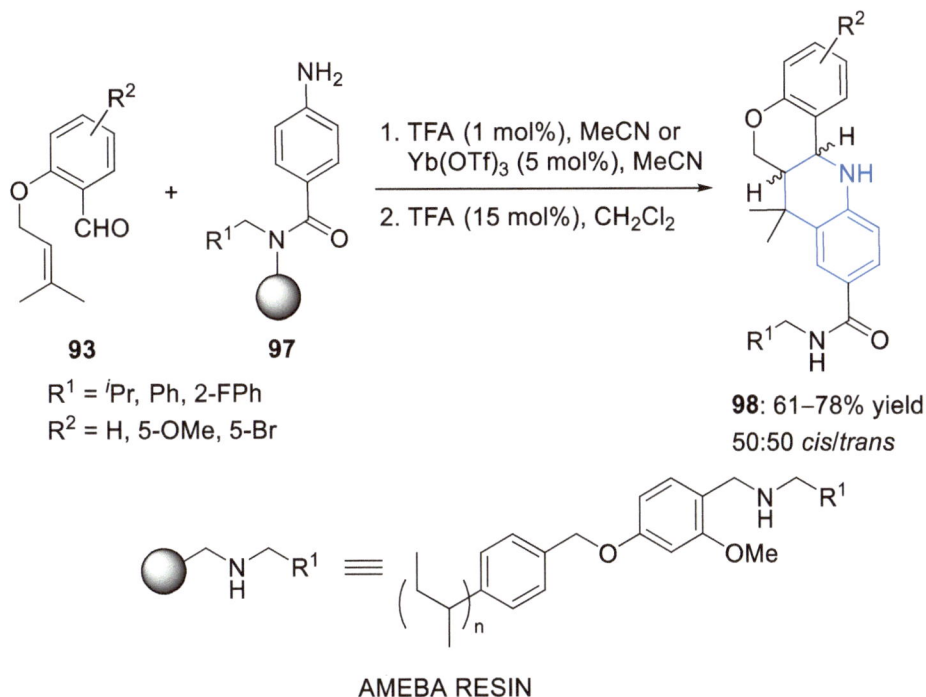

Scheme (33). Solid-phase synthesis of tetrahydrochromeno[4,3-*b*]quinolines.

Not only anilines but diamines have also been used in the intramolecular Povarov reaction for the preparation of tetrahydrochromeno[4,3-*b*]quinoline derivatives. When diamines with two aromatic systems linked through an oxygen atom or a methylene group react with salicylaldehydes **93** (Scheme 33) in the presence of TPP or [bmim]BF$_4$ as catalysts, bis-tetrahydrochromeno[4,3-*b*]quinolines as a mixture of three isomers *cis/cis*, *cis/trans* and *trans/trans* in good yields in nearly a 1:1:1 ratio have been obtained [94]. However, when using 4,4'-oxydianiline and [bmim]BF$_4$ as the catalyst, the product was achieved exclusively as *cis/trans*-bis-adduct under similar reaction conditions [95].

The chromenoquinoline skeleton has been synthesized under mild reaction conditions involving the intramolecular aza-Diels-Alder reaction/aromatization cascade using a strategy of combination of Lewis-acid-catalysis and visible-light photoredox [85]. The *in situ* generated imines **100**, derived from arylamines **8** and salicylaldehydes **99** bearing an alkene-tethered partner, undertake an intramolecular aza-Diels-Alder cycloaddition promoted by 10 mol% of BF$_3$·Et$_2$O as Lewis acid (Scheme **34**). Oxidative aromatization of the obtained tetrahydrochromeno[4,3-*b*]quinolines **101** using Ru(bpy)$_3$(PF$_6$)$_2$ as photosensitizer in acetonitrile under aerobic conditions with the irradiation of visible light affords chromeno[4,3-*b*]quinolines **102**.

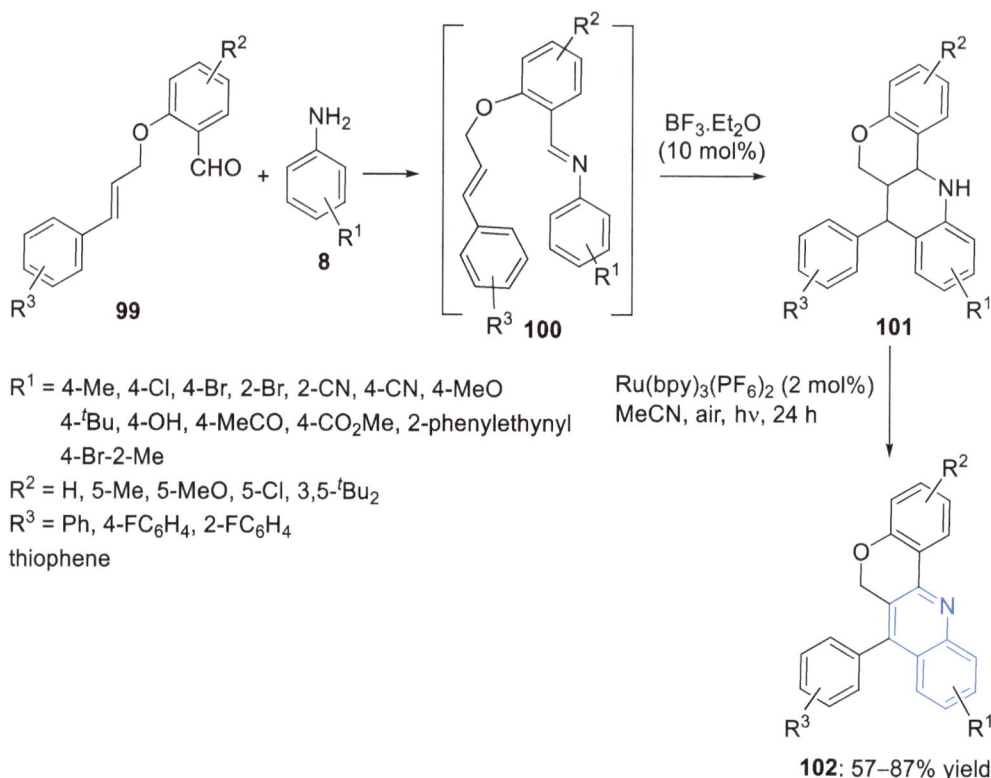

R^1 = 4-Me, 4-Cl, 4-Br, 2-Br, 2-CN, 4-CN, 4-MeO
      4-^{t}Bu, 4-OH, 4-MeCO, 4-CO$_2$Me, 2-phenylethynyl
      4-Br-2-Me
R^2 = H, 5-Me, 5-MeO, 5-Cl, 3,5-^{t}Bu$_2$
R^3 = Ph, 4-FC$_6$H$_4$, 2-FC$_6$H$_4$
thiophene

102: 57–87% yield

Scheme (34).   Preparation of chromeno[4,3-*b*]quinolines *via* intramolecular aza-Diels-Alder reaction/aromatization cascade.

Terminal alkyne C-H bond activation of *O*-propargyl substituted salicylaldehyde ethers is another strategy for the preparation of chromenoquinoline derivatives. In this way, Nagarajan's group [98] described a straightforward approach to chromenoquinolines using a mixture of copper (I) iodide and lanthanum triflate as an efficient catalyst in the intramolecular Povarov reaction. Thus, intermediate aldimines **104**, derived from the reaction of aromatic amines **8** with *O*-propargylated salicyladehydes **103** undergo an intramolecular Povarov reaction affording good yields of 6*H*-chromeno[4,3-*b*]quinolines **105** (Scheme **35**). Substitution at *O*-propargylated salicyladehyde ring seems not to affect the reaction. However, aromatic amines **8** with ring-activating groups in *ortho*, *meta* or *para*-positions participated in this reaction giving chromenoquinolines **105** with remarkably comparable yields (Scheme **35**). Aromatic amines **8** with electron-withdrawing groups did not afford the expected products **105**.

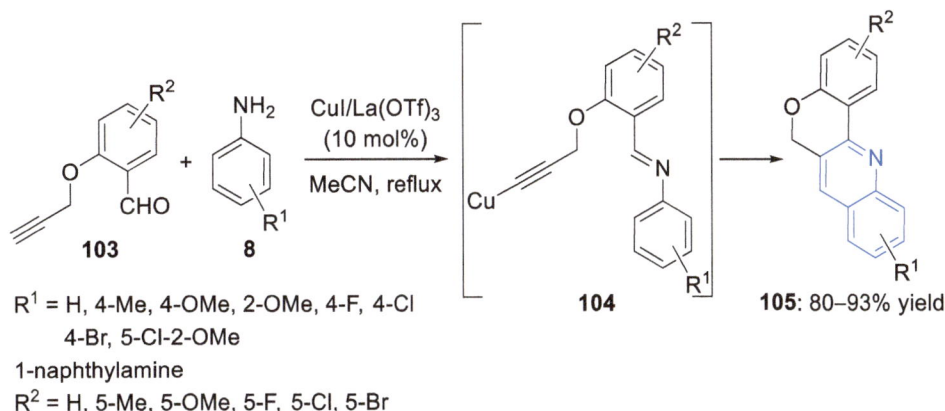

R¹ = H, 4-Me, 4-OMe, 2-OMe, 4-F, 4-Cl
    4-Br, 5-Cl-2-OMe
1-naphthylamine
R² = H, 5-Me, 5-OMe, 5-F, 5-Cl, 5-Br

Scheme (35). Efficient intramolecular Povarov reaction of 6*H*-chromeno[4,3-*b*]quinolines promoted by CuI/La(OTf)₃.

More recently, a green and simple intramolecular domino condensation/aza-Diel--Alder reaction between electron-rich anilines and *O*-propargylated salicylaldehydes using CuI as a catalyst has been used to obtain 6*H*-chromeno[4,3-*b*]quinolines in chemical yields ranging from 75-83% [99]. The best yield was obtained using only highly electron-rich anilines. The main advantages of this method are the simplicity of the starting materials, good yields of the products, and the use of green, cheap and nontoxic solvents.

Chromenoquinolines have also been developed through Cu-catalyzed aza-Diels–Alder reaction/halogenation cascade reaction. Cu₂O worked both as a Lewis acid and transition-metal catalyst in the aza-Diels–Alder reaction and halogenation reaction, respectively. Chlorination or bromination was achieved using chloranil or bromanil as halogen sources. Both perform dual functions, that is, as a halogen source and oxidant [100].

Quinolines fused with sulfur-containing heterocyclic compounds will be mentioned in this section. The preparation of hexahydrothiopyranoquinoline derivatives **108** can be performed through an intramolecular imino-Diels-Alder reaction of *N*-aryl imines **107** generated *in situ* from anilines **8** and *S*-allyl-1*H*-pyrazole-4-carbaldehyde derivatives **106** (Scheme **36**) [101]. This process is highly diastereoselective by the exclusive isolation of the *cis*-cycloadduct by using 5 mol% of BiCl₃ as the catalyst of this transformation. The synthesis of tetracyclic compounds **108** has also been reported by Raghunathan *et al.* [102] *via* InCl₃-promoted intramolecular Povarov reaction of aldehydes **106** with substituted anilines **8** (Scheme **36**). Cycloadducts **108** were attained with diastereoselectivities higher than 94:6 in favor of the *cis*-quinoline derivatives and chemical yields ranging from 85 to 96%.

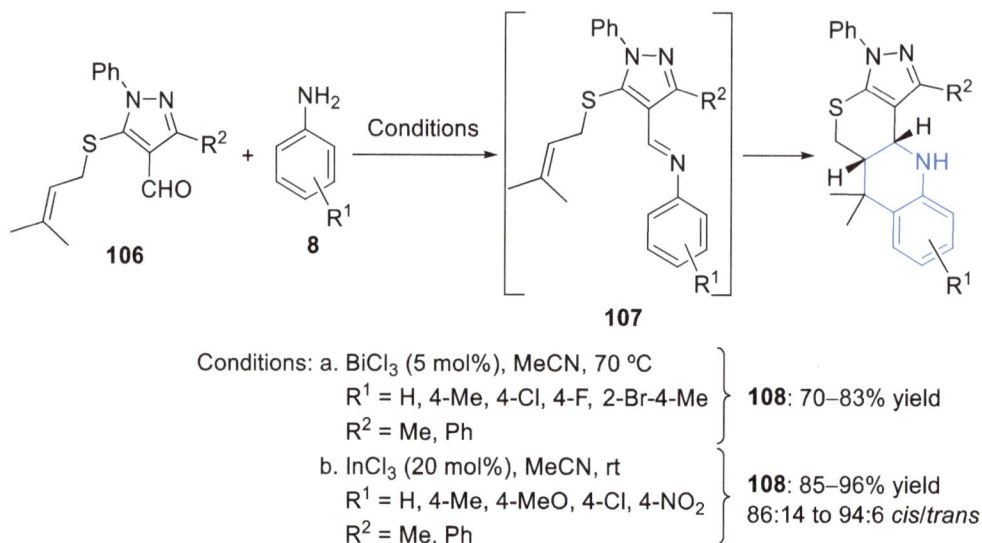

Scheme (36). Quinoline annulated sulfur heterocycles by intramolecular Povarov reaction.

Bis-anilines also react with *S*-allyl-1*H*-pyrazole-4-carbaldehyde derivatives **106** affording intermediate aldimines, whose subsequent bis-intramolecular Povarov reaction, catalyzed by 40 mol% of InCl$_3$, underwent bis-tetrahydropyrazol--thiopyrano[4,3-*b*]quinoline derivatives **109** as a mixture of three inseparable isomers *cis*/*cis*, *cis*/*trans*, and *trans*/*trans* in favor of the *cis*/*cis*-isomer (Fig. **2**) [102].

Masson's methodology applied to the asymmetric synthesis of chromeno fused quinoline as well as dibenzo fused naphthyridine derivatives has been widened to the preparation of optically active tetrahydrothiochromeno[4,3-*b*]quinolin-6-ones **111** (Scheme **37**) [87]. These new fused nitrogen and sulfur-containing tetraheterocycles were attained in excellent diastereo- and enantioselectivities (93-96% *ee*). Since the reaction conversion is not complete, a slight increase of the catalyst loading to 2 mol% was accomplished to afford the corresponding cycloadducts in excellent yields (91-97%).

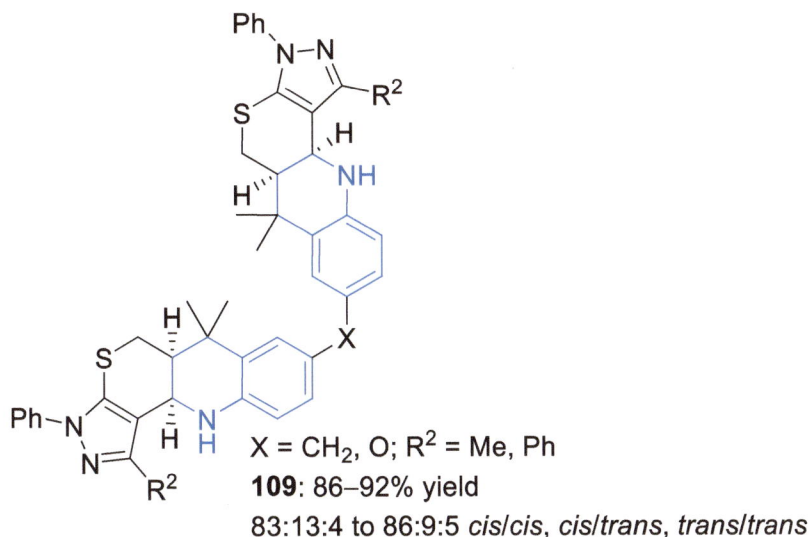

Fig. (2). Bis-quinoline annulated sulfur heterocycles by bis-intramolecular Povarov reaction catalyzed by InCl₃.

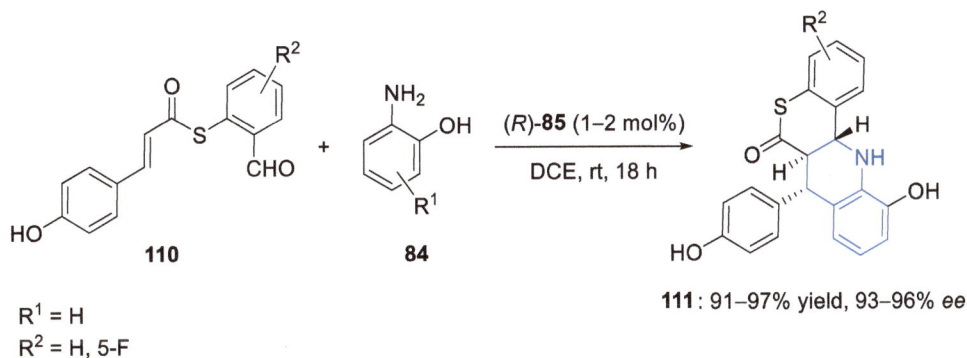

Scheme (37). Enantioselective organocatalytic intramolecular Povarov reaction for the preparation of tetrahydrothiochromeno[4,3-*b*]quinolin-6-ones.

The 7-halogenated thiochromenoquinolines **113** have been obtained *via* Cu-catalyzed aza-Diels–Alder reaction/halogenation cascade [100]. Cu₂O functioned both as a Lewis acid in the aza-Diels–Alder reaction and transition-metal catalyst in the halogenation reaction. Chloranil and bromanil used in chlorination and bromination reactions, respectively, also worked as halogen source and oxidant. The 7-halogenated thiochromenoquinolines **113** have been obtained in moderate yields, however, this method is highly useful in organic synthesis because of mild reaction conditions and experimental simplicity (Scheme **38**).

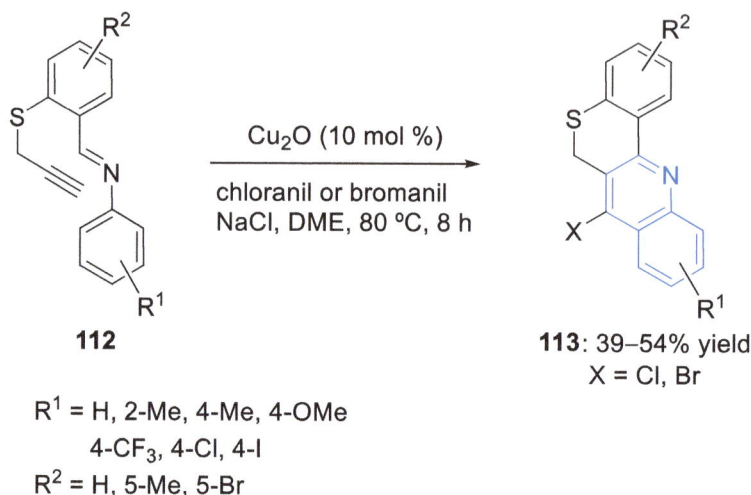

112

R^1 = H, 2-Me, 4-Me, 4-OMe
    4-CF$_3$, 4-Cl, 4-I
R^2 = H, 5-Me, 5-Br

Scheme (38). Synthesis of halogenated thiochromenoquinolines *via* Cu-catalyzed aza-Diels–Alder reaction/ halogenation cascade.

3.4. Formation of Naphthyridines Fused with Heterocycles Containing Nitrogen Atoms

Hybrid substituted quinolino[4,3-*b*] [1, 5]naphthyridine derivatives **119** and **121** and quinolino[4,3-*b*] [1, 5]naphthyridin-6(5*H*)-one derivatives **120** and **122** were achieved by an intramolecular Povarov [4+2] cycloaddition reaction using BF$_3$·Et$_2$O as Lewis acid when 3-aminopyridines are used instead of aromatic amines (Scheme **39**) [103]. 5-Tosyl functionalized aldehydes **114** (X = CH$_2$) or **115** (X = CO) reacted with 3-aminopyridines **116** affording aldimines **117** or **118**. Intramolecular cyclization in refluxing chloroform promoted by BF$_3$·Et$_2$O and subsequent prototropic tautomerization gave the corresponding cycloadducts **119** or **120** regio- and stereospecifically. 4 Equivalents of MnO$_2$ in toluene at 111 °C for 48 h was necessary for the dehydrogenation reaction of cycloadducts **119** or **120**, yielding dihydroquinolino[4,3-*b*] [1, 5]naphthyridines **121** and quino-lino[4,3-*b*] [1, 5]naphthyridin-6(5*H*)-one **122** in almost quantitative yield (Scheme **39**). *N*-Tosyl deprotection in **121** or **122** was accomplished using magnesium under acidic conditions. All these new 1,5-naphthyridine derivatives have been tested as topoisomerase I inhibitors. Likewise, the synthesis of substituted dihydroquinolino[4,3-*b*] [1, 5]naphthyridines **121** may also be obtained in good to excellent yields by using *N*-propargyl substituted aldehyde derivative. The corresponding aldimines undergo an intramolecular cycloaddition in the presence of BF$_3$·Et$_2$O to afford dihydroquinolino[4,3-*b*] [1, 5]naphthyridines **121** [103].

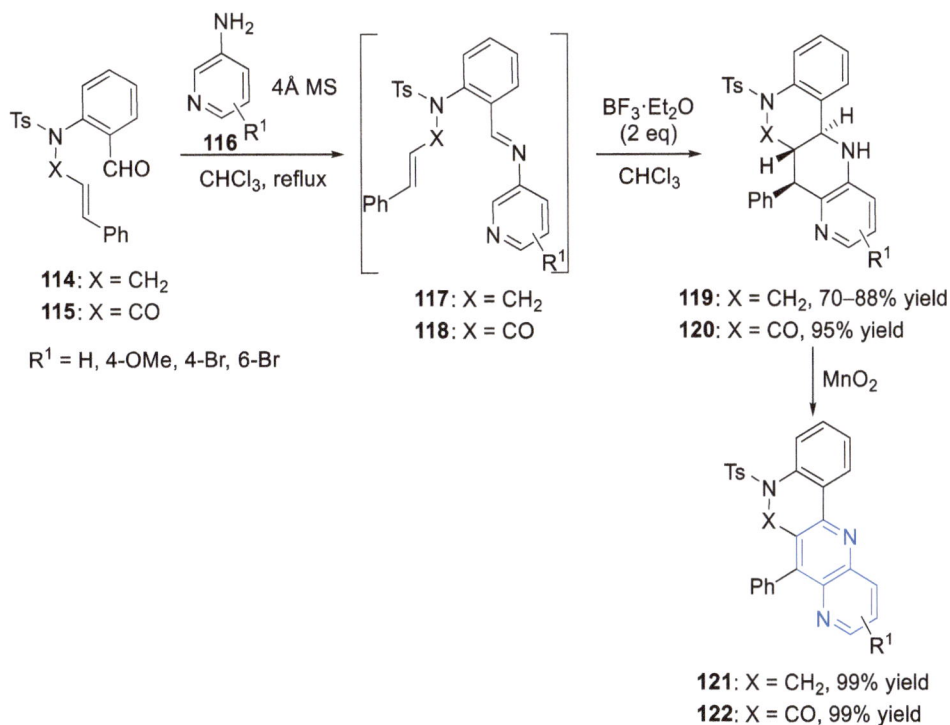

Scheme (39). Intramolecular Povarov reaction in the preparation of dihydroquinolino[4,3-*b*] [1, 5]naphthyridines and quinolino[4,3-*b*] [1, 5]naphthyridin-6(5*H*)-one.

A similar strategy has been applied to the synthesis of dihydroquinolino[4,3-*b*] [1, 8]naphthyridines **125** and **126** starting from 2-aminopyridines **123** [104]. As before, the intermediate imines **124** generated from the condensation of **123** with 5-tosyl functionalized aldehydes **114** suffer a Lewis catalyzed intramolecular Povarov reaction followed by prototropic tautomerization yielding hexahydroquinolino[4,3-*b*] [1, 8]naphthyridine derivatives **125** (Scheme 40). Aromatization step was accomplished using MnO_2 in refluxing toluene for 48h. The final dihydroquinolino[4,3-*b*] [1, 8]naphthyridines **126** were attained in 56-99% yield, whose *N*-tosyl deprotection was accomplished with magnesium under acidic conditions. Likewise, dihydroquinolino[4,3-*b*] [1, 8]naphthyridines **126**, may also be directly obtained in 81-86% yield when 2-aminopyridines **123** and the corresponding *N*-propargyl substituted aldehyde was used in the $BF_3 \cdot Et_2O$ catalyzed intramolecular cyclization [104].

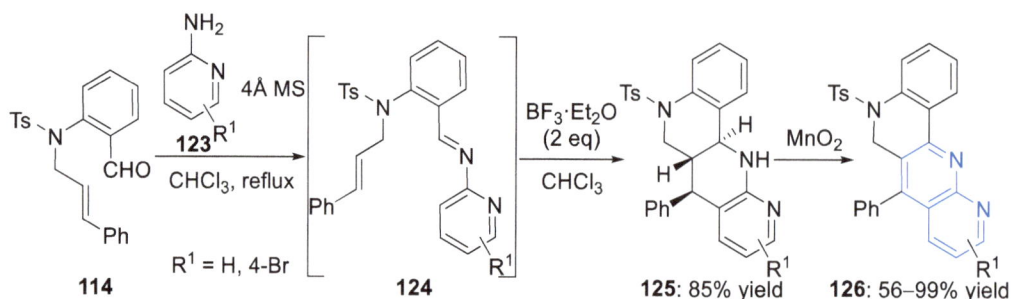

Scheme (40). Intramolecular Povarov reaction in the synthesis of dihydroquinolino[4,3-*b*] [1, 8] naphthyridines.

3.5. Formation of Naphthyridines Fused with Heterocycles Containing Oxygen Atoms

1,5-Naphthyridines annulated with oxygen-containing heterocycles have also been described [105]. The synthesis entails an intramolecular Povarov reaction promoted by $BF_3 \cdot Et_2O$ as Lewis acid of functionalized aldimines **129** or **130** achieved by the condensation of 3-aminopyridine derivatives **116** with aldehydes **127** or **128** and subsequent prototropic tautomerization. This stereoselective protocol allows the generation of three stereogenic centers with the formation of *endo*-tetrahydrochromeno[4,3-*b*] [1, 5]naphthyridines **131** or **132** (Scheme **41**). Dehydrogenation reaction of derivatives **131** or **132** takes place using 1 equivalent of DDQ as an oxidant in toluene at 120 °C under microwave irradiation for 2h, affording the corresponding tetracyclic chromeno[4,3-*b*] [1, 5]naphthyridines **133** or chromeno[4,3-*b*] [1, 5]naphthyridin-6-ones **134** in good to excellent yields (Scheme **41**). It is noteworthy that using propargyl oxyaldehydes as carbonyl components entails straightforward access to the corresponding aromatic 1,5-naphthyridine derivatives **133** in 71-85% yield.

Following the same strategy, 1,8-naphthyridine derivatives fused with other oxygen-containing heterocycles, such as chromene or chromen-2-one have been prepared. In this case, aldimines **137** or **138** derived from the condensation of 2-aminopyridine derivatives **123** with aldehydes containing a double carbon-carbon bond in *ortho* position **135** or **136** (Scheme **42**) [104]. As reported before, this strategy is stereoselective affording exclusively *trans*-tetrahydrochromeno[4,3-*b*] [1, 8]naphthyridines **139** or *trans*-tetrahydrochromeno[4,3-*b*] [1, 5]naphthyridin-6-ones **140**. The dehydrogenation step was performed with 4 equivalents of MnO_2 in toluene at 111 °C for 48 h to yield aromatic 1,5-naphthyridine derivatives **141** or **142** in quantitative yield (Scheme **42**). The preparation of chromeno[4,3-*b*] [1, 8]naphthyridines **141** in a single step and without the need to previously obtain their corresponding dehydrogenated adducts may also be performed. When *O*-

propargylated aldehyde and 2-aminopyridine derivatives **123** were used in the presence of 2 equivalents of $BF_3 \cdot Et_2O$, 1,8-naphthyridine derivatives **141** were isolated in 72-79% chemical yield.

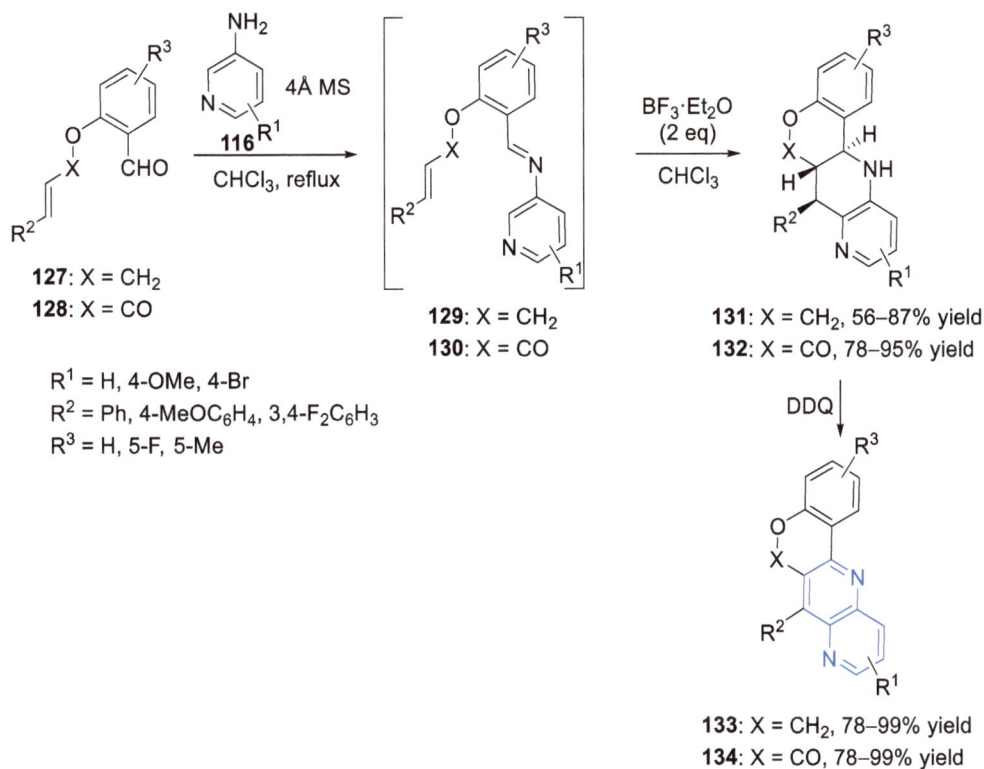

127: X = CH$_2$
128: X = CO

129: X = CH$_2$
130: X = CO

131: X = CH$_2$, 56–87% yield
132: X = CO, 78–95% yield

R^1 = H, 4-OMe, 4-Br
R^2 = Ph, 4-MeOC$_6$H$_4$, 3,4-F$_2$C$_6$H$_3$
R^3 = H, 5-F, 5-Me

133: X = CH$_2$, 78–99% yield
134: X = CO, 78–99% yield

Scheme (41). Intramolecular Povarov cyclization in the preparation of new chromeno[4,3-*b*] [1, 5]naphthyridine and chromeno[4,3-*b*] [1, 5]naphthyridin-6-one derivatives.

3.6. Formation of other Tetracyclic Heterocycles

Intermediate imines **144**, derived from the condensation of 2-amino pyrrole derivatives **44** and commercially available 2-hydroxybenzaldehyde **143** and subsequent *O*-alkylation reaction, underwent a catalyst-free intramolecular [4+2] cycloaddition reaction conducted in water under microwave irradiation (Scheme 43) [70]. This process is highly diastereoselective, yielding *trans*-tetra-hydropyrrolo[2,3-*b*]pyridine annulated hetereocycles **145**. Although the *cis*-isomer can also be isolated using a non-polar solvent such as *p*-xylene.

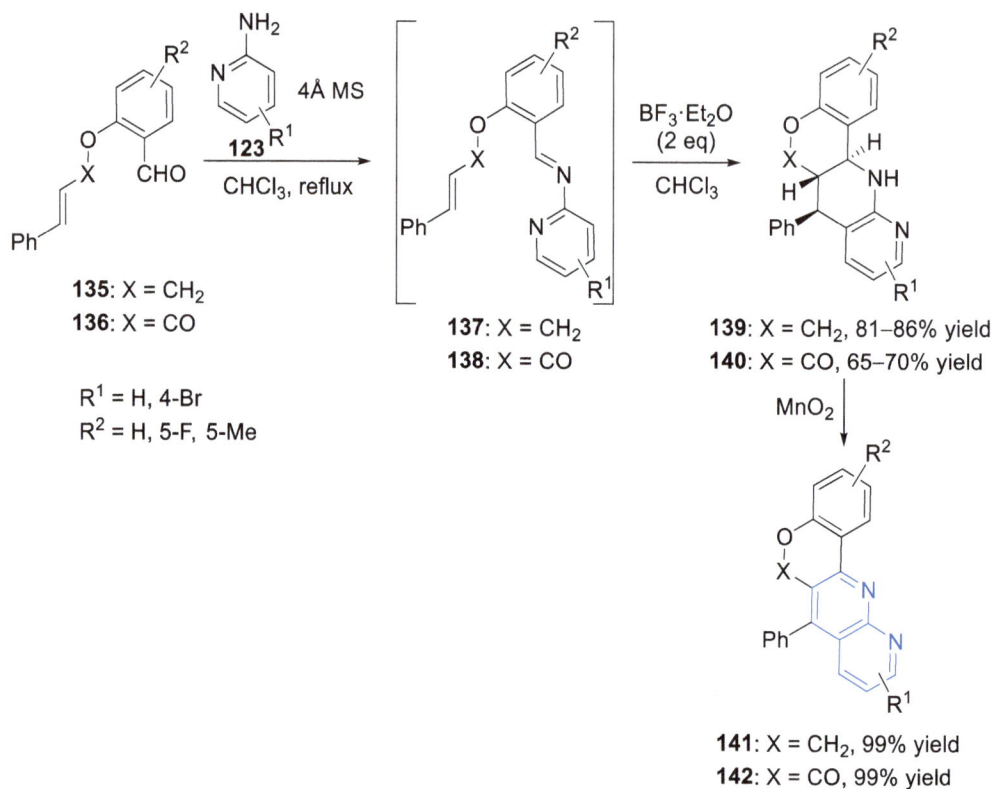

Scheme (42). Intramolecular Povarov cyclization in the preparation of new chromeno[4,3-*b*] [1, 8] naphthyridine and chromeno[4,3-*b*] [1, 8]naphthyridin-6-one derivatives.

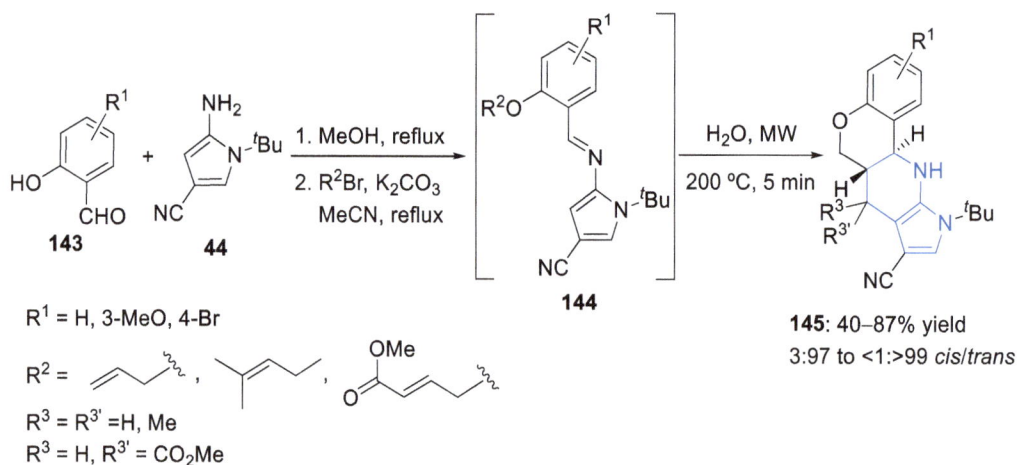

Scheme (43). Tetrahydropyrrolo[2,3-*b*]pyridine annulated hetereocycles *via* intramolecular Povarov reaction.

When propargyl bromide was used as an alkylating reagent in the *O*-alkylation reaction, intermediate imine **146** underwent an intramolecular Povarov reaction affording pyrrolo[2,3-*b*]pyridine annulated hetereocycles **147** in good yields *via* spontaneous aromatization of the corresponding cycloadducts (Scheme **44**) [70].

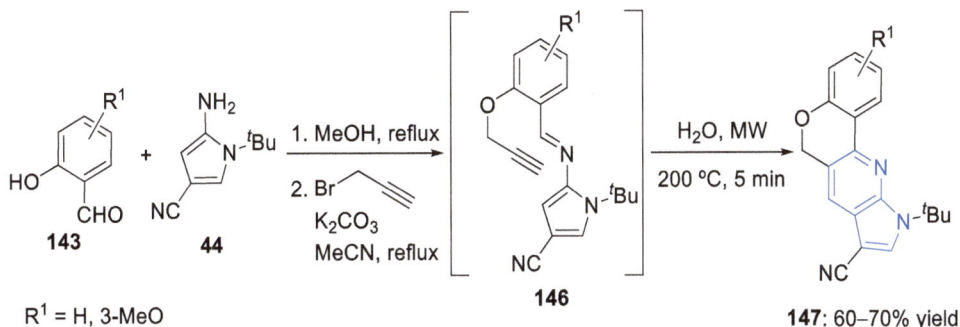

Scheme (44). Intramolecular Povarov reaction using 2-amino pyrrole and alkyne-tethered aldehydes.

Vilches-Herrera *et al.* [70] applied the same methodology to the preparation of tetrahydropyrazolopyridine annulated heterocycles. In this case, the intramolecular aza-Diels-Alder cycloaddition took place using 2-aminopyrazole derivatives **148** and the tetracyclic tetrahydropyrazolo[3,4-*b*]pyridines **150** were attained in high diastereoselectivity and good yields (Scheme **45**). This green process is catalyst-free, carried out in the water and under microwave irradiation, which fulfills all the requirements for sustainable chemistry. Concerning the stereoselectivity, this is highly solvent-dependent. Only the *trans*-isomer is attained when the reaction is performed in water. Conversely, the *cis*-isomer is produced using *p*-xylene as a non-polar solvent. As reported previously, for the preparation of pyrrolo[2,3-*b*]pyridine annulated hetereocycles **147**, when propargyl bromide was used as alkylating reagent, pyrazolo[3,4-*b*]pyridine annulated heterocycles can be obtained after intramolecular Povarov reaction of the intermediate imine.

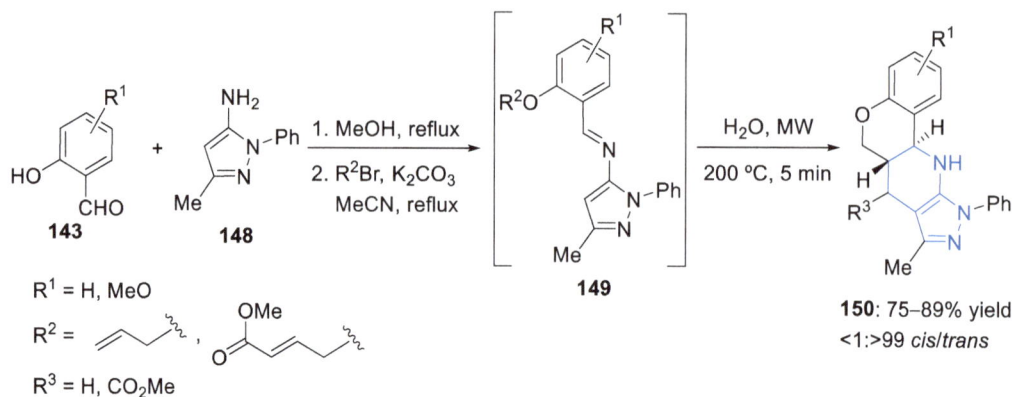

Scheme (45). Intramolecular Povarov reaction for the synthesis of tetracyclic tetrahydropyrazolo[3,4-*b*] pyridines.

In 1992 Tietze *et al.* reported the intramolecular Povarov reaction under thermal conditions using functionalized heteroaromatic amines [106]. Condensation of benzaldehydes **151** and aminoisoxazole **152** generated the corresponding imines **153**, which selectively cyclized to form the *cis*-fused cycloadduct **154** or the *trans*-fused tetrahydropyridine **155** (Scheme **46**). The selectivity of these reactions seems to be steric and electronic effects dependent. Since the reaction of **151** (R^1 = R^2 = Cl, R^3 = R^4 = Me) with **152** only yielded the *cis*-fused compound **154**, the reaction of **151** (R^1 = R^2 = R^3 = H, R^4 = H or CO_2Me) afforded only the *trans*-annulated tetrahydropyridine **155**. Surprisingly, in the reaction of **151** (R^1 = R^2 = H, R^3 = H or Me, R^4 = Me) with aminoisoxazole **152** a mixture of *cis*-**154** and *trans*-annulated heterocycles **155** has been obtained in diasteroselectivities ranging from 1:1.4 to 1:5.

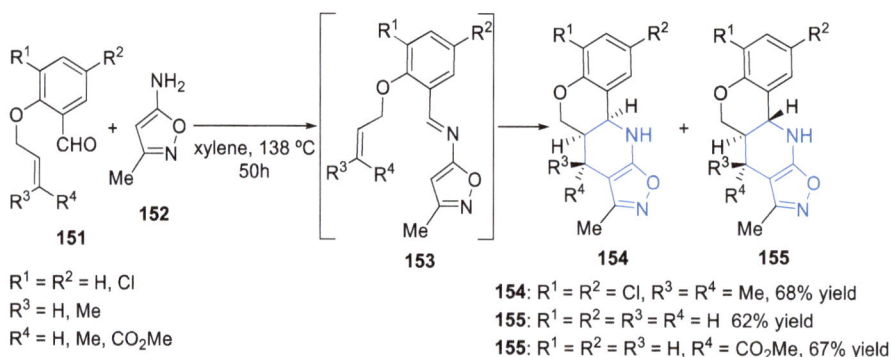

Scheme (46). Annulated tetrahydroisoxazolo[5,4-*b*]pyridine derivatives through hetero-Diels-Alder reaction using 5-amino-3-methylisoxazole.

Pyrido[3,2-*d*]pyrimidine derivatives have been synthesized by Majumdar *et al.* [107] by means of a Lewis acid catalytic intramolecular Povarov reaction. The reaction takes place between *O*-propargylated salicylaldehydes **103** and 5-amin-1,3-dimethyl uracil **156** (Scheme **47**). After a deep screening using different Brønsted (TFA) and Lewis acids (CuBr, CuI, BF$_3$·OEt$_2$, and Yb(OTf)$_3$), and solvents (DMF, DMSO, THF, MeCN, EtOH, and toluene); 10 mol% of BF$_3$·OEt$_2$ as the catalyst in toluene provides the best results. Under these reaction conditions, pyrido[3,2-*d*]pyrimidines **157** attained good chemical yields (Scheme **47**).

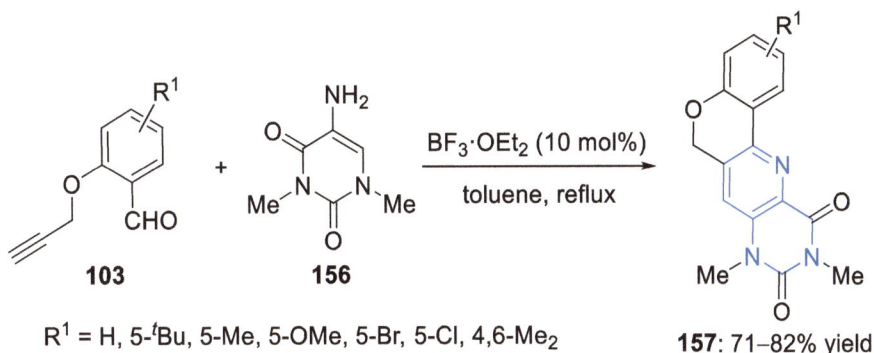

R^1 = H, 5-^{t}Bu, 5-Me, 5-OMe, 5-Br, 5-Cl, 4,6-Me$_2$ **157**: 71–82% yield

Scheme (47). Intramolecular Povarov reaction involving 5-amino-1,3-dimethyl uracil for the preparation of pyrido[3,2-*d*]pyrimidine annulated heterocycles.

4. SYNTHESIS OF PENTACYCLIC FUSED HETEROCYCLES

4.1. Formation of Quinolines Fused with Carbocycles

Pentacyclic fused heterocycles could also be achieved through intramolecular Povarov reaction. In this way, the reaction involving 2′-alkynylbiaryl-2-carbaldehydes **158** and arylamines **8** with tandem oxidation using catalytic FeCl$_3$, through the formation of a quinoline ring, represents an efficient general synthesis of dibenzo[*a,c*]acridines **160** with moderate to high yields. The optimum reaction conditions for the general synthesis of dibenzo[*a,c*]acridine derivatives **160** were obtained with a 10 mol% FeCl$_3$ in toluene at 100 °C in the open air (Scheme **48**) [108].

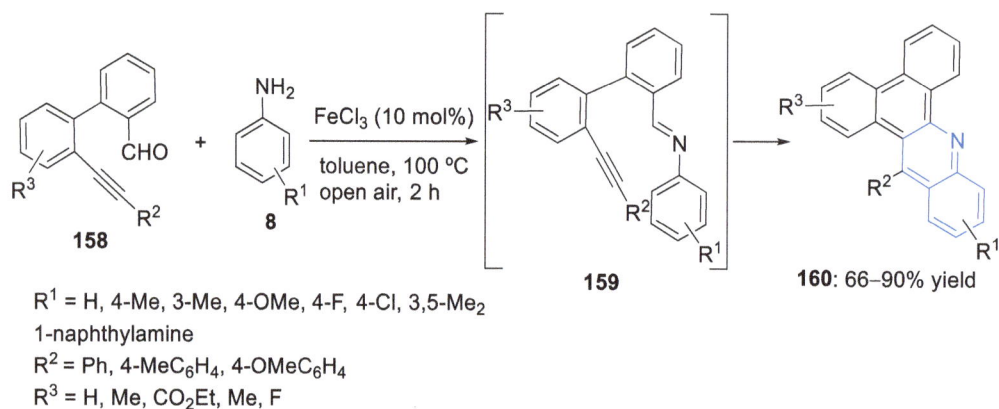

R^1 = H, 4-Me, 3-Me, 4-OMe, 4-F, 4-Cl, 3,5-Me$_2$
1-naphthylamine
R^2 = Ph, 4-MeC$_6$H$_4$, 4-OMeC$_6$H$_4$
R^3 = H, Me, CO$_2$Et, Me, F

Scheme (48). Synthesis of dibenzo[*a,c*]acridines using FeCl$_3$.

4.2. Formation of Quinolines Fused with Heterocycles Containing Nitrogen Atoms

Obtaining quinoline rings through the intramolecular imino Diels-Alder reaction also allows the formation of other different polycyclic heterocycles such as indolopyrroloquinolines. The synthesis of indolo-annulated pyrroloquinoline *via* the imino-Diels-Alder reaction has been described by Nagarajan *et al.* [109]. In this case, Lewis acid-catalyzed intramolecular imino-Diels-Alder reaction of *N*-prenylated-2-formyl-3-chloroindoles **161** (R^2 = R^3 = Me, R^4 = Cl) and substituted anilines **8** produced indolopyrroloquinolines **162** in moderate to excellent chemical yields and high *cis*-diastereoselectivity (Scheme **49**). An array of Lewis acid catalysts has been tested, and among them, La(OTf)$_3$, Sc(OTf)$_3$, and Yb(OTf)$_3$ gave better diastereoselectivities. Only the *cis*-isomer was observed in the presence of La(OTf)$_3$, when the reaction was performed at 130-140 °C. Similarly, when *N*-alkenyl indole-2-carbaldehydes **161** (R^4 = H) reacted with various *p*-substituted anilines **8** in the presence of InCl$_3$ as Lewis acid catalyst indolo[2,1-*a*]pyrrolo[4',3':2,3]-7a,8,13,13b-tetrahydroquinolines **162** were also obtained (Scheme **49**) [110]. The best overall yields were obtained when 20 mol% of InCl$_3$ was used, and under these reaction conditions, the corresponding cycloadducts **162** were obtained with *cis*-diastereoselectivities ranging from 80:20 to 96:4.

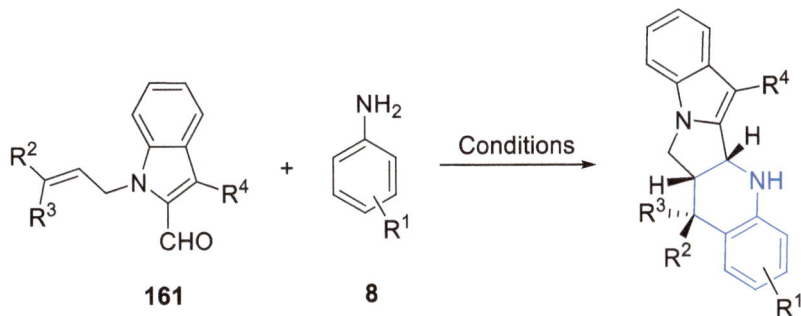

Conditions: a. La(OTf)$_3$ (10 mol%), 1,4-dioxane, 130–140 °C
        R^1 = H, 4-Me, 4-MeO, 4-Cl, 4-Br
        1-naphthylamine, 8-aminonaphthalen-2-ol
        R^2 = R^3 = Me; R^4 = Cl

      b. InCl$_3$ (20 mol%), MeCN, rt
        R^1 = H, 4-Me, 4-MeO, 4-Cl, 4-NO$_2$
        R^2 = Me, H; R^3 = Me, Ph; R^4 = H

162: 52–92% yield

162: 76–92% yield
80:20 to 96:4 *cis/trans*

Scheme (49). Lewis acid-catalyzed intramolecular Povarov reaction for the synthesis of indolo-annulated pyrroloquinolines.

A small library of A- and D-ring modified luotonin-inspired heterocyclic systems was synthesized in moderate to good yields following a six-step route that starts from phenylalanine. The key step of this total synthesis consists in an intramolecular Povarov reaction of imines **164** obtained *in situ* from a tetrahydroquinoline-derived alkynyl aldehyde **163** and various arylamines **8** in the presence of 4 Å molecular sieves (Scheme **50**). Without isolation, subsequent treatment of *N*-arylimines **164** with 1.5 equivalents of BF$_3$·Et$_2$O afforded the target pentacyclic heterocycles **165** in yields that were approximately in the 40-50% range [111].

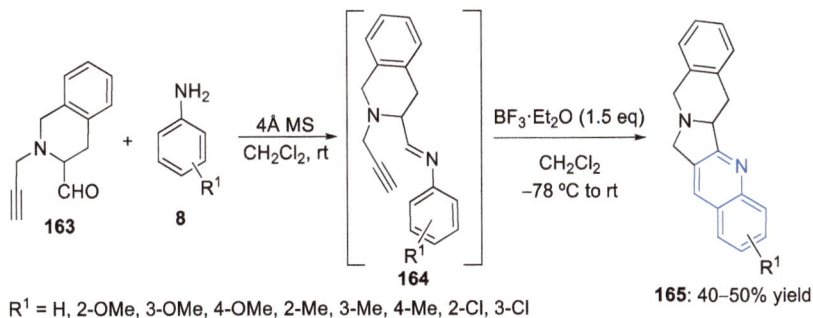

R^1 = H, 2-OMe, 3-OMe, 4-OMe, 2-Me, 3-Me, 4-Me, 2-Cl, 3-Cl

165: 40–50% yield

Scheme (50). Synthesis of luotonin A analogues *via* intramolecular Povarov reaction.

On the other hand, fused 1,7-naphthyridine scaffolds were also obtained *via* intramolecular Povarov reaction. The Lewis acid-catalyzed cyclization of *N*-arylimines **167** obtained from aldehydes **166** derived from L-phenylalanine catalyzed by EtAlCl$_2$ was studied (Scheme **51**) [112]. In this way, the *cis*-octahydro-5*H*-benz[*b*]isoquinolino[2,3-*h*] [1, 7]naphthyridines **168** were obtained. The starting aldehyde, (*S*)-*N*-(4-methyl-3-pentenyl)-1,2,3,4-tetrahydroisoquinoline -3-carboxaldehyde **166**, prepared from L-phenylalanine, was treated with various arylamines in the presence of molecular sieves giving rise to the corresponding imines **167** which were immediately cyclized in the presence of EtAlCl$_2$ to the 1,7-naphthyridine **168**. The formal hetero-Diels-Alder reaction of **167** proceeded with high diastereoselectivity in favor of the all-*cis*-configured product. The amino-substituent into a rigid pentacyclic system like **168** resulted in a good cytotoxic activity against human brain tumor cell lines.

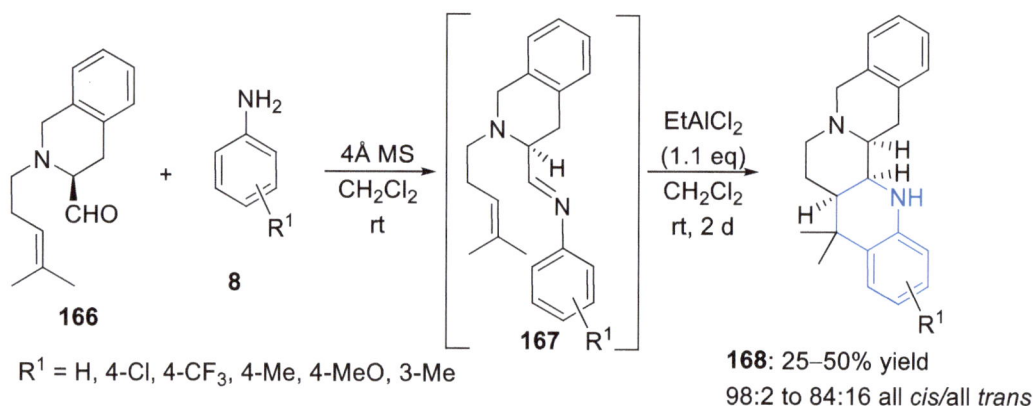

Scheme (51). Synthesis of fused benzo[*g*]quinolino[2,3-*a*]quinolidines.

The same group also developed a similar strategy by using enantiomeric aldehyde **169** derived from the D-phenylalanine. Thus, the corresponding imine intermediate **170**, obtained by condensation of aldehyde **169** with ethyl 4-aminobenzoate **8** (R^1 = 4-CO$_2$Et) in the presence of molecular sieves, directly treated with EtAlCl$_2$ gave the pentacyclic benzo[*b*]isoquinolino[2,3-*h*] [1, 7]naphthyridine **171** with high *cis*-diastereoselectivity (Scheme **52**). Whereas, when the same imine was treated with SnCl$_4$ the pentacyclic *trans*-diastereomer **172** was obtained in a dr of 99.5/0.5 [113]. To improve the ring planarity in benzo[*b*]isoquinolino[2,3-*h*] [1, 7]naphthyridines an intramolecular Diels-Alder starting from an aldehyde tethered to an alkyne instead of an alkene was performed.

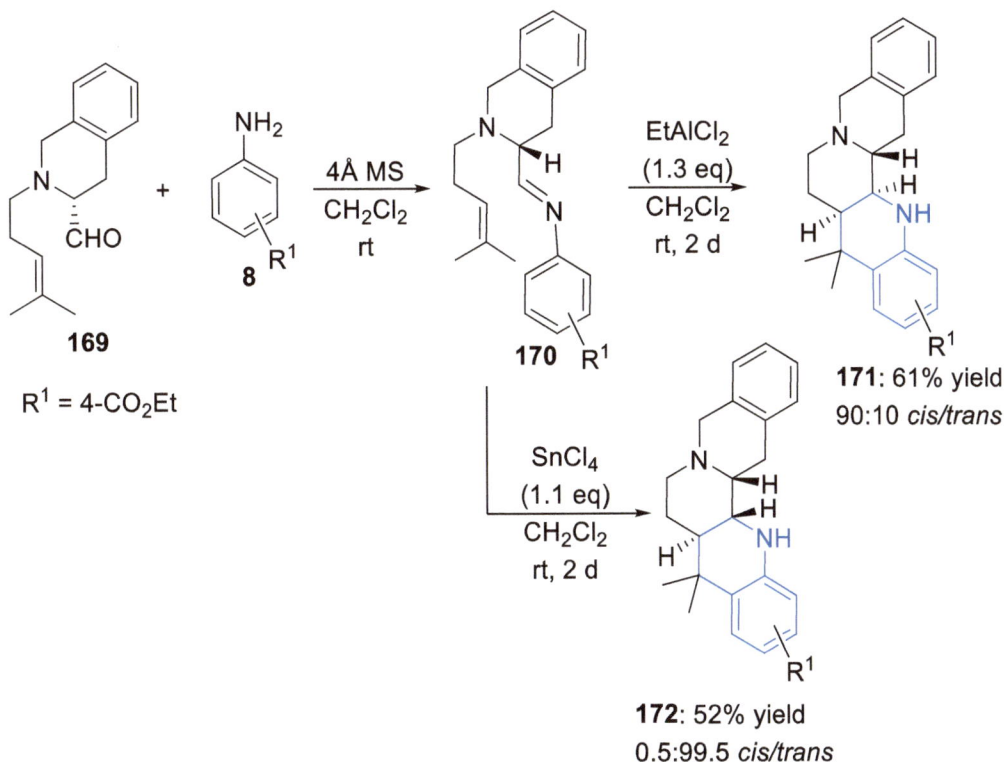

Scheme (52). Synthesis of fused benzo[*b*]isoquinolino[2,3-*h*] [1, 7]naphthyridines.

Quinoline rings with a higher level of complexity can also be prepared. In another approach, the intramolecular Povarov reaction allowed the synthesis of the quinolino[2',3':3,4]pyrrolo[2,1-*b*]quinazolin-11(13*H*)-one **174**, known as luotonin A. The reaction of *N*-propargylic substituted aldehyde derived from quinazoline **173** and aniline **8** (R^1 = H) in the presence of 10 mol% Dy(OTf)$_3$ in acetonitrile gave luotonin A in 51% yield (Scheme **53**) [78].

2,7-Naphthyridine regioisomers can also be obtained. In 2008, Zhang's group developed a new intramolecular Povarov reaction for the preparation of luotonin A analogues [114]. This approach entails the *in situ* formation of imidates through activation of corresponding chemically stable amides. Thus, bis(triphenyl)oxodiphosphonium trifluoromethanesulfonate, formed *in situ* from Ph$_3$PO and Tf$_2$O, used for the total synthesis of camptothecin [115] and luotonin A [116], works as an amide-activating reagent to convert the amide moiety to its corresponding imidate under mild reaction conditions, and also to promote the subsequent intramolecular Povarov reaction in the desired direction. Using these catalytic conditions, cyclization of *N*-allyl naphthyridones **175** to afford luotonin

A analogues **178** through the corresponding imidates **176** was attained in 64-78% yield (Scheme **54**). The formation of compound **178** may be rationalized by the stability of aromatic system of **178**, driven by the catalytic system acidity. In the same way, luotonin A analogues [114] were also synthesized using bis(tri-phenyl)oxodiphosphonium trifluoromethanesulfonate, as an amide-activating reagent, from *N*-propargyl naphthyridones **177** yielding luotonin A analogues **178** in moderate yield (Scheme **54**).

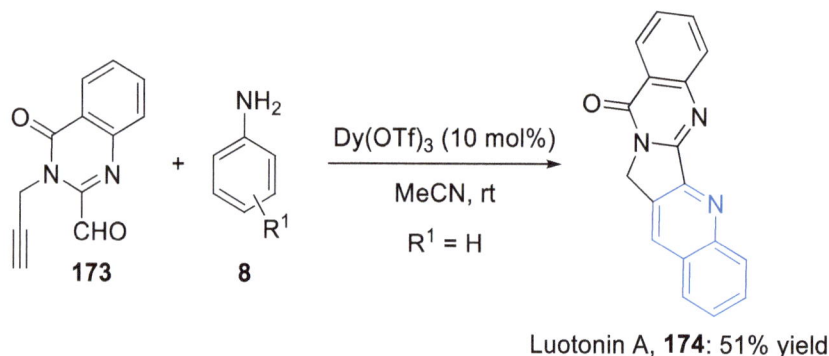

Scheme (53). Intramolecular Povarov reaction in the total synthesis of luotonin A.

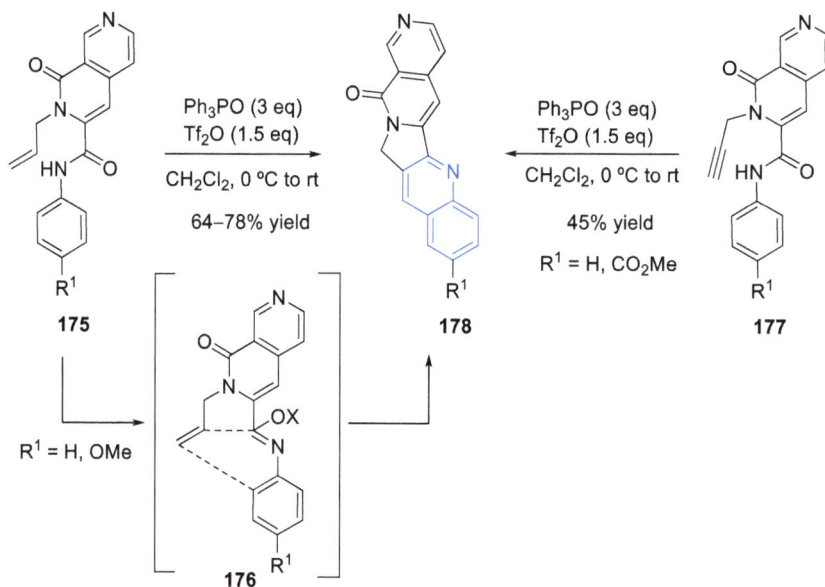

Scheme (54). Cyclization of *N*-alkenyl or *N*-propargyl naphthyridones into luotonin A analogues using amide-activating reagents.

More recently, a small library of benzimidazole-fused pyrrolo[3,4-*b*]quinolines, considered decarbonyl analogues of the anticancer alkaloid luotonin A, synthesized from readily available benzimidazole 2-carbaldehyde **179** and various substituted arylamines **8** was reported [117]. The cycloadducts **181** could only be obtained through the treatment of aldimines **180** with 20 mol% of BF₃·OEt₂ in DCE at 80 °C in good yields (Scheme **55**). In addition, cycloadducts **181** were achieved in 65-80% yield *via* one-pot intramolecular Povarov reaction when substituted anilines **8** reacted with benzimidazole 2-carbaldehyde **179** in the presence of BF₃·OEt₂.

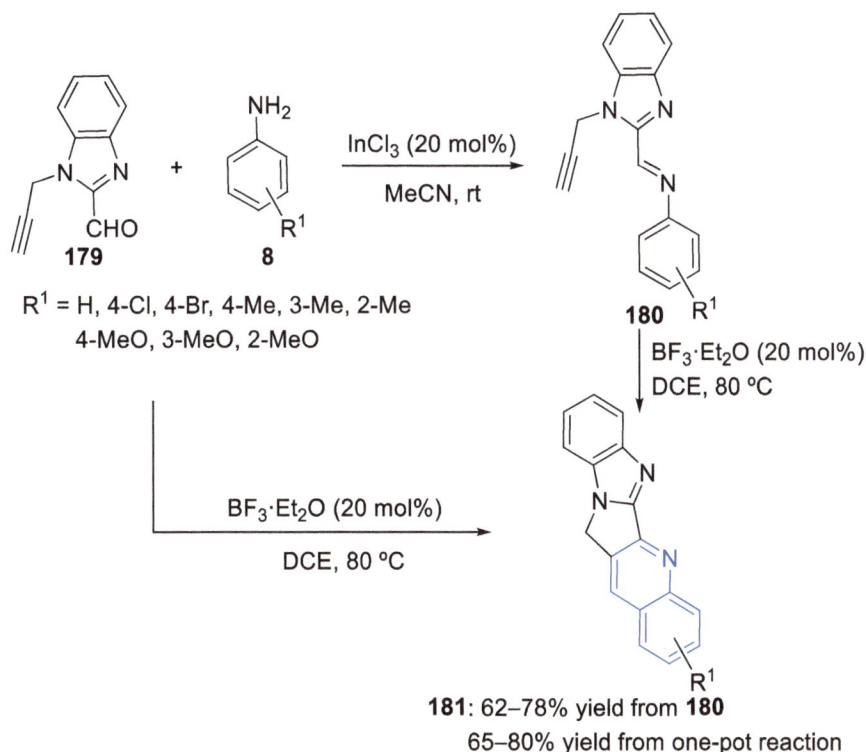

Scheme (55). Synthesis of decarbonyl analogues of the anticancer alkaloid luotonin A by BF₃·OEt₂ assisted-intramolecular Povarov reaction.

N-Alkenyl pyrrolopyrimidine-6-carbaldehydes **182** have been used as carbonyl components for the preparation of uracil annulated quinoline derivatives **183** in good yields and good to excellent stereoselectivities (Scheme **56**) [20, 76]. Indium trichloride proved to be the most efficient Lewis acid catalyst, with overall yields higher compared to other tested Lewis acid catalysts [20]. The cyclization pathway proceeds by a stepwise mechanism. Tetrahydroquinoline annulated heterocycles **183** (R = Cl) were evaluated for their antibacterial activity against six different bacterial strains, being as active as the antibiotic ciprofloxacin and presenting a MIC value of 2.5 mg/mL against *Escherichia coli* [76].

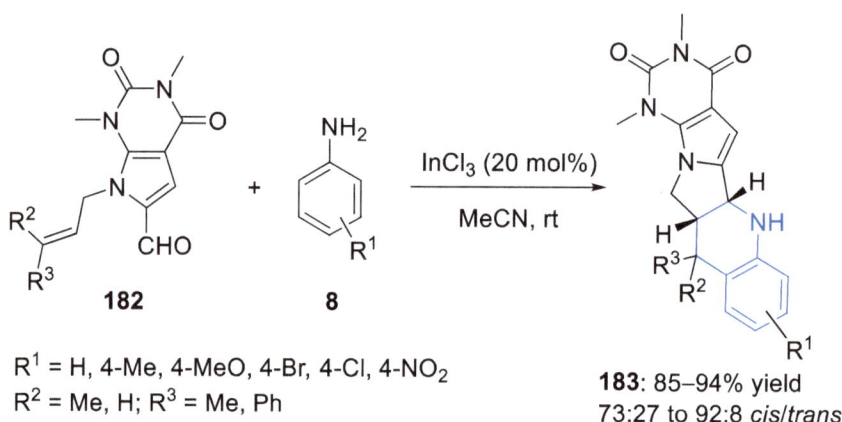

R^1 = H, 4-Me, 4-MeO, 4-Br, 4-Cl, 4-NO$_2$
R^2 = Me, H; R^3 = Me, Ph

183: 85–94% yield
73:27 to 92:8 *cis/trans*

Scheme (56). Synthesis of quinoline annulated heterocycles by InCl$_3$ assisted-intramolecular Povarov reaction.

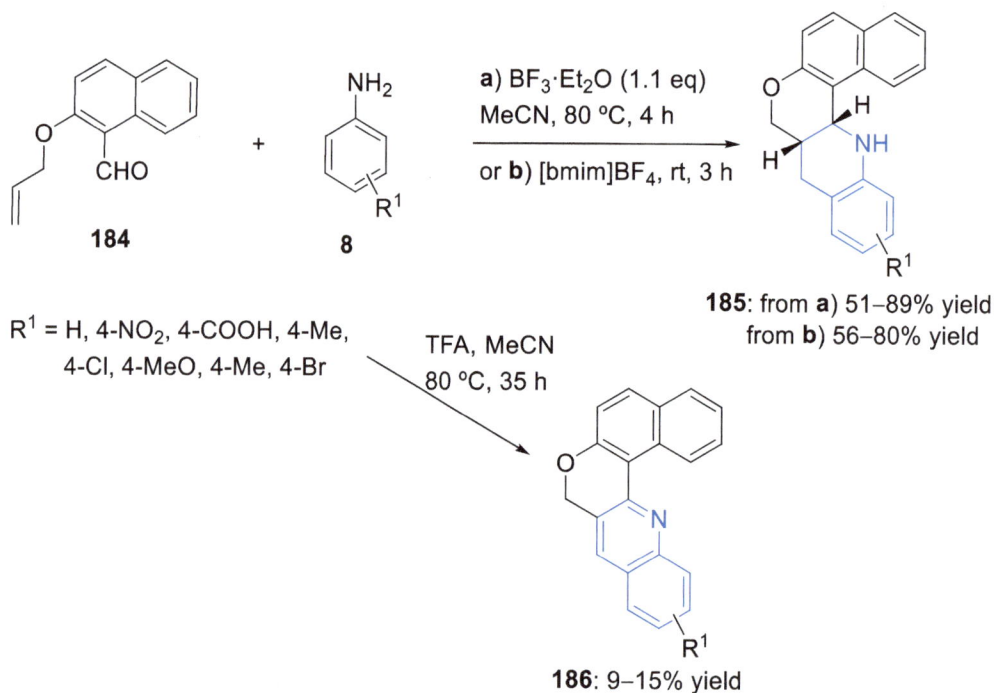

R^1 = H, 4-NO$_2$, 4-COOH, 4-Me,
4-Cl, 4-MeO, 4-Me, 4-Br

TFA, MeCN
80 °C, 35 h

185: from **a**) 51–89% yield
from **b**) 56–80% yield

186: 9–15% yield

Scheme (57). Synthesis of 8a,9,14,14a-tetrahydro-8*H*-benzo [5, 6]chromeno[4,3-*b*]quinolines.

4.3. Formation of Quinolines Fused with Heterocycles Containing Oxygen Atoms

Other pentacyclic heterocycles derived from quinoline containing one oxygen atom may be formed. The condensation of 2-allyloxynaphthalene-1-carbaldehyde **184** with substituted anilines **8** and subsequent intramolecular cyclization in the presence of $BF_3 \cdot Et_2O$ yielded 8a,9,14,14a-tetrahydro-8*H*-benzo [5, 6]chromeno[4,3-*b*]quinolines **185** (Scheme **57**) [118]. When the reaction was carried out with TFA, the obtained products were not those expected but their dehydrogenated derivatives **186**. In a similar manner, *cis*-compounds **185** could also be obtained performing the intramolecular aza-Diels-Alder reaction in [bmim]BF_4 ionic medium [95], being the last one a green protocol that offers significant advantages over reported methods.

Pentacyclic polyaromatic chromenoacridine derivatives can be synthesized in a single-pot intramolecular Povarov reaction [119]. This protocol involves the formation of the intermediate imine **188** between *para*- and *ortho*-substituted aromatic amines **8** and alkene-tethered chromene-3-carbaldehyde **187**, followed by the $BF_3 \cdot OEt_2$-induced cycloaddition. Under these reaction conditions, a set of chromeno[2,3-*c*]acridines **189** were obtained in 66-88% yield with high *trans*-stereoselectivity (Scheme **58**). Other Lewis acids and Brønsted acids were ineffective for this transformation in terms of both yield and selectivity. All attempts to extend this protocol to diamines such as 1,5-diaminonaphthalene did not furnish the desired product.

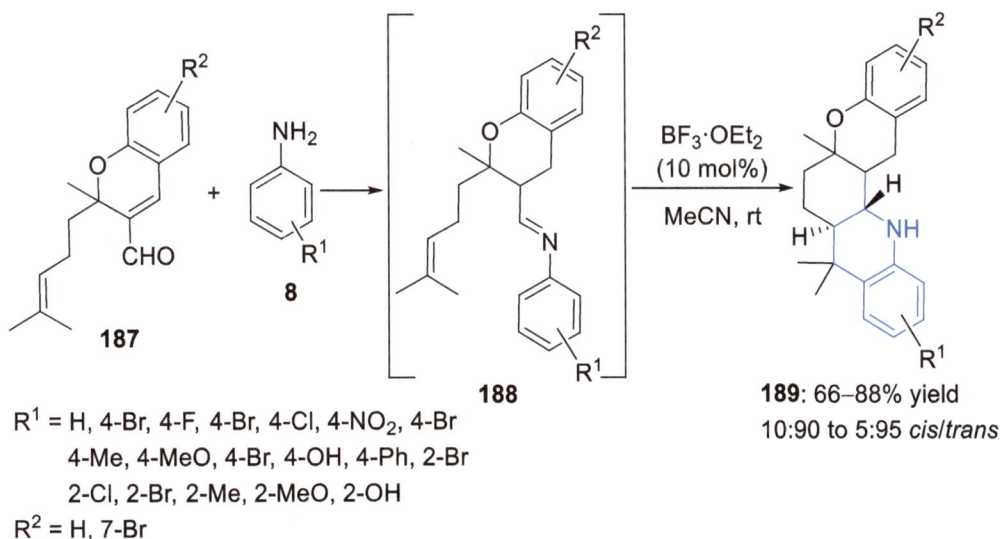

R^1 = H, 4-Br, 4-F, 4-Br, 4-Cl, 4-NO$_2$, 4-Br
4-Me, 4-MeO, 4-Br, 4-OH, 4-Ph, 2-Br
2-Cl, 2-Br, 2-Me, 2-MeO, 2-OH
R^2 = H, 7-Br

189: 66–88% yield
10:90 to 5:95 *cis/trans*

Scheme (58). Synthesis of chromenoacridine derivatives through intramolecular Povarov reaction.

The use of *O*-prenylated compounds derived from 8-formyl chromenones allows the synthesis of pyranochromeno[4,3-*b*]quinoline derivatives. In this sense, diastereoselective synthesis of cycloaducts **192** by intramolecular Povarov reaction of formal 2-azadienes derived *in situ* from aromatic amines **8** and *O*-prenylated compounds **190** derived from 8-formyl chromenones was reported (Scheme **59**) [120]. Several Lewis and Brønsted acids catalysts were tested in this reaction and among them Yb(OTf)$_3$ and Sc(OTf)$_3$ were found to be almost equal efficient according to reaction yields, times, and diastereoselectivities. In all cases, exclusive formation of *cis*-tetrahydrochromenoquinolines **192** was obtained, which may be due to the steric effect of the chromenone moiety. This method allows the use of aromatic amines **8** with electron-withdrawing or electron-donating groups, giving compounds **192** in very good yields. Some of this synthesized tetrahydropyranochromeno [4,3-*b*]quinolines **192** exhibited significant antiproliferative activity against MCF-7 breast cancer cell line and low inhibitory activity against MDA-MB-231 breast cancer cell line. Similarly, *O*-propargylated compounds derived from 8-formyl chromenones have been used as carbonyl compounds in the preparation of pyranochromeno[4,3-*b*]quinolines [121]. Several copper catalysts were analyzed for the activation of the terminal alkyne C-H bond, and CuFe$_2$O$_4$ nanoparticles were found to be the best catalyst for this transformation. Due to their magnetic properties, they can be easily separated from the reaction mixture and reused without loss of activity. Kinfe's group reported the preparation of benzopyran-fused pyranoquinolines through a domino Ferrier rearrangement intramolecular Povarov-like reaction [122]. The process involves a three-component condensation of a glycal with a variety of anilines and 2-hydroxybenzaldehydes under Lewis acid catalysis.

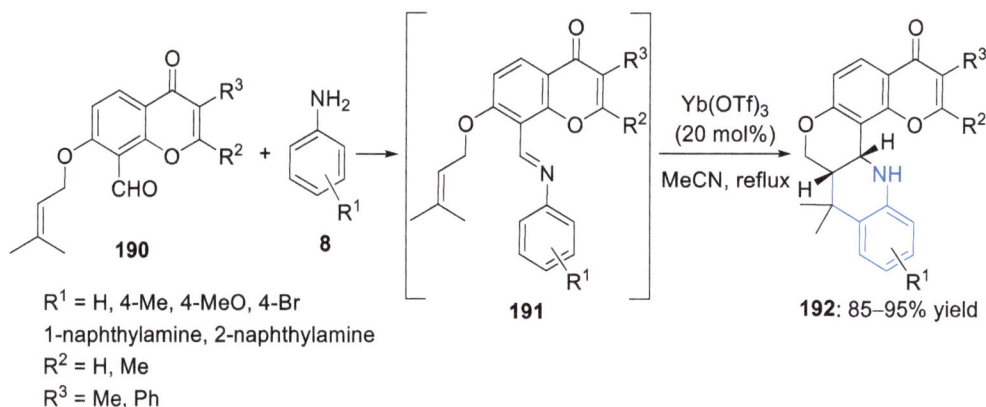

R^1 = H, 4-Me, 4-MeO, 4-Br
1-naphthylamine, 2-naphthylamine
R^2 = H, Me
R^3 = Me, Ph

Scheme (59).   Intramolecular Povarov reaction in the preparation of antiproliferative tetrahydro-pyranochromeno[4,3-*b*]quinolines.

Aldehydes derived from natural carbohydrates present a certain absolute stereochemistry and are interesting substrates as chiral auxiliaries or chiral building blocks. Intramolecular hetero-Diels-Alder reactions of carbohydrate-derived aldehydes have been performed by Sabitha *et al.* [123]. In this work, octahydro-3b*H* [1, 3]dioxolo[4",5":4',5']furo-[2',3':5,6]pyrano[4,3-*b*]quinolines **195** have been prepared in a highly efficient and stereoselective way (Scheme **60**). Aldimines **194** generated *in situ* from aromatic amines **8** and the *O*-allyl derivative of the D-glucose aldehyde **193** were treated in acetonitrile in the presence of a catalytic amount of BiCl$_3$. This Lewis acid is the most suitable, since it can be used in substoichiometric amounts (10 mol%), while other Lewis acids are needed in stoichiometric amounts. Cycloadducts **195** were obtained with high selectivity and good to excellent yields. In general, the reactions led to the formation of *trans*-isomers as major products, although small amounts of *cis*-isomers were observed. However, when a bulky group, such as *tert*-butyl (R^1 = ^{t}Bu), was present in the *ortho* position of the amine **8**, only the *trans*-adduct **195** was exclusively obtained.

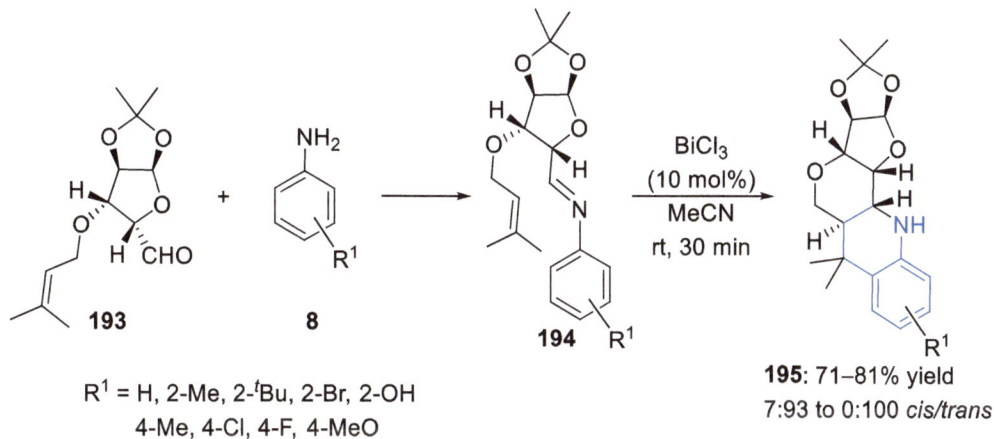

R^1 = H, 2-Me, 2-^{t}Bu, 2-Br, 2-OH
4-Me, 4-Cl, 4-F, 4-MeO

195: 71–81% yield
7:93 to 0:100 *cis/trans*

Scheme 60. Synthesis of pyrano[4,3-*b*]quinolines from *O*-allyl carbohydrate-derived aldehydes.

4.4. Formation of Quinolines Fused with Heterocycles Containing Nitrogen and Oxygen Atoms

Based on the protocol mentioned in the previous section, the same group has performed the preparation of pentacyclic furo[3,2-*h*] [1, 6]naphthyridine derivatives from anilines and an amino sugar derivatives [124]. In this case, an *N*-prenylated sugar aldehyde **196** and different aromatic amines **8** were used in the condensation reaction to give the imines **197** (Scheme **61**). Afterwards, the intramolecular hetero-Diels-Alder reaction proceeded in the presence of bismuth(III) chloride as catalyst, under very mild conditions at room temperature, and was completed within 30 min to give exclusively the corresponding *trans-*fused products **198** stereoselectively in good to excellent yields.

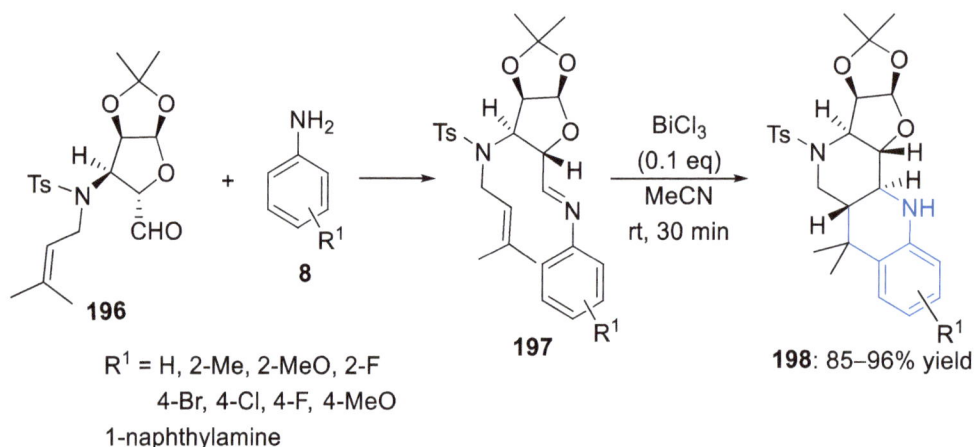

Scheme (61). Synthesis of furo[3,2-*h*] [1, 6]naphthyridines from *N*-prenylated carbohydrate-derived aldehydes.

More recently, the combination of visible-light-photoredox and acid catalysis has also been applied to the formal synthesis of pentacyclic derivative **200**, a precursor of 10-hydroxycamptothecin and irinotecan. The intramolecular Povarov cycloaddition/dehydrogenation aromatization cascade of pyridone carbaldehyde **199** and aromatic amine **8** in the presence of a photocatalyst (Ru(bpy)$_3$Cl$_2$·6H$_2$O) and a Brønsted acid catalyst, *p*-toluenesulfonic acid (*p*-TsOH), yielded the compound **200** in 92% yield (Scheme **62**) [77].

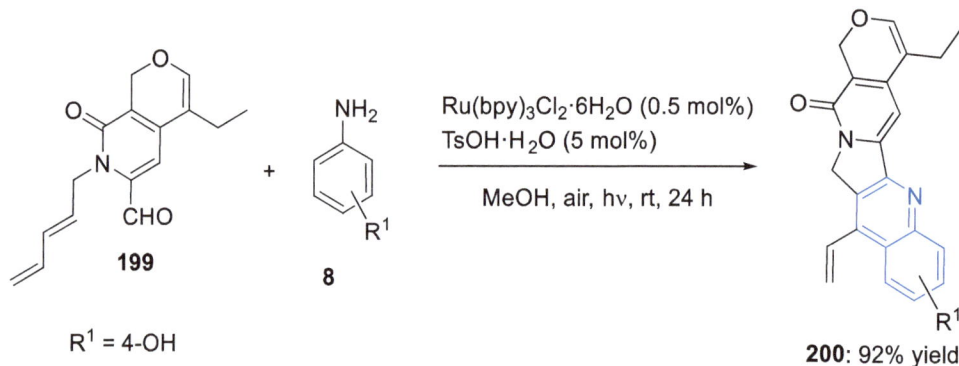

Scheme (62). Tandem acid-catalyzed intramolecular Povarov reaction/visible-light-photoredox for the synthesis of the precursor of 10-hydroxycamptothecin and irinotecan.

Aldehydes derived from alkene-tethered aminochromenes have been employed for the intramolecular inverse electron demand [4+2] cycloaddition reaction [125]. 2-(*N*-Alkenyl-*N*-aryl)aminochromene-3-carbaldehyde **201** underwent intramolecular Povarov reaction with aromatic amines **8** in the presence of Lewis acids to furnish octahydro-14*H*-benzo[*b*]chromeno[2,3-*h*] [1, 6]naphthyridin-1--one **202** or **203** (Scheme **63**) [126]. Thus, the reaction of 4-substituted aromatic amines **8** with **201** in the presence of 40 mol% of TPP [94], afforded *cis*-**202** or *trans*-chromenonaphthyridines **203**. *cis*-Adduct **202** is favored when R^5 = Me, while *trans*-chromenonaphthyridines **203** were observed, when R^5 = Ph.

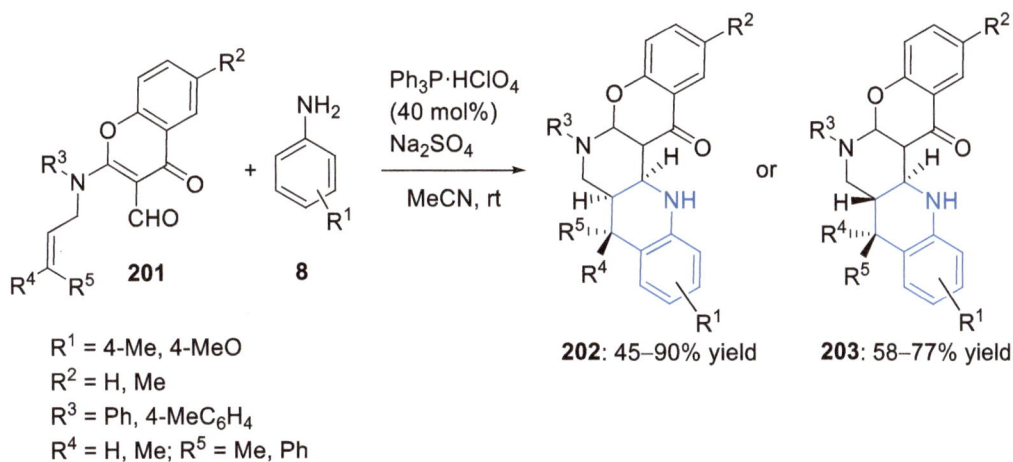

R^1 = 4-Me, 4-MeO
R^2 = H, Me
R^3 = Ph, 4-MeC$_6$H$_4$
R^4 = H, Me; R^5 = Me, Ph

Scheme (63). Synthesis of chromenonaphthyridines *via* TPP-induced intramolecular Povarov reaction.

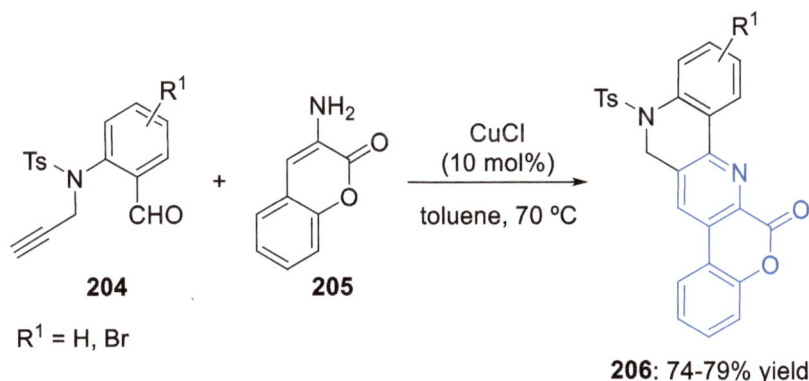

R^1 = H, Br

Scheme (64). Synthesis of 12,13-dihydro-6*H*benzo[*h*]chromeno[3,4-*b*] [1, 6]naphthyridin-6-ones *via* copper catalyzed reaction.

On the other hand, 12,13-dihydro-6*H*benzo[*h*]chromeno[3,4-*b*] [1, 6]naphthyridin-6-one derivatives **206** can also be prepared when 3-amino-2*H*-chromen-2-one **205** reacted with *N*-propargyl derivated aldehydes **204**, which possess an alkyne moiety, in the presence of CuCl as a catalyst in high yields (Scheme **64**) [84].

Raghunathan *et al.* [127] reported a simple procedure for the synthesis of pyrano derivatives using indium trichloride supported in silica gel. Therefore, the reaction of aromatic amines **8** with *O*-alkenyl quinoline-3-carbaldehyde **207** in the presence of InCl$_3$ in acetonitrile furnished a mixture of *cis*- and *trans*-pyrano **209** with diastereoselectivities ranging from 65:35 to 84:26 by intramolecular Povarov reaction of the intermediate imine **208** generated in the one-pot reaction (Scheme **65**). As a further extension of this work, the same reaction was carried out using InCl$_3$ impregnated in silica gel as Lewis acid catalyst under microwave irradiation. These reaction conditions dramatically increase the overall yields from 55-82% to 75-97%, retaining nearly the same diastereoselectivity ratios. This eco-friendly protocol avoids the use of organic solvents, has general applicability, and notably enhances reaction rates and chemical yields.

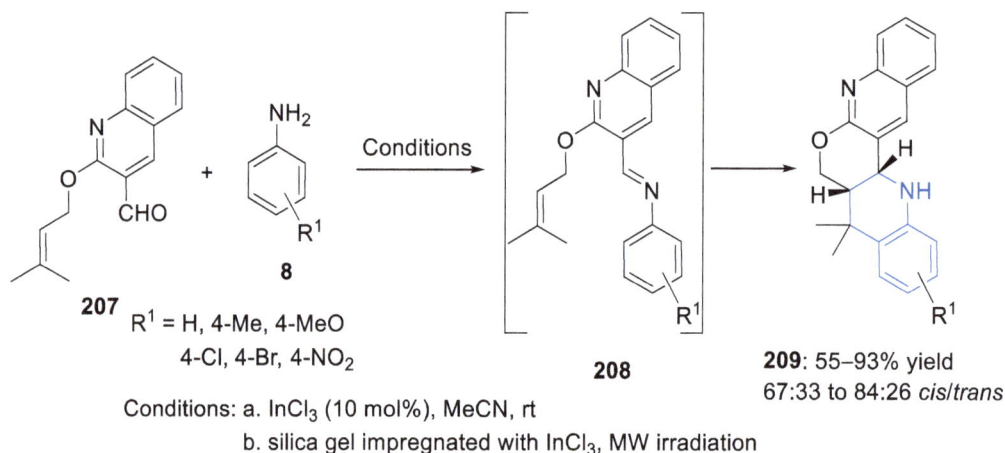

Scheme (65). Synthesis of pyranoquinolines through InCl$_3$ catalyzed intramolecular Povarov reaction.

4.5. Formation of Quinolines Fused with Heterocycles Containing Nitrogen and Sulfur Atoms

As an extension of the methodology applied in the previous section, thiopyranoquinolines **211** can also be obtained if *S*-alkenyl quinoline--carbaldehyde **210** was used instead of the corresponding *O*-alkenylcarbaldehydes (Scheme **66**) [127].

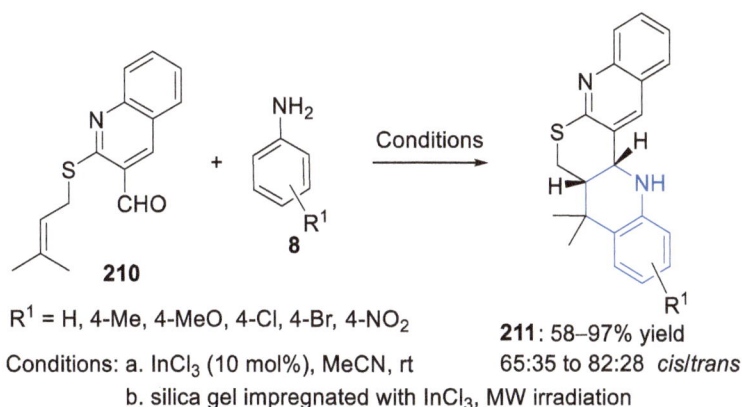

R^1 = H, 4-Me, 4-MeO, 4-Cl, 4-Br, 4-NO$_2$

Conditions: a. InCl$_3$ (10 mol%), MeCN, rt
b. silica gel impregnated with InCl$_3$, MW irradiation

211: 58–97% yield
65:35 to 82:28 *cis/trans*

Scheme (66). Synthesis of pyrano and thiopyranoquinolines through InCl$_3$ catalyzed intramolecular Povarov reaction.

4.6. Formation of other Pentacyclic Heterocycles

The preparation of a series of pentacyclic heterocyclic systems **213** involving the intramolecular Povarov reactions between 3-aminocoumarins **212** and *O*-cinnamylsalicylaldehydes **135** (Scheme **67**) was reported by Bodwell *et al.* [128]. This intramolecular version provides three points of diversity, *i.e.,* the salicylaldehyde component that links the dienophile to diene, the 3-aminocoumarin unit and the cinnamyl unit (dienophile). The Povarov adducts **213** were obtained with high selectivity for the *trans,trans* relative stereochemistry in the newly formed fused ring system. 3-Aminocoumarins **212** have also been used in the preparation of pyrido-coumarin derivatives through intramolecular Povarov reaction with inactivated alkynes. Therefore, treatment of 3-aminocoumarins **212** with 2-(propargyloxy)benzaldehydes in the presence of Lewis [Yb(OTf)$_3$] [128] or Brønsted acid catalysts (triflic acid) [129] in refluxing acetonitrile gave to the formation of pyrido[2,3-*c*]coumarins. In both cases, no co-oxidant was required for the aromatization of the desired products.

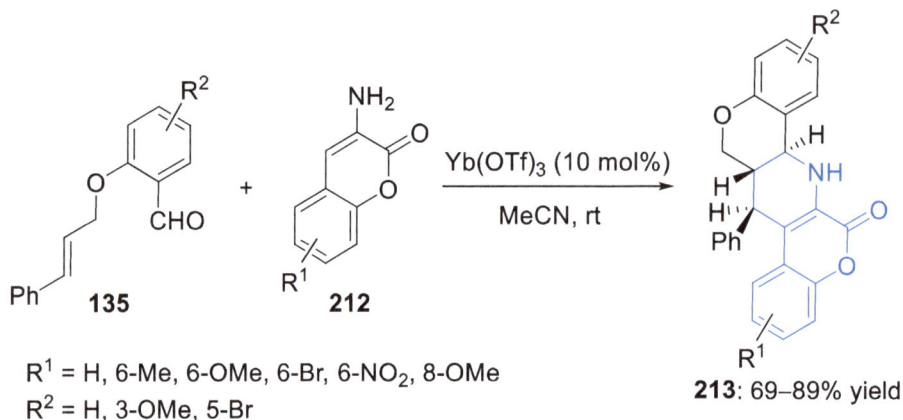

Scheme (67). Intramolecular Povarov reactions involving 3-aminocoumarins.

A $BF_3 \cdot Et_2O$ catalyzed intramolecular Povarov reaction, followed by oxidation with DDQ, was used to synthesize chromenopyridine fused thiazolino-2-pyridone peptidomimetics with the ability to bind α-synuclein and amyloid-β fibrils *in vitro*. The reaction works with several *O*-alkylated salicylaldehydes **127** and amino functionalized thiazolino-2-pyridones **214**, to generate polyheterocycles **215** with diverse substitution in moderate to excellent yields (Scheme **68**). On the contrary, attempts to synthesize C-7 unsubstituted molecules **216** (R^2 = H) through intramolecular Povarov reaction, using *O*-allylsalicylaldehyde transcurred in very low yields, but the use of a vinyl ester moiety as electron-donating auxiliary **127** (R^2 = OCOPh), allowed to achieve the C-7 unsubstituted compounds in reasonable reaction times and moderate yields [130].

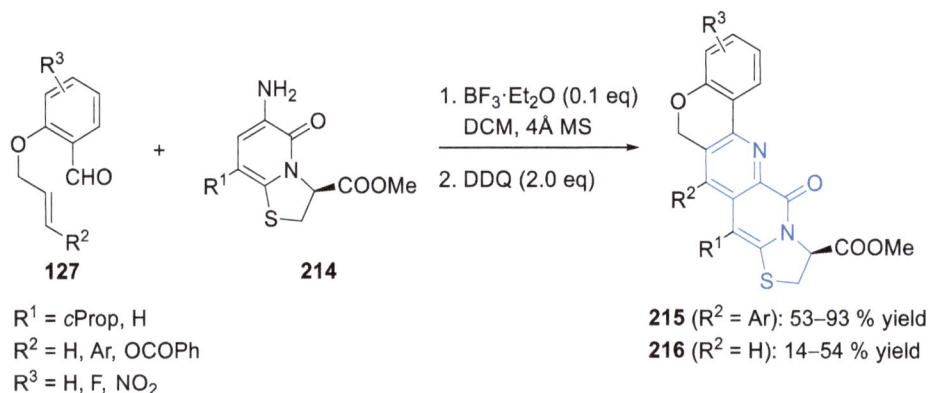

Scheme (68). Synthesis of 2-pyridone-based polyheterocycles.

The synthesis of indoloacridine **219** in the intramolecular Povarov reaction of imine derived from aliphatic aldehyde **43** as dienophile and a heteroaromatic amine such as 3-aminocarbazole **217** has been performed (Scheme **69**) [131]. The reaction proceeded very smoothly in the presence of Lewis acid La(OTf)$_3$. Although the diastereoselectivity is highly temperature-dependent, the *trans*-isomer is the major product isolated in this reaction, and better diastereoselectivities (10:90) were attained at lower reaction temperatures.

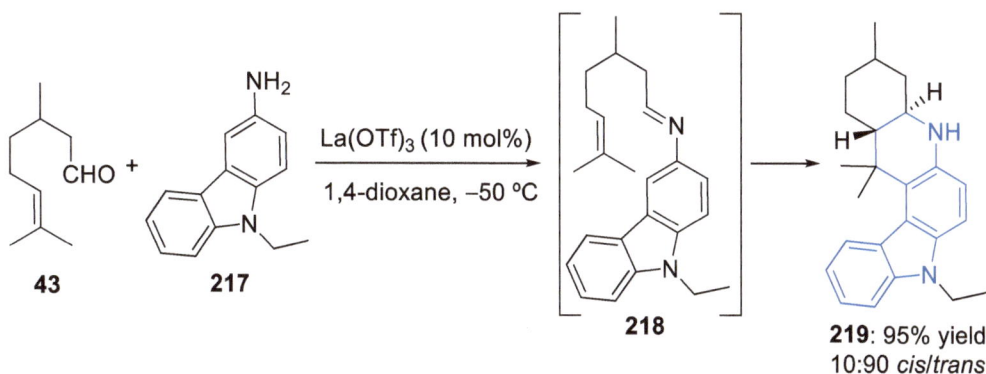

Scheme (69). Intramolecular Povarov reaction of aliphatic aldehydes and 3-aminocarbazol.

5. SYNTHESIS OF HEXA- HEPTA- AND OCTACYCLIC FUSED HETEROCYCLES

Hexacyclic fused heterocycles containing a tetrahydroquinoline core in their structure have also been prepared by intramolecular Povarov reaction. Thus, after the reaction of estrone-derived aldehyde **220** with different anilines **8**, treatment with BF$_3$·Et$_2$O afforded two different cyclic products, hybrid derivatives of tetrahydroquinolines condensed to a steroid skeleton **223** and another product **224** apparently obtained by an intramolecular Prins reaction (Scheme **70**) [132, 133]. Although compound **223** is the formal Diels-Alder adduct, the authors indicated that the compound might be obtained in a two-step mechanism from the initially formed iminium ion **221**, which led to the carbocation **222** and then undergoes an electrophilic aromatic substitution to give **223**.

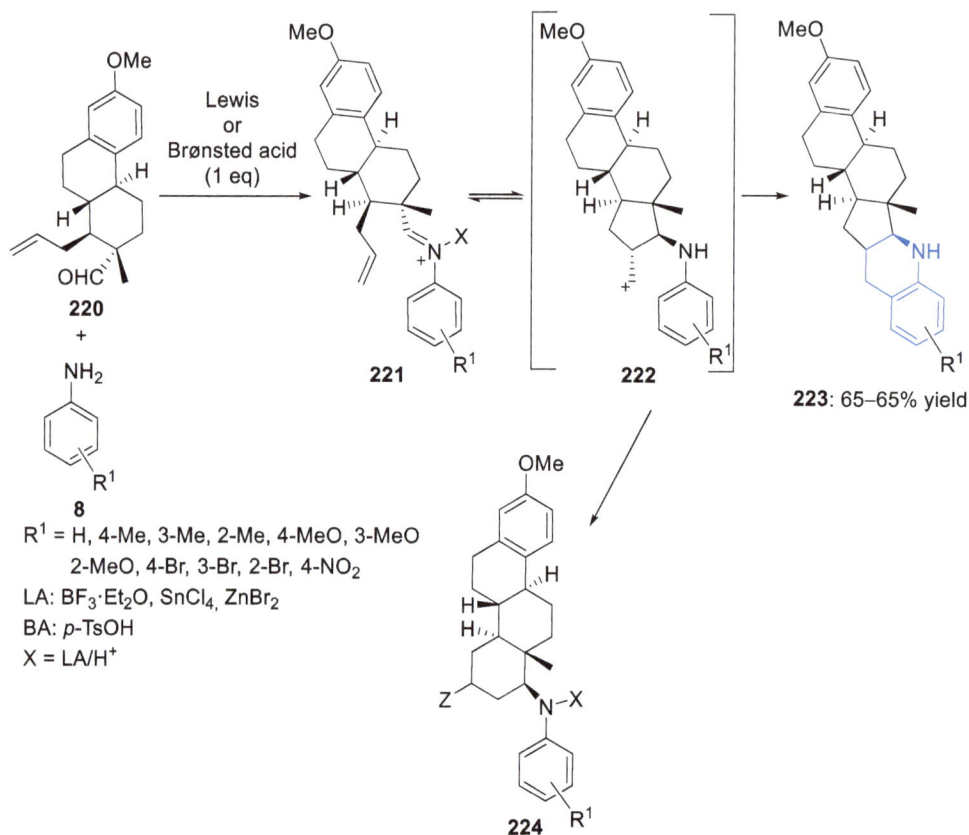

Scheme (70). Synthesis of octahydroacridines condensed to a steroid skeleton.

Similarly, arylimino steroids **226**, generated from the condensation of various anilines **8** with steroid-derived aldehydes **225**, undergo a Lewis acid-catalyzed intramolecular cyclization in the presence of $BF_3 \cdot Et_2O$ to afford tetrahydroquinolines fused heterocycles **227** in good yields (Scheme **71**) [134]. The reactivity of these arylimino steroids **226** is dependent on aniline substitution, since different tetrahydroquinoline derivatives were formed on unsubstituted ($R^1 = H$), brominated ($R^1 = 4$-Br) or methoxy- ($R^1 = 4$-OMe) substituted anilines **8**. Conversely, nitroaniline ($R^1 = 4$-NO$_2$) gave to the formation of a fluoro-D-homosteroid derivative, apparently through an intramolecular Prins reaction.

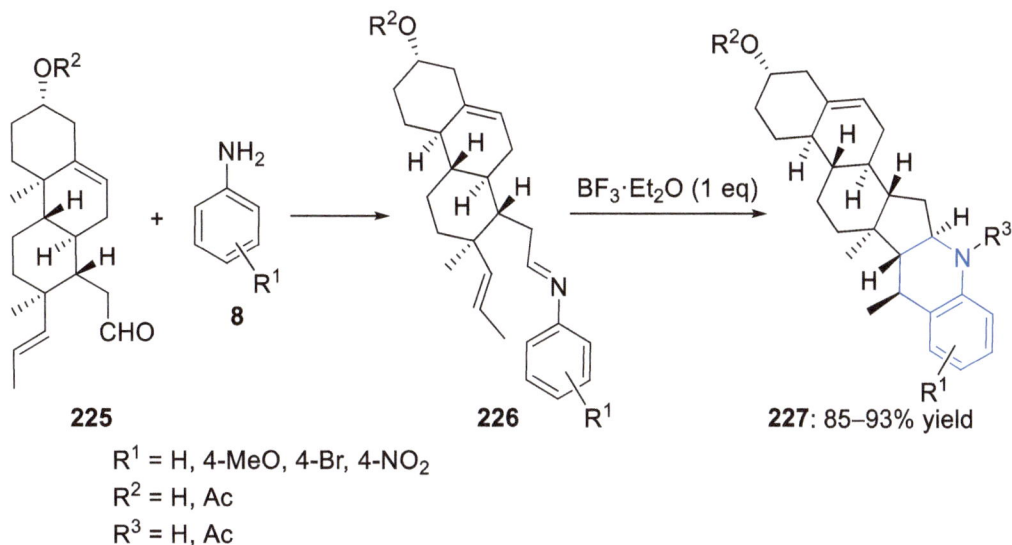

Scheme (71). Synthesis of hexacyclic fused heterocycles from steroid-derived aldehydes.

Nagarajan *et al.* [131] reported the synthesis of ellipticine derivatives **229** through the intramolecular Povarov reaction of imines **228** derived from aromatic aldehydes **93** (*O*-prenylated salicylaldehydes) and 3-aminocarbazol **217** (Scheme **72**). The cyclization, which was catalyzed by La(OTf)$_3$ as Lewis acid, yielded isomeric ellipticine derivatives **229** in 85-92% yield and very good diastereoselectivities (95:5 to 98:2) in favor of *cis*-isomer (Scheme **72**).

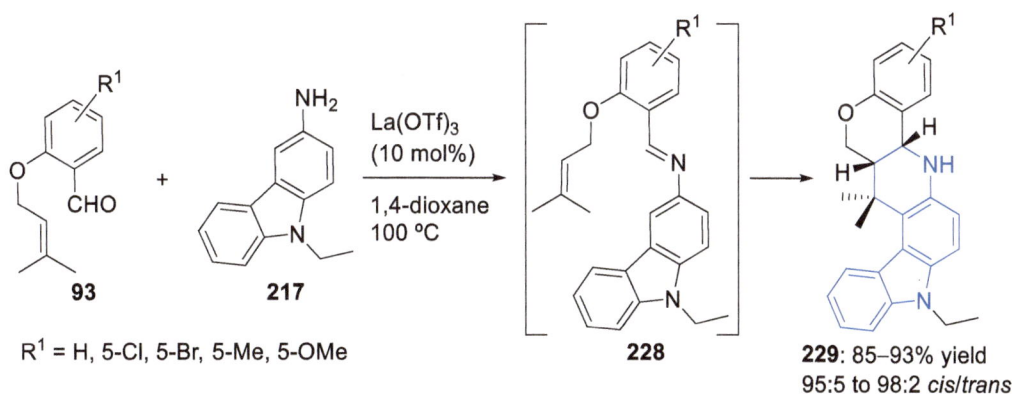

Scheme (72). Synthesis of ellipticine derivatives through intramolecular Povarov reaction of aromatic aldehydes and 3-aminocarbazole.

The same group reported the synthesis of isomeric isoellipticine derivatives through a straightforward CuI/La(OTf)$_3$ catalyzed tandem reaction in ionic liquid

[bmim]BF$_4$ [135]. The reaction of carbazole-derived amine **217** with *O*-propargylated salicylaldehyde **230** (R^1 = H) in the presence of CuI/La(OTf)$_3$ and ionic liquid afforded isoellipticine fused with dihydro chromene derivatives **231**. In this case, the intramolecular Povarov occurred through C-2 of the carbazole ring instead of C-4 position of the carbazole ring in the intermolecular case. Nevertheless, when using bromo, fluor, chloro, methyl, or methoxy substituted *O*-propargylic aldehydes **230** under the optimized reaction conditions, the intramolecular cyclization occurred through the C-4 position of the carbazole ring to yield isoellipticine fused with dihydro chromene derivatives **231** in 80-96% chemical yield (Scheme **73**).

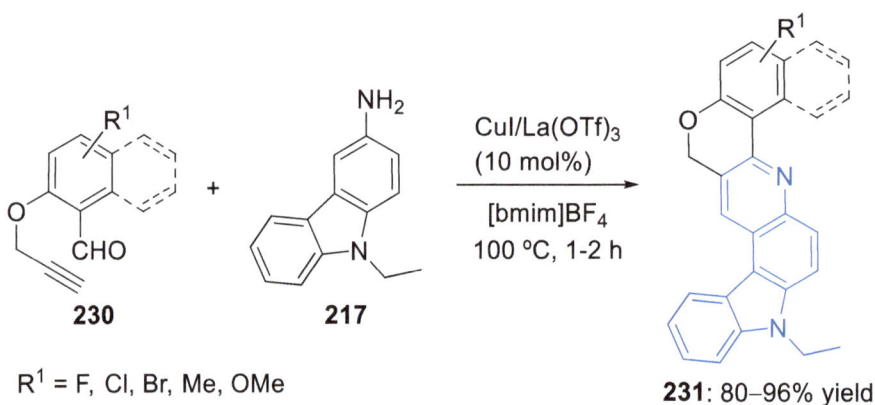

Scheme (73). Intramolecular Povarov reactions in the synthesis of isoellipticine fused with dihydrochromene derivatives.

Imines resulting from the condensation of *N*-prenylated indole-2-carbaldehydes **232** with aminocarbazoles **233**, undergo an intramolecular Povarov reaction catalyzed by Lewis or Brønsted acids [131]. The best catalytic conditions were found for La(OTf)$_3$ (10 mol%) in 1,4-dioxane at 150-160 °C, yielding heptacyclic isomeric ellipticine ring system derivatives **235** (R^3 = R^4 = H) in good yields and excellent diatereoselectivities (Scheme **74**). The process diastereoselectivity is highly influenced by the nature of the catalyst. The intramolecular Povarov reaction occurred through C-4 of the carbazole ring. Conversely, when amine **233** with substitution at C-1 and C-4 (R^3 = R^4 = Me) reacted with aldehyde **232**, the intramolecular cyclization occurred through the C-2 position of the carbazole ring, affording the corresponding product **237** in 51% yield (Scheme **74**).

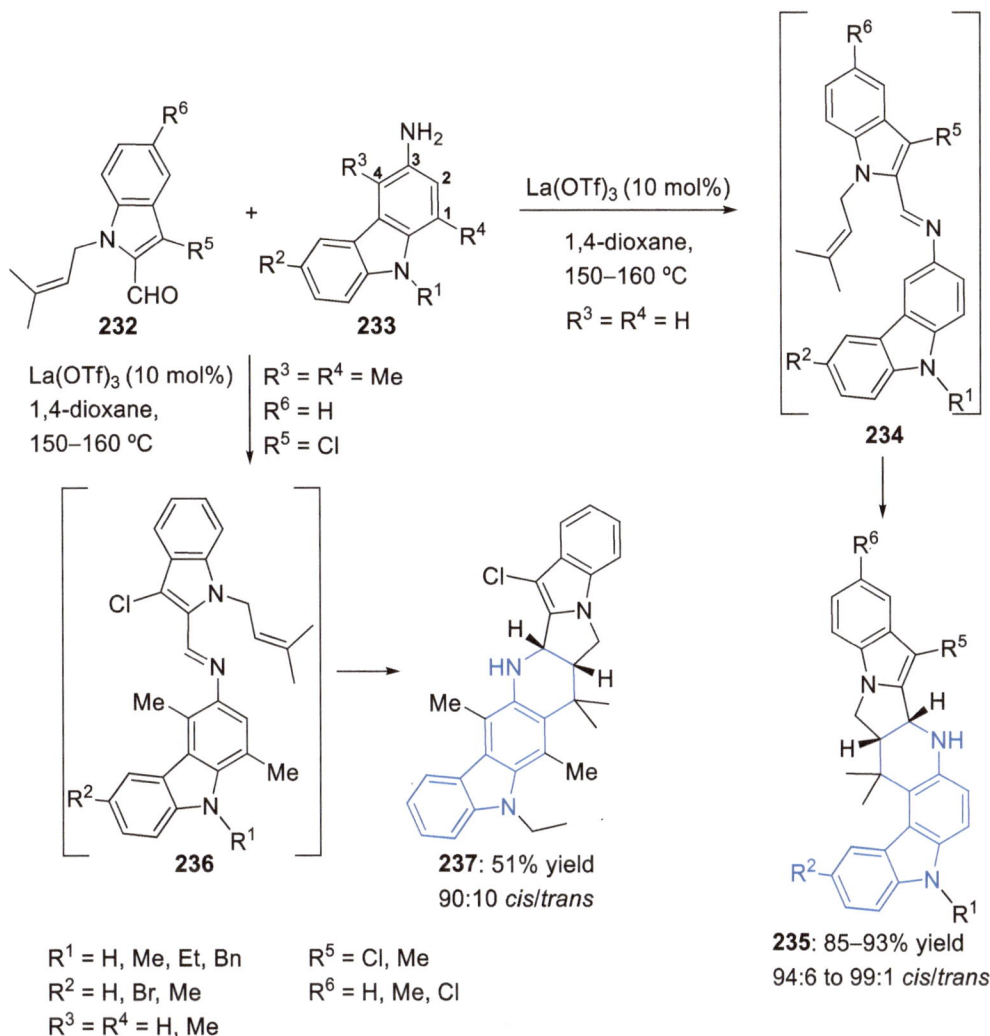

Scheme (74). Synthesis of heptacyclic isomeric ellipticine derivatives by intramolecular Povarov reaction.

Laschat's group studied the Povarov reaction of bis(imine) **238** generated from the condensation of proline-derived aldehyde **236** with 1,5-diaminonaphthalene **237** for the generation of octacyclic heterocycles (Scheme **75**) [82]. The bis-cyclization of **238** proceeded with high diastereoselectivity. The all-*cis*-configured rigid indolizino[7',8':2,3]quinolino[8,7-*h*]quinoline **239** was attained when EtAlCl$_2$ was used in the bis-cyclization reaction, while the corresponding all-*trans*-product **240** was isolated from the SnCl$_4$-promoted reaction (Scheme **75**).

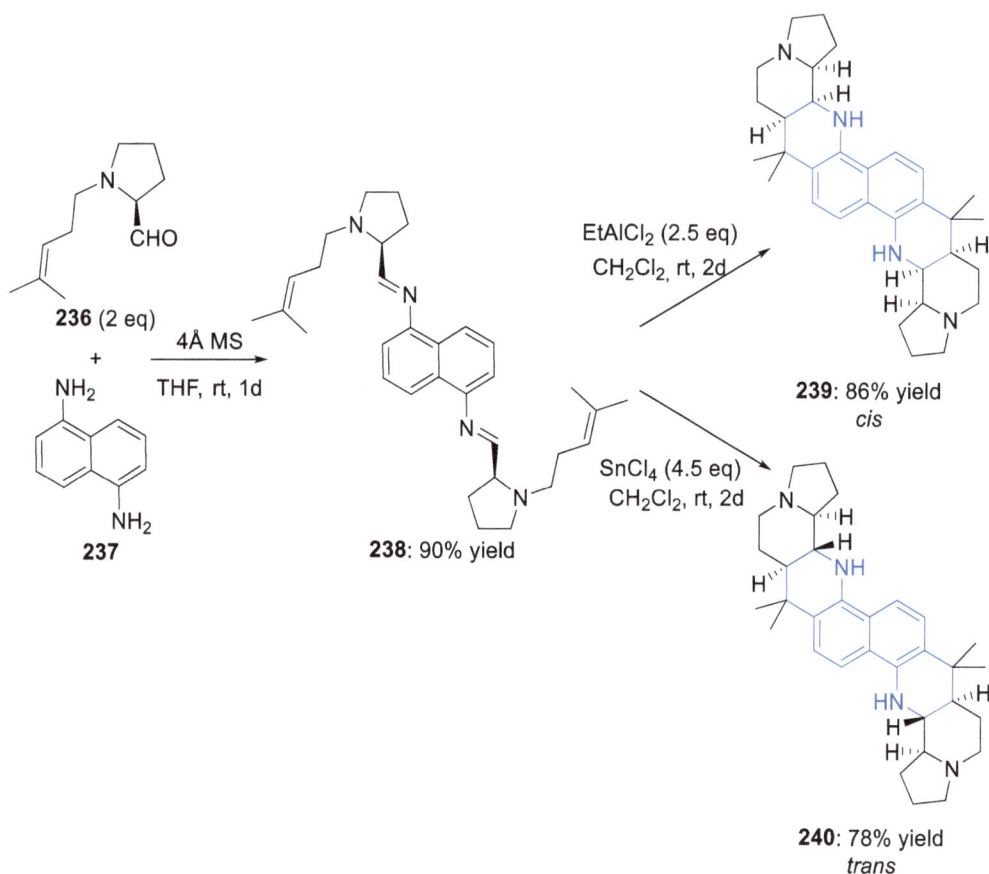

Scheme (75). Synthesis of fused bis(benzo[*b*]pyrrolo[1,2-*h*] [1, 7]naphthyridines.

CONCLUDING REMARKS

The intramolecular Povarov reaction, a type of intramolecular aza-Diels-Alder reaction, is an appropriate method for the preparation of a wide variety of fused heterocyclic compounds. This methodology is a direct route to the formation of *N*-heterocycles in a single step from imines derived from functionalized aldehydes with double or triple bonds and aromatic or heteroaromatic amines. Through the direct formation of a 6-membered nitrogen cycle (direct adduct of the [4+2] cycloaddition reaction), at least one other fused cycle is obtained from the amine to which other cycles are fused depending on the structure of the aldehyde used. In other words, the products that can be obtained range from simple quinoline derivatives to polycyclic systems as octacyclic fused heterocycles.

Polycyclic ring compounds have important applications in pharmaceuticals, biology and materials chemistry. Thus, the availability of simple tools for the preparation of complex molecules, like in this case, the intramolecular Povarov reaction, holds great potential.

Different synthetic methods using the intramolecular Povarov cycloaddition as the key step for the preparation of several drugs or alkaloids have been described. For example, the total synthesis of alkaloids such as camptothecin, luotonin A and 22-hydroxyacuminatine, as well as attempts to the preparation of isoschizozygane alkaloid core, the synthesis of uncialamycin natural product core, or the preparation of drug precursors such as irinotecan or ellipticine have been included.

As indicated throughout the text, the reaction conditions, including the type of Lewis and Brønsted acids used, have been modulated to obtain the target compounds in chemo-, regio- and stereoselective way. However, there is still much to be discovered in the future in this field, for example, the improvement of the methods described so far, the discovery of new synthetic strategies for the preparation of new molecules as well as the search for efficient catalytic asymmetric protocols to obtain compounds in an enantioselective way. In addition, as in the case of the intermolecular Povarov reaction, the development of new crystalline porous aromatic frameworks (covalent organic frameworks, COFs) from the intramolecular Povarov reaction of imine covalent organic frameworks could represent an interesting application of this methodology.

ACKNOWLEDGEMENTS

Financial support from the *Ministerio de Ciencia, Innovación y Universidades (MCIU), Agencia Estatal de Investigación (AEI) y Fondo Europeo de Desarrollo Regional (FEDER;* RTI2018-101818-B-I00, UE) and by *Gobierno Vasco, Universidad del País Vasco* (GV, IT 992-16; UPV) is gratefully acknowledged.

LIST OF ABBREVIATIONS

BA	Brønsted acid
bmim	1-butyl-3-methylimidazolium
bmim	1-hexyl-3-methylimidazolium
CAN	Ceric ammonium nitrate
DCE	dichloroethane
DCM	dichloromethane
DDQ	2,3-dichloro-5,6-dicyano-*p*-benzoquinone

DME	dimethoxyethane
DMF	dimethylformamide
DMSO	dimethylsulfoxide
eq	Equivalents
MCR	Multicomponent reaction
MDA-MB-231	Breast cancer cell line
MIC	Minimum inhibitory concentration
MS	Molecular sieves
MW	Microwave
LA	Lewis acid
LPDE	Lithium perchlorate in diethyl ether
Octmim	1-octyl-3-methylimidazolium
OHA	Octahydroacridine
rt	Room temperature
TFA	trifluoroacetic acid
TFE	trifluoroethanol
THF	tetrahydrofurane
TRIP	(*R*)-3,3'-bis(2,4,6-triisopropylphenyl)-1,1'-binaphthyl-2,2'-diyl hydrogen phosphate
TPP	triphenylphosphonium perchlorate
Ts	tosylate
TS	Transition state

REFERENCES

[1]   Cabrele, C.; Reiser, O. The modern face of synthetic heterocyclic chemistry. *J. Org. Chem.,* **2016**, *81*(21), 10109-10125.
[http://dx.doi.org/10.1021/acs.joc.6b02034] [PMID: 27680573]

[2]   Sridharan, V.; Suryavanshi, P.A.; Menéndez, J.C. Advances in the chemistry of tetrahydroquinolines. *Chem. Rev.,* **2011**, *111*(11), 7157-7259.
[http://dx.doi.org/10.1021/cr100307m] [PMID: 21830756]

[3]   Carey, J.S.; Laffan, D.; Thomson, C.; Williams, M.T. Analysis of the reactions used for the preparation of drug candidate molecules. *Org. Biomol. Chem.,* **2006**, *4*(12), 2337-2347.
[http://dx.doi.org/10.1039/b602413k] [PMID: 16763676]

[4]   Schreiber, S.L. Target-oriented and diversity-oriented organic synthesis in drug discovery. *Science,* **2000**, *287*(5460), 1964-1969.
[http://dx.doi.org/10.1126/science.287.5460.1964] [PMID: 10720315]

[5]   Muthukrishnan, I.; Sridharan, V.; Menéndez, J.C. Progress in the chemistry of tetrahydroquinolines. *Chem. Rev.,* **2019**, *119*(8), 5057-5191.
[http://dx.doi.org/10.1021/acs.chemrev.8b00567] [PMID: 30963764]

[6]   Ghashghaei, O.; Masdeu, C.; Alonso, C.; Palacios, F.; Lavilla, R. Recent advances of the Povarov reaction in medicinal chemistry. *Drug Discov. Today. Technol.,* **2018**, *29*, 71-79.

[http://dx.doi.org/10.1016/j.ddtec.2018.08.004] [PMID: 30471676]

[7]   Bello Forero, J.S.; Jones, J., Jr; da Silva, F.M. The Povarov reaction as a versatile strategy for the preparation of 1,2,3,4-tetrahydroquinoline derivatives: an overview. *Curr. Org. Synth., **2016**, *13*, 157-175.
        [http://dx.doi.org/10.2174/1570179412666150706183906]

[8]   Bello, D.; Ramón, R.; Lavilla, R. Mechanistic variations of the Povarov multicomponent reaction and related processes. *Curr. Org. Chem., **2010**, *14*(4), 332-356.
        [http://dx.doi.org/10.2174/138527210790231883]

[9]   Kouznetsov, V.V. Recent synthetic developments in a powerful imino Diels-Alder reaction (Povarov reaction): application to the synthesis of *N*-polyheterocycles and related alkaloids. *Tetrahedron*, **2009**, *65*(14), 2721-2750.
        [http://dx.doi.org/10.1016/j.tet.2008.12.059]

[10]   Povarov, L.S.; Mikhailov, B.M. *A new type of Diels-Alder reaction*; Izv. Akad. Nauk SSR, **1963**, pp. 955-956.

[11]   Kobayashi, S.; Ishitani, H.; Nagayama, S. Lanthanide triflate catalyzed imino Diels-Alder reactions; convenient syntheses of pyridine and quinoline derivatives. *Synthesis*, **1995**, *1995*(9), 1195-1202.
        [http://dx.doi.org/10.1055/s-1995-4066]

[12]   Kudale, A.A.; Kendall, J.; Miller, D.O.; Collins, J.L.; Bodwell, G.J. Povarov Reactions Involving 3-Aminocoumarins: Synthesis of 1,2,3,4-Tetrahydropyrido[2,3- *c*]coumarins and Pyrido[2,3- *c*] coumarins. *J. Org. Chem., **2008**, *73*(21), 8437-8447.
        [http://dx.doi.org/10.1021/jo801411p] [PMID: 18821803]

[13]   Palacios, F.; Alonso, C.; Arrieta, A.; Cossío, F.P.; Ezpeleta, J.M.; Fuertes, M.; Rubiales, G. Lewis acid activated aza Diels–Alder reaction of *N*-(3-pyridyl)aldimines: an experimental and computational study. *Eur. J. Org. Chem., **2010**, *2010*(11), 2091-2099.
        [http://dx.doi.org/10.1002/ejoc.200901325]

[14]   Povarov, L.S. αβ-unsaturated ethers and their analogues in reactions of diene synthesis. *Russ. Chem. Rev., **1967**, *36*(9), 656-670.
        [http://dx.doi.org/10.1070/RC1967v036n09ABEH001680]

[15]   Ma, Y.; Qian, C.; Xie, M.; Sun, J. Lanthanide chloride catalyzed imino Diels–Alder reaction. One-pot synthesis of pyrano[3,2-*c*]- and furo[3,2-*c*]quinolines. *J. Org. Chem., **1999**, *64*(17), 6462-6467.
        [http://dx.doi.org/10.1021/jo982220p]

[16]   Gaddam, V.; Meesala, R.; Nagarajan, R. A Rapid intramolecular imino Diels-Alder reaction of aminoanthraquinones with citronellal or prenylated salicylaldehydes: substituent effect on changing the reaction pathway from Diels-Alder to ene-type cyclization. *Synthesis*, **2007**, *16*, 2503-2512.

[17]   Vidari, G.; Ferrino, S.; Grieco, P.A. Quassinoids: total synthesis of dl-quassin. *J. Am. Chem. Soc.,* **1984**, *106*(12), 3539-3548.
        [http://dx.doi.org/10.1021/ja00324a023]

[18]   Laurent-Robert, H.; Garrigues, R.; Dubac, J. Bismuth(III) chloride and triflate: new efficient catalysts for the aza-Diels-Alder reaction. *Synlett,* **2000**, 1160-1162.
        [http://dx.doi.org/10.1055/s-2000-6738]

[19]   Hirashita, T.; Kawai, D.; Araki, S. GaCl₃-catalyzed [4+2] annulations of allyltrimethylsilane and trimethyl(propargyl)silane with aldimines. *Tetrahedron Lett.,* **2007**, *48*(31), 5421-5424.
        [http://dx.doi.org/10.1016/j.tetlet.2007.06.013]

[20]   Ramesh, E.; Raghunathan, R. Indium chloride catalyzed intramolecular cyclization of *N*-aryl imines: synthesis of pyrrolo[2,3-d]pyrimidine annulated tetrahydroquinoline derivatives. *Tetrahedron Lett.,* **2008**, *49*(16), 2583-2587.
        [http://dx.doi.org/10.1016/j.tetlet.2008.02.105]

[21]   Sridharan, V.; Avendaño, C.; Menéndez, J.C. New findings on the cerium(IV) ammonium nitrate

catalyzed Povarov reaction: stereoselective synthesis of 4-alkoxy-2-aryl-1,2,3,4-tetrahydroquinoline derivatives. *Synthesis,* **2008**, *7*, 1039-1044.
[http://dx.doi.org/10.1055/s-2008-1032126]

[22]   Khadem, S.; Udachin, K.A.; Enright, G.D.; Prakesch, M.; Arya, P. One-pot construction of isoindolo[2,1-a]quinoline system. *Tetrahedron Lett.,* **2009**, *50*(48), 6661-6664.
[http://dx.doi.org/10.1016/j.tetlet.2009.09.075]

[23]   Baudelle, R.; Melnyk, P.; Déprez, B.; Tartar, A. Parallel synthesis of polysubstituted tetrahydroquinolines. *Tetrahedron,* **1998**, *54*(16), 4125-4140.
[http://dx.doi.org/10.1016/S0040-4020(98)00140-9]

[24]   Dai, W.; Jiang, X.L.; Tao, J.Y.; Shi, F. Application of 3-methyl-2-vinylindoles in catalytic asymmetric Povarov reaction: diastereo- and enantioselective synthesis of indole-derived tetrahydroquinolines. *J. Org. Chem.,* **2016**, *81*(1), 185-192.
[http://dx.doi.org/10.1021/acs.joc.5b02476] [PMID: 26652222]

[25]   Akiyama, T.; Morita, H.; Fuchibe, K. Chiral Brønsted acid-catalyzed inverse electron-demand aza Diels-Alder reaction. *J. Am. Chem. Soc.,* **2006**, *128*(40), 13070-13071.
[http://dx.doi.org/10.1021/ja064676r] [PMID: 17017784]

[26]   Luo, H.X.; Niu, Y.; Jin, X.; Cao, X.P.; Yao, X.; Ye, X.S. Indolo-quinoline boron difluoride dyes: synthesis and spectroscopic properties. *Org. Biomol. Chem.,* **2016**, *14*(18), 4185-4188.
[http://dx.doi.org/10.1039/C6OB00623J] [PMID: 27098049]

[27]   Zhu, Z.B.; Shao, L.X.; Shi, M. Brønsted acid or solid acid catalyzed aza-Diels-Alder reactions of methylenecyclopropanes with ethyl (arylimino)acetates. *Eur. J. Org. Chem.,* **2009**, *2009*(15), 2576-2580.
[http://dx.doi.org/10.1002/ejoc.200900050]

[28]   Abranches, P.A.S.; de Paiva, W.F.; de Fátima, Â.; Martins, F.T.; Fernandes, S.A. Calix[n]arene-catalyzed three-component Povarov reaction: microwave-assisted synthesis of julolidines and mechanistic insights. *J. Org. Chem.,* **2018**, *83*(4), 1761-1771.
[http://dx.doi.org/10.1021/acs.joc.7b02532] [PMID: 29337547]

[29]   Zhu, J.; Wang, Q.; Wang, M., Eds. *Multicomponent Reactions in Organic Synthesis*; Wiley: Chichester, **2014**.
[http://dx.doi.org/10.1002/9783527678174]

[30]   Müller, T.J.J., Ed. Science of Synthesis; Thieme: Stuttgart, 2014, Vol. 1 and 2.

[31]   Wang, Q.; Zhu, J. *Multicomponent Domino Process: Rational Design and Serendipity*; Tietze, L.F., Ed.; Willey: Chichester, **2014**, pp. 579-610.

[32]   Brauch, S.; van Berkel, S.S.; Westermann, B. Higher-order multicomponent reactions: beyond four reactants. *Chem. Soc. Rev.,* **2013**, *42*(12), 4948-4962.
[http://dx.doi.org/10.1039/c3cs35505e] [PMID: 23426583]

[33]   Feng, X.; Song, Y.; Lin, W. Dimensional reduction of Lewis acidic metal−organic frameworks for multicomponent reactions. *J. Am. Chem. Soc.,* **2021**, *143*(21), 8184-8192.
[http://dx.doi.org/10.1021/jacs.1c03561] [PMID: 34018731]

[34]   Maujean, T.; Chataigner, I.; Girard, N.; Gulea, M.; Bonnet, D. Endocyclic enamides derived from aza-diketopiperazines as olefin partners in Povarov reaction: an access to tetracyclic *N*-heterocycles. *Eur. J. Org. Chem.,* **2020**, *2020*(47), 7385-7395.
[http://dx.doi.org/10.1002/ejoc.202001339]

[35]   Wang, S.J.; Wang, Z.; Tang, Y.; Chen, J.; Zhou, L. Asymmetric synthesis of quinoline-naphthalene atropisomers by central-to-axial chirality conversion. *Org. Lett.,* **2020**, *22*(22), 8894-8898.
[http://dx.doi.org/10.1021/acs.orglett.0c03285] [PMID: 33124830]

[36]   Dayal, N.; Wang, M.; Sintim, H.O. HSD1787, a tetrahydro-3*H*-pyrazolo[4,3-*f*]quinoline compound synthesized *via* Povarov reaction, potently inhibits proliferation of cancer cell lines at nanomolar

concentrations. *ACS Omega,* **2020**, *5*(37), 23799-23807.
[http://dx.doi.org/10.1021/acsomega.0c03001] [PMID: 32984700]

[37]   Tejería, A.; Pérez-Pertejo, Y.; Reguera, R.M.; Carbajo-Andrés, R.; Balaña-Fouce, R.; Alonso, C.; Martin-Encinas, E.; Selas, A.; Rubiales, G.; Palacios, F. Antileishmanial activity of new hybrid tetrahydroquinoline and quinoline derivatives with phosphorus substituents. *Eur. J. Med. Chem.,* **2019**, *162*, 18-31.
[http://dx.doi.org/10.1016/j.ejmech.2018.10.065] [PMID: 30408746]

[38]   Alonso, C.; Fuertes, M.; Martín-Encinas, E.; Selas, A.; Rubiales, G.; Tesauro, C.; Knudssen, B.K.; Palacios, F. Novel topoisomerase I inhibitors. Syntheses and biological evaluation of phosphorus substituted quinoline derivates with antiproliferative activity. *Eur. J. Med. Chem.,* **2018**, *149*, 225-237.
[http://dx.doi.org/10.1016/j.ejmech.2018.02.058] [PMID: 29501943]

[39]   Alonso, C.; Martín-Encinas, E.; Rubiales, G.; Palacios, F. Reliable synthesis of phosphino- and phosphine sulfide-1,2,3,4-tetrahydroquinolines and phosphine sulfide quinolines. *Eur. J. Org. Chem.,* **2017**, *2017*(20), 2916-2924.
[http://dx.doi.org/10.1002/ejoc.201700258]

[40]   Tejería, A.; Pérez-Pertejo, Y.; Reguera, R.M.; Balaña-Fouce, R.; Alonso, C.; Fuertes, M.; González, M.; Rubiales, G.; Palacios, F. Antileishmanial effect of new indeno-1,5-naphthyridines, selective inhibitors of Leishmania infantum type IB DNA topoisomerase. *Eur. J. Med. Chem.,* **2016**, *124*, 740-749.
[http://dx.doi.org/10.1016/j.ejmech.2016.09.017] [PMID: 27639365]

[41]   Alonso, C.; Fuertes, M.; González, M.; Rubiales, G.; Tesauro, C.; Knudsen, B.R.; Palacios, F. Synthesis and biological evaluation of indeno[1,5]naphthyridines as topoisomerase I (TopI) inhibitors with antiproliferative activity. *Eur. J. Med. Chem.,* **2016**, *115*, 179-190.
[http://dx.doi.org/10.1016/j.ejmech.2016.03.031] [PMID: 27017547]

[42]   Mazaheripour, A.; Dibble, D. J.; Umerani, M. J.; Park, Y. S.; Lopez, R.; Laidlaw, D.; Vargas, E.; Ziller, J. W.; Gorodetsky, A. A. An aza-Diels-Alder approach to crowded benzoquinolines. *Org. Lett,* **2016**, *18*, 156-159.
[http://dx.doi.org/10.1021/acs.orglett.5b02939]

[43]   Mi, X.; Chen, J.; Xu, L. FeCl$_3$-catalyzed SF5-containing quinoline synthesis: three-component coupling reactions of SF5-anilines, aldehydes and alkynes. *Eur. J. Org. Chem.,* **2015**, *2015*(7), 1415-1418.
[http://dx.doi.org/10.1002/ejoc.201403613]

[44]   Meyet, C.E.; Larsen, C.H. One-step catalytic synthesis of alkyl-substituted quinolines. *J. Org. Chem.,* **2014**, *79*(20), 9835-9841.
[http://dx.doi.org/10.1021/jo5015883] [PMID: 25229642]

[45]   Selas, A.; Ramírez, G.; Palacios, F.; Alonso, C. Design, synthesis and cytotoxic evaluation of diphenyl(quinolin-8-yl)phosphine oxides. *Tetrahedron Lett.,* **2021**, *70*, 153019.
[http://dx.doi.org/10.1016/j.tetlet.2021.153019]

[46]   Alonso, C.; González, M.; Palacios, F.; Rubiales, G. Study of the hetero-[4+2]-cycloaddition reaction of aldimines and alkynes. Synthesis of 1,5- naphthyridine and isoindolone derivatives. *J. Org. Chem.,* **2017**, *82*(12), 6379-6387.
[http://dx.doi.org/10.1021/acs.joc.7b00977] [PMID: 28537387]

[47]   Gensicka-Kowalewska, M.; Cholewiński, G.; Dzierzbicka, K. Recent developments in the synthesis and biological activity of acridine/acridone analogues. *RSC Advances,* **2017**, *7*(26), 15776-15804.
[http://dx.doi.org/10.1039/C7RA01026E]

[48]   Cholewiński, G.; Dzierzbicka, K.; Kołodziejczyk, A.M. Natural and synthetic acridines/acridones as antitumor agents: their biological activities and methods of synthesis. *Pharmacol. Rep.,* **2011**, *63*(2), 305-336.
[http://dx.doi.org/10.1016/S1734-1140(11)70499-6] [PMID: 21602588]

[49]   Laschat, S.; Lauterwein, J. Intramolecular hetero-Diels-Alder reaction of *N*-arylimines. Applications to the synthesis of octahydroacridine derivatives. *J. Org. Chem.*, **1993**, *58*(10), 2856-2861.
[http://dx.doi.org/10.1021/jo00062a033]

[50]   Temme, O.; Laschat, S. Effect of molecular sieves on the formation and acid-catalysed mono- and bis-cyclization of N-arylimines: easy entry to polycyclic ring systems by a novel cascade reaction. *J. Chem. Soc., Perkin Trans. 1,* **1995**, (2), 125-131.
[http://dx.doi.org/10.1039/p19950000125]

[51]   Schulte, J.L.; Laschat, S.; Kotila, S.; Hecht, J.; Frohlich, R. Wibbeling, Birgit. Synthesis of +-(octahydroacrydine)-chromium tricarbonyl complexes with non-polar tails *via* molecular sieves-catalyzed cyclization of A'-arylimines and subsequent diastereoselective complexation. *Heterocycles,* **1996**, *43*, 2713-2724.
[http://dx.doi.org/10.3987/COM-96-7622]

[52]   Sabitha, G.; Reddy, E.V.; Yadav, J.S. Bismuth(III) chloride: an efficient catalyst for the one-pot stereoselective synthesis of octahydroacridines. *Synthesis,* **2002**, *2002*(3), 409-412.
[http://dx.doi.org/10.1055/s-2002-20045]

[53]   Jacob, R.G.; Perin, G.; Botteselle, G.V.; Lenardão, E.J. Clean and atom-economic synthesis of octahydroacridines: application to essential oil of citronella. *Tetrahedron Lett.,* **2003**, *44*(36), 6809-6812.
[http://dx.doi.org/10.1016/S0040-4039(03)01749-0]

[54]   Jacob, R.G.; Silva, M.S.; Mendes, S.R.; Borges, E.L.; Lenardão, E.J.; Perin, G. Atom-economic synthesis of functionalized octahydroacridines from citronellal or 3-(phenylthio)-citronellal. *Synth. Commun.,* **2009**, *39*(15), 2747-2762.
[http://dx.doi.org/10.1080/00397910802663469]

[55]   Zaccheria, F.; Santoro, F.; Iftitah, E.; Ravasio, N. Brønsted and Lewis solid acid catalysts in the valorization of citronellal. *Catalysts,* **2018**, *8*(10), 410.
[http://dx.doi.org/10.3390/catal8100410]

[56]   Kiselyov, A.S.; Smith, L., II; Armstrong, R.W. Solid support synthesis of polysubstituted tetrahydroquinolines *via* three-component condensation catalyzed by Yb(OTf)3. *Tetrahedron,* **1998**, *54*(20), 5089-5096.
[http://dx.doi.org/10.1016/S0040-4020(98)00248-8]

[57]   Mayekar, N.V.; Nayak, S.K.; Chattopadhyay, S. Two convenient one-pot strategies for the synthesis of octahydroacridines. *Synth. Commun.,* **2004**, *34*(17), 3111-3119.
[http://dx.doi.org/10.1081/SCC-200028567]

[58]   Lenardão, E.J.; Mendes, S.R.; Ferreira, P.C.; Perin, G.; Silveira, C.C.; Jacob, R.G. Selenium- and tellurium-based ionic liquids and their use in the synthesis of octahydroacridines. *Tetrahedron Lett.,* **2006**, *47*(42), 7439-7442.
[http://dx.doi.org/10.1016/j.tetlet.2006.08.049]

[59]   Yadav, J.S.; Reddy, B.V.S.; Chetia, L.; Srinivasulu, G.; Kunwar, A.C. Ionic liquid accelerated intramolecular hetero-Diels-Alder reactions: a protocol for the synthesis of octahydroacridines. *Tetrahedron Lett.,* **2005**, *46*(6), 1039-1044.
[http://dx.doi.org/10.1016/j.tetlet.2004.11.137]

[60]   Dickmeiss, G.; Jensen, K.L.; Worgull, D.; Franke, P.T.; Jørgensen, K.A. An asymmetric organocatalytic one-pot strategy to octahydroacridines. *Angew. Chem. Int. Ed.,* **2011**, *50*(7), 1580-1583.
[http://dx.doi.org/10.1002/anie.201006608] [PMID: 21308909]

[61]   Magomedov, N.A. Efficient construction of cyclopenta[b]quinoline core of isoschizozygane alkaloids *via* intramolecular formal hetero-Diels-Alder reaction. *Org. Lett.,* **2003**, *5*(14), 2509-2512.
[http://dx.doi.org/10.1021/ol034776r] [PMID: 12841767]

[62] Desrat, S.; van de Weghe, P. Intramolecular imino Diels-Alder reaction: progress toward the synthesis of uncialamycin. *J. Org. Chem.,* **2009**, *74*(17), 6728-6734.
[http://dx.doi.org/10.1021/jo901291t] [PMID: 19637842]

[63] Spaller, M.R.; Thielemann, W.T.; Brennan, P.E.; Bartlett, P.A. Combinatorial synthetic design. Solution and polymer-supported synthesis of heterocycles *via* intramolecular aza Diels-Alder and imino alcohol cyclizations. *J. Comb. Chem.,* **2002**, *4*(5), 516-522.
[http://dx.doi.org/10.1021/cc020027+] [PMID: 12217025]

[64] More, D.A.; Shinde, G.H.; Shaikh, A.C.; Muthukrishnan, M. Oxone promoted dehydrogenative Povarov cyclization of *N*-aryl glycine derivatives: an approach towards quinoline fused lactones and lactams. *RSC Advances,* **2019**, *9*(52), 30277-30291.
[http://dx.doi.org/10.1039/C9RA06212B] [PMID: 35530246]

[65] Jayagobi, M.; Poornachandran, M.; Raghunathan, R. A novel heterotricyclic assembly through intramolecular imino Diels-Alder reaction: synthesis of pyrrolo[3,4-b]quinolines. *Tetrahedron Lett.,* **2009**, *50*(6), 648-650.
[http://dx.doi.org/10.1016/j.tetlet.2008.11.092]

[66] Jayagobi, M.; Raghunathan, R.; Sainath, S.; Raghunathan, M. Synthesis and antibacterial property of pyrrolopyrano quinolinones and pyrroloquinolines. *Eur. J. Med. Chem.,* **2011**, *46*(6), 2075-2082.
[http://dx.doi.org/10.1016/j.ejmech.2011.02.060] [PMID: 21444131]

[67] Jayagobi, M.; Raghunathan, R. Novel diastereoselective synthesis of *trans*-fused pyrrolo[3,4-*b*]quinolines through intramolecular imino-Diels-Alder reaction. *Synth. Commun.,* **2012**, *42*(19), 2917-2930.
[http://dx.doi.org/10.1080/00397911.2011.572218]

[68] Almansour, A.I.; Arumugam, N.; Suresh Kumar, R.; Carlos Menéndez, J.; Ghabbour, H.A.; Fun, H.K.; Ranjith Kumar, R. Straightforward synthesis of pyrrolo[3,4-*b*]quinolines through intramolecular Povarov reactions. *Tetrahedron Lett.,* **2015**, *56*(49), 6900-6903.
[http://dx.doi.org/10.1016/j.tetlet.2015.10.107]

[69] Mishra, A.; Rastogi, N.; Batra, S. 2-(*N*-Allylaminomethyl)cinnamaldehydes as substrates for syntheses of aza-polycycles *via* intramolecular cycloaddition reactions. *Tetrahedron,* **2012**, *68*(9), 2146-2154.
[http://dx.doi.org/10.1016/j.tet.2012.01.016]

[70] Lezana, N.; Matus-Pérez, M.; Galdámez, A.; Lühr, S.; Vilches-Herrera, M. Highly stereoselective and catalyst-free synthesis of annulated tetrahydropyridines by intramolecular imino-Diels-Alder reaction under microwave irradiation in water. *Green Chem.,* **2016**, *18*(13), 3712-3717.
[http://dx.doi.org/10.1039/C6GC00912C]

[71] Chen, S.Y.; Li, Q.; Liu, X.G.; Wu, J.Q.; Zhang, S.S.; Wang, H. Polycyclization enabled by relay catalysis: one-pot manganese-catalyzed C-H allylation and silver-catalyzed Povarov reaction. *ChemSusChem,* **2017**, *10*(11), 2360-2364.
[http://dx.doi.org/10.1002/cssc.201700452] [PMID: 28471522]

[72] Chen, M.; Sun, N.; Liu, Y. Environmentally benign synthesis of indeno[1,2-*b*]quinolines *via* an intramolecular Povarov reaction. *Org. Lett.,* **2013**, *15*(21), 5574-5577.
[http://dx.doi.org/10.1021/ol402775k] [PMID: 24128093]

[73] Linkert, F.; Laschat, S.; Knickmeier, M. *Chelation control by a second nitrogen atom in formal hetero-Diels-Alder reactions of N-arylimines*; Liebigs Annal, **1995**, pp. 985-993.

[74] Chen, Y.; Ramanathan, M.; Liu, S.T. Intramolecular aza-Diels-Alder cyclization of a dimerized citral with anilines catalyzed by InCl$_3$. *J. Chin. Chem. Soc. (Taipei),* **2015**, *62*(9), 761-765.
[http://dx.doi.org/10.1002/jccs.201500112]

[75] Sabitha, G.; Shankaraiah, K.; Sindhu, K.; Latha, B. Bismuth(III) chloride catalyzed intramolecular hetero-Diels-Alder reaction: Access to *cis*-fused angular hexahydrobenzo[*c*]acridines. *Synthesis,* **2014**, *47*(1), 124-128.

[http://dx.doi.org/10.1055/s-0034-1379022]

[76]   Ramesh, E.; Manian, R.D.R.S.; Raghunathan, R.; Sainath, S.; Raghunathan, M. Synthesis and antibacterial property of quinolines with potent DNA gyrase activity. *Bioorg. Med. Chem.,* **2009**, *17*(2), 660-666.
[http://dx.doi.org/10.1016/j.bmc.2008.11.058] [PMID: 19097914]

[77]   Dong, W.; Yuan, Y.; Hu, B.; Gao, X.; Gao, H.; Xie, X.; Zhang, Z. Combining visible-light-photoredox and Lewis acid catalysis for the synthesis of indolizino[1,2-*b*]quinolin-9(11*H*)-ones and irinotecan precursor. *Org. Lett.,* **2018**, *20*(1), 80-83.
[http://dx.doi.org/10.1021/acs.orglett.7b03395] [PMID: 29215891]

[78]   Twin, H.; Batey, R.A. Intramolecular hetero Diels-Alder (Povarov) approach to the synthesis of the alkaloids luotonin A and camptothecin. *Org. Lett.,* **2004**, *6*(26), 4913-4916.
[http://dx.doi.org/10.1021/ol0479848] [PMID: 15606098]

[79]   Linkert, F.; Laschat, S. Intramolecular hetero-Diels-Alder reaction of prolinal-derived *N*-arylimines. Lewis acid-dependent reversal of the diastereoselectivity. *Synlett,* **1994**, *1994*(2), 125-126.
[http://dx.doi.org/10.1055/s-1994-22764]

[80]   Linkert, F.; Laschat, S.; Kotila, S.; Fox, T. Evidence for a stepwise mechanism in formal hetero-Diel-
-Alder reactions of *N*-arylimines. *Tetrahedron,* **1996**, *52*(3), 955-970.
[http://dx.doi.org/10.1016/0040-4020(95)00947-7]

[81]   Dickner, T.; Laschat, S. Stereoselective synthesis and binding properties of novel concave-shaped indolizino[3,4-*b*]quinolines. *J. Prakt. Chem.,* **2000**, *342*(8), 804-811.
[http://dx.doi.org/10.1002/1521-3897(200010)342:8<804::AID-PRAC804>3.0.CO;2-F]

[82]   Temme, O.; Dickner, T.; Laschat, S.; Fröhlich, R.; Kotila, S.; Bergander, K. Synthesis of aza polycyclic systems based on the indolizino[3,4-*b*]quinoline skeleton. A diastereoselective entry to potential oligodentate artificial receptors. *Eur. J. Org. Chem.,* **1998**, *1998*(4), 651-659.
[http://dx.doi.org/10.1002/(SICI)1099-0690(199804)1998:4<651::AID-EJOC651>3.0.CO;2-Z]

[83]   Sabitha, G.; Venkata Reddy, E.; Maruthi, C.; Yadav, J.S. Bismuth(III) chloride-catalyzed intramolecular hetero-Diels-Alder reactions: a novel synthesis of hexahydrodibenzo [*b,h*] [1,6]naphthyridines. *Tetrahedron Lett.,* **2002**, *43*(8), 1573-1575.
[http://dx.doi.org/10.1016/S0040-4039(02)00018-7]

[84]   Muthukrishnan, I.; Vinoth, P.; Vivekanand, T.; Nagarajan, S.; Maheswari, C.U.; Menéndez, J.C.; Sridharan, V. Synthesis of 5,6-dihydrodibenzo[*b,h*][1,6]naphthyridines *via* copper bromide catalyzed intramolecular [4+2] hetero-Diels−Alder reactions. *J. Org. Chem.,* **2016**, *81*(3), 1116-1124.
[http://dx.doi.org/10.1021/acs.joc.5b02669] [PMID: 26694659]

[85]   Dong, W.; Yuan, Y.; Gao, X.; Hu, B.; Xie, X.; Zhang, Z. Merging visible-light photoredox and Lewis acid catalysis for the intramolecular aza-Diels-Alder reaction: synthesis of substituted chromeno[4,3-*b*]quinolines and [1,6]naphthyridines. *ChemCatChem,* **2018**, *10*(13), 2878-2886.
[http://dx.doi.org/10.1002/cctc.201800192]

[86]   Jarrige, L.; Blanchard, F.; Masson, G. Enantioselective organocatalytic intramolecular aza-Diels-Alder reaction. *Angew. Chem. Int. Ed.,* **2017**, *56*(35), 10573-10576.
[http://dx.doi.org/10.1002/anie.201705746] [PMID: 28661020]

[87]   Jarrige, L.; Gandon, V.; Masson, G. Enantioselective synthesis of complex fused heterocycles through chiral phosphoric acid catalyzed intramolecular inverse-electron-demand aza-Diels-Alder reactions. *Chemistry,* **2020**, *26*(6), 1406-1413.
[http://dx.doi.org/10.1002/chem.201904902] [PMID: 31663177]

[88]   Yang, F.; Zheng, L.; Xiang, J.; Dang, Q.; Bai, X. Synthesis of Hexahydrobenzo[*b*]pyrimido[4,5- *h*][1,6]naphthyridines *via* an Intramolecular Hetero-Diels−Alder Reaction. *J. Comb. Chem.,* **2010**, *12*(4), 476-481.
[http://dx.doi.org/10.1021/cc100018b] [PMID: 20550172]

[89]   Jones, W.; Kiselyov, A.S. Intramolecular cyclization of aromatic imines: an approach to tetrahydrochromano[4,3-b]quinolines. *Tetrahedron Lett.,* **2000**, *41*(14), 2309-2312.
[http://dx.doi.org/10.1016/S0040-4039(00)00156-8]

[90]   Rajagopal, N.; Magesh, C.J.; Perumal, P.T. Inter- and intramolecular imino-Diels-Alder reactions catalyzed by sulfamic acid: a mild and efficient catalyst for a one-pot synthesis of tetrahydroquinolines. *Synthesis,* **2004**, *1*, 69-74.

[91]   Elamparuthi, E.; Anniyappan, M.; Muralidharan, D.; Perumal, P.T. InCl₃ as an efficient catalyst for intramolecular imino-Diels-Alder reactions: synthesis of tetrahydrochromenoquinolines. *ARKIVOC,* **2005**, *11*, 6-16.

[92]   Sabitha, G.; Reddy, E.V.; Yadav, J.S. Bismuth(III) chloride-catalyzed intramolecular hetero-Diel-Alder reaction: application to the synthesis of tetrahydrochromeno[4,3-*b*]quinoline derivatives. *Synthesis,* **2001**, *2001*(13), 1979-1984.
[http://dx.doi.org/10.1055/s-2001-17709]

[93]   Yadav, J.S.; Reddy, B.V.S.; Rao, C.V.; Srinivas, R. LPDE-catalyzed intramolecular cyclization of arylimines: a facile synthesis of tetrahydrochromenoquinolines. *Synlett,* **2002**, *2002*(6), 0993-0995.
[http://dx.doi.org/10.1055/s-2002-31902]

[94]   Anniyappan, M.; Muralidharan, D.; Perumal, P.T. Triphenylphosphonium perchlorate as an efficient catalyst for mono- and bis-intramolecular imino Diels-Alder reactions: synthesis of tetrahydrochromanoquinolines. *Tetrahedron Lett.,* **2003**, *44*(18), 3653-3657.
[http://dx.doi.org/10.1016/S0040-4039(03)00707-X]

[95]   Yadav, J.S.; Reddy, B.V.S.; Kondaji, G.; Sowjanya, S.; Nagaiah, K. Intramolecular imino-Diels-Alder reactions in [bmim]BF4 ionic medium: Green protocol for the synthesis of tetrahydrochromanoquinolines. *J. Mol. Catal. Chem.,* **2006**, *258*(1-2), 361-366.
[http://dx.doi.org/10.1016/j.molcata.2006.07.016]

[96]   Imrich, H.G.; Conrad, J.; Beifuss, U. The First domino reduction/imine formation/intramolecular aza-Diels-Alder reaction for the diastereoselective preparation of tetrahydrochromeno[4,3-*b*]quinolines. *Eur. J. Org. Chem.,* **2016**, *2016*(34), 5706-5715.
[http://dx.doi.org/10.1002/ejoc.201600976]

[97]   Zhang, D.; Kiselyov, A.S. A solid-phase approach to tetrahydrochromeno[4,3-*b*]quinolines. *Synlett,* **2001**, *2001*(7), 1173-1175.
[http://dx.doi.org/10.1055/s-2001-15159]

[98]   Ramesh, S.; Gaddam, V.; Nagarajan, R. A flexible approach to the chromenoquinolines under copper/Lewis acid catalysis. *Synlett,* **2010**, 757-760.
[http://dx.doi.org/10.1055/s-0029-1219364]

[99]   Rahimzadeh, G.; Soheilizad, M.; Kianmehr, E.; Larijani, B.; Mahdavi, M. Copper-catalyzed intramolecular domino synthesis of 6*H*-chromeno[4,3-*b*]quinolines in green condition. *ARKIVOC,* **2018**, *2018*(5), 20-28.
[http://dx.doi.org/10.24820/ark.5550190.p010.406]

[100]   Yu, X.; Wang, J.; Xu, Z.; Yamamoto, Y.; Bao, M. Copper-catalyzed aza-Diels−Alder reaction and halogenation: an approach to synthesize 7-halogenated chromenoquinolines. *Org. Lett.,* **2016**, *18*(10), 2491-2494.
[http://dx.doi.org/10.1021/acs.orglett.6b01065] [PMID: 27145113]

[101]   Sabitha, G.; Reddy, Ch.S.; Maruthi, Ch.; Reddy, E.V.; Yadav, J.S. BiCl₃-catalyzed diastereoselective intramolecular [4+2] cycloaddition reactions leading to pyrazole annulated new sulfur heterocycles. *Synth. Commun.,* **2003**, *33*, 3063-3070.
[http://dx.doi.org/10.1081/SCC-120022482]

[102]   Manian, R.D.R.S.; Jayashankaran, J.; Ramesh, R.; Raghunathan, R. Rapid synthesis of tetrahydroquinolines by indium trichloride catalyzed mono- and bis-intramolecular imino Diels-Alder

reactions. *Tetrahedron Lett.,* **2006**, *47*(43), 7571-7574.
[http://dx.doi.org/10.1016/j.tetlet.2006.08.088]

[103] Martín-Encinas, E.; Selas, A.; Tesauro, C.; Rubiales, G.; Knudsen, B.R.; Palacios, F.; Alonso, C. Synthesis of novel hybrid quinolino[4,3-*b*][1,5]naphthyridines and quinolino[4,3-*b*][1,5]naphthyridin-6(5*H*)-one derivatives and biological evaluation as topoisomerase I inhibitors and antiproliferatives. *Eur. J. Med. Chem.,* **2020**, *195*, 112292.
[http://dx.doi.org/10.1016/j.ejmech.2020.112292] [PMID: 32279049]

[104] Martín-Encinas, E.; Rubiales, G.; Knudsen, B.R.; Palacios, F.; Alonso, C. Fused chromeno and quinolino[1,8]naphthyridines: Synthesis and biological evaluation as topoisomerase I inhibitors and antiproliferative agents. *Bioorg. Med. Chem.,* **2021**, *40*, 116177.
[http://dx.doi.org/10.1016/j.bmc.2021.116177] [PMID: 33962152]

[105] Martín-Encinas, E.; Rubiales, G.; Knudssen, B.R.; Palacios, F.; Alonso, C. Straightforward synthesis and biological evaluation as topoisomerase I inhibitors and antiproliferative agents of hybrid Chromeno[4,3-*b*][1,5]Naphthyridines and Chromeno[4,3-*b*][1,5]Naphthyridin-6-ones. *Eur. J. Med. Chem.,* **2019**, *178*, 752-766.
[http://dx.doi.org/10.1016/j.ejmech.2019.06.032] [PMID: 31229877]

[106] Tietze, L.F.; Utecht, J. Inter- and intramolecular hetero diels-alder reactions, 41. Unusual stereocontrol in intramolecular hetero diels-alder reactions of 2-Aza-1,3-butadienes. A stereoselective sequential synthesis of annulated tetrahydropyridines. *Chem. Ber.,* **1992**, *125*(10), 2259-2263.
[http://dx.doi.org/10.1002/cber.19921251014]

[107] Majumdar, K.; Ponra, S.; Ganai, S. Copper(I)-catalyst-free approach for the synthesis of chromene-fused pyrido[3,2-*d*]pyrimidines by Lewis acid catalyzed aza-Diels-Alder reaction. *Synlett,* **2010**, *2010*(17), 2575-2578.
[http://dx.doi.org/10.1055/s-0030-1258768]

[108] Chakraborty, B.; Kar, A.; Chanda, R.; Jana, U. Application of the Povarov reaction in biaryls under iron catalysis for the general synthesis of dibenzo[*a,c*]acridines. *J. Org. Chem.,* **2020**, *85*(14), 9281-9289.
[http://dx.doi.org/10.1021/acs.joc.0c01300] [PMID: 32588630]

[109] Gaddam, V.; Nagarajan, R. Highly diastereoselective synthesis of new indolopyrroloquinolines through intramolecular imino Diels-Alder reactions. *Tetrahedron Lett.,* **2007**, *48*(41), 7335-7338.
[http://dx.doi.org/10.1016/j.tetlet.2007.08.016]

[110] Manian, R.D.R.S.; Jayashankaran, J.; Raghunathan, R. Indium trichloride catalyzed one-pot synthesis of indolo[2,1-*a*]pyrrolo[4′,3′:2,3]-7a,8,13,13b-tetrahydroquinolines through intramolecular imino Diels-Alder reactions. *Tetrahedron Lett.,* **2007**, *48*(23), 4139-4142.
[http://dx.doi.org/10.1016/j.tetlet.2007.03.170]

[111] Almansour, A.I.; Suresh Kumar, R.; Arumugam, N.; Bianchini, G.; Menéndez, J.C.; Al-thamili, D.M.; Periyasami, G.; Altaf, M. Design and synthesis of A- and D ring-modified analogues of luotonin A with reduced planarity. *Tetrahedron Lett.,* **2019**, *60*(23), 1514-1517.
[http://dx.doi.org/10.1016/j.tetlet.2019.05.010]

[112] Monsees, A.; Laschat, S.; Hotfilder, M.; Jones, P.G. Diastereoselective synthesis of octahydro-14*H*-benzo[*g*]quinolino-[2,3-*a*]quinolidines. Improved cytotoxic activity against human brain tumor cell lines as a result of the increased rigidity of the molecular backbone. *Bioorg. Med. Chem. Lett.,* **1998**, *8*(20), 2881-2884.
[http://dx.doi.org/10.1016/S0960-894X(98)00506-X] [PMID: 9873641]

[113] Koepler, O.; Mazzini, S.; Bellucci, M.C.; Mondelli, R.; Baro, A.; Laschat, S.; Hotfilder, M.; Viseur, C.; Frey, W. Synthesis and DNA binding properties of novel benzo[*b*]isoquino[2,3-*h*]-naphthyridines. *Org. Biomol. Chem.,* **2005**, *3*(15), 2848-2858.
[http://dx.doi.org/10.1039/b503281d] [PMID: 16032363]

[114] Dai, X.; Cheng, C.; Ding, C.; Yao, Q.; Zhang, A. Synthesis of 2,7-naphthyridine-containing analogues

of luotonin A. *Synlett,* **2008,** 2989-2992.
[http://dx.doi.org/10.1055/s-0028-1087299]

[115] Zhou, H.B.; Liu, G.S.; Yao, Z.J. Highly efficient and mild cascade reactions triggered by bis(triphenyl)oxodiphosphonium trifluoromethanesulfonate and a concise total synthesis of camptothecin. *Org. Lett.,* **2007,** *9*(10), 2003-2006.
[http://dx.doi.org/10.1021/ol0706307] [PMID: 17432868]

[116] Zhou, H.B.; Liu, G.S.; Yao, Z.J. Short and efficient total synthesis of luotonin A and 22-hydroxyacuminatine using a common cascade strategy. *J. Org. Chem.,* **2007,** *72*(16), 6270-6272.
[http://dx.doi.org/10.1021/jo070837d] [PMID: 17608538]

[117] Almansour, A.I.; Arumugam, N.; Suresh Kumar, R.; Mahalingam, S.M.; Sau, S.; Bianchini, G.; Menéndez, J.C.; Altaf, M.; Ghabbour, H.A. Design, synthesis and antiproliferative activity of decarbonyl luotonin analogues. *Eur. J. Med. Chem.,* **2017,** *138,* 932-941.
[http://dx.doi.org/10.1016/j.ejmech.2017.07.027] [PMID: 28753517]

[118] Tomashevskaya, M.M.; Tomashenko, O.A.; Tomashevskii, A.A.; Sokolov, V.V.; Potekhin, A.A. New one-step procedure for the synthesis of 6*H*-chromeno[4,3-*b*]quinolines and 8a,9,14,14a-tetrahydro-8*H*-benzo[5,6]chromeno[4,3-*b*]quinolines. *Russ. J. Org. Chem.,* **2007,** *43*(1), 77-82.
[http://dx.doi.org/10.1134/S1070428007010095]

[119] Reddy, B.V.S.; Antony, A.; Yadav, J.S. Novel intramolecular aza-Diels-Alder reaction: a facile synthesis of *trans*-fused 5*H*-chromeno[2,3-*c*]acridine derivatives. *Tetrahedron Lett.,* **2010,** *51*(23), 3071-3074.
[http://dx.doi.org/10.1016/j.tetlet.2010.04.018]

[120] Nagaiah, K.; Venkatesham, A.; Srinivasa Rao, R.; Saddanapu, V.; Yadav, J.S.; Basha, S.J.; Sarma, A.V.S.; Sridhar, B.; Addlagatta, A. Synthesis of new *cis*-fused tetrahydrochromeno[4,3-*b*]quinolines and their antiproliferative activity studies against MDA-MB-231 and MCF-7 breast cancer cell lines. *Bioorg. Med. Chem. Lett.,* **2010,** *20*(11), 3259-3264.
[http://dx.doi.org/10.1016/j.bmcl.2010.04.061] [PMID: 20451380]

[121] Kumar, B.N.; Venkatesham, A.; Nagaiah, K.; Babu, N.J. Synthesis of New Pyrano[2',3': 5,6]chromeno[4,3-*b*]quinolin-4-ones *via* Aza-*Diels-Alder* Reaction. *Helv. Chim. Acta,* **2015,** *98*(3), 417-426.
[http://dx.doi.org/10.1002/hlca.201400286]

[122] Moshapo, P.T.; Sokamisa, M.; Mmutlane, E.M.; Mampa, R.M.; Kinfe, H.H. A convenient domino Ferrier rearrangement-intramolecular cyclization for the synthesis of novel benzopyran-fused pyranoquinolines. *Org. Biomol. Chem.,* **2016,** *14*(24), 5627-5638.
[http://dx.doi.org/10.1039/C5OB02536B] [PMID: 26806268]

[123] Sabitha, G.; Reddy, V. E.; Yadav, J. S.; Rama Krishna, K. V. S.; Ravi Sankar, A. Stereoselective synthesis of octahydro-3b*H*-[1,3]dioxolo[4",5":4',5']furo[2',3':5,6]pyrano[4,3-*b*]quinolines *via* intramolecular hetero-Diels-Alder reactions catalyzed by bismuth(III) chloride. *Tetrahedron Lett.,* **2002,** *43,* 4029-4032.
[http://dx.doi.org/10.1016/S0040-4039(02)00704-9]

[124] Sabitha, G.; Maruthi, C.; Reddy, E.V.; Srinivas, C.; Yadav, J.S.; Dutta, S.K.; Kunwar, A.C. Intramolecular hetero-Diels-Alder reactions catalyzed by BiCl. Stereoselective synthesis of benzoannelated decahydrofuro[3,2-*h*][1,6]-naphthyridine derivatives. *Helv. Chim. Acta,* **2006,** *89*(11), 2728-2731.
[http://dx.doi.org/10.1002/hlca.200690243]

[125] Maiti, S.; Panja, S.K.; Bandyopadhyay, C. Substituent-controlled domino-Knoevenagel-hetero Diels-Alder reaction—a one-pot synthesis of polycyclic heterocycles. *Tetrahedron,* **2010,** *66*(38), 7625-7632.
[http://dx.doi.org/10.1016/j.tet.2010.07.028]

[126] Maiti, S.; Panja, S.K.; Sadhukhan, K.; Ghosh, J.; Bandyopadhyay, C. Effects of substituent and catalyst on the intramolecular Povarov reaction—synthesis of chromenonaphthyridines. *Tetrahedron Lett.,* **2012**, *53*(6), 694-696.
[http://dx.doi.org/10.1016/j.tetlet.2011.11.130]

[127] Ramesh, E.; Sree Vidhya, T.K.; Raghunathan, R. Indium chloride/silica gel supported synthesis of pyrano/thiopyranoquinolines through intramolecular imino Diels-Alder reaction using microwave irradiation. *Tetrahedron Lett.,* **2008**, *49*(17), 2810-2814.
[http://dx.doi.org/10.1016/j.tetlet.2008.02.128]

[128] Kudale, A.A.; Miller, D.O.; Dawe, L.N.; Bodwell, G.J. Intramolecular Povarov reactions involving 3-aminocoumarins. *Org. Biomol. Chem.,* **2011**, *9*(20), 7196-7206.
[http://dx.doi.org/10.1039/c1ob05867c] [PMID: 21858320]

[129] Khan, A.; Belal, M.; Das, D. Synthesis of pyrido[2,3-*c*]coumarin derivatives by an intramolecular Povarov reaction. *Synthesis,* **2015**, *47*(8), 1109-1116.
[http://dx.doi.org/10.1055/s-0034-1380131]

[130] Adolfsson, D.E.; Tyagi, M.; Singh, P.; Deuschmann, A.; Ådén, J.; Gharibyan, A.L.; Jayaweera, S.W.; Lindgren, A.E.G.; Olofsson, A.; Almqvist, F. Intramolecular Povarov reactions for the synthesis of chromenopyridine fused 2-pyridone polyheterocycles binding to α-synuclein and amyloid-β fibrils. *J. Org. Chem.,* **2020**, *85*(21), 14174-14189.
[http://dx.doi.org/10.1021/acs.joc.0c01699] [PMID: 33099999]

[131] Gaddam, V.; Nagarajan, R. An Efficient, one-pot synthesis of isomeric ellipticine derivatives through intramolecular imino-Diels-Alder reaction. *Org. Lett.,* **2008**, *10*(10), 1975-1978.
[http://dx.doi.org/10.1021/ol800497u] [PMID: 18412351]

[132] Wölfling, J.; Frank, E.; Schneider, G.; Bes, M.T.; Tietze, L.F. Synthesis of azasteroids and D-homosteroids by intramolecular cyclization reactions of steroid arylimines. *Synlett,* **1998**, *1998*(11), 1205-1206.
[http://dx.doi.org/10.1055/s-1998-3142]

[133] Wölfling, J.; Frank, É.; Schneider, G.; Tietze, L.F. Synthesis of novel steroid alkaloids by cyclization of arylimines from estrone. *Eur. J. Org. Chem.,* **1999**, *1999*(11), 3013-3020.
[http://dx.doi.org/10.1002/(SICI)1099-0690(199911)1999:11<3013::AID-EJOC3013>3.0.CO;2-H]

[134] Magyar, A.; Wölfling, J.; Kubas, M.; Cuesta Seijo, J.A.; Sevvana, M.; Herbst-Irmer, R.; Forgó, P.; Schneider, G. Synthesis of novel steroid-tetrahydroquinoline hybrid molecules and d-homosteroids by intramolecular cyclization reactions. *Steroids,* **2004**, *69*(5), 301-312.
[http://dx.doi.org/10.1016/j.steroids.2004.01.004] [PMID: 15219408]

[135] Gaddam, V.; Ramesh, S.; Nagarajan, R. CuI/La(OTf)₃ catalyzed, one-pot synthesis of isomeric ellipticine derivatives in ionic liquid. *Tetrahedron,* **2010**, *66*(23), 4218-4222.
[http://dx.doi.org/10.1016/j.tet.2010.03.095]

CHAPTER 3

Use of Barbituric Acid as a Precursor for the Synthesis of Bioactive Compound

Sundaram Singh[1,*] and **Savita Kumari**[1]

[1] *Department of Chemistry, Indian Institute of Technology (BHU), Varanasi 221005, Uttar Pradesh, India*

Abstract: Barbituric acid is an organic compound containing a pyrimidine heterocyclic skeleton. It is a water-soluble and odorless compound. Barbituric acid served as a starting material for many barbiturate drugs. The variable properties of the products achieved from barbituric acid motivate organic chemists to investigate its chemistry and current developments have suggested it by multicomponent reactions (MCR). Barbituric acid and its derivatives, commonly known as barbiturates, are important in pharmaceutical chemistry because they are fascinating building blocks for synthesizing biologically active compounds. The first barbiturate to be prepared was Barbital (5, 5-diethyl barbituric acid), and it is hypnotic and sedative and was used as an anxiolytic and sleeping aid. Barbituric acid derivatives act on the central nervous system and are used as sedatives, anxiolytics, anticonvulsants, and hypnotics. Recent investigations show that barbituric acid derivatives may have applications in matrix metalloproteinases, inhibiting collagen-ase-3 (MMP-3), anti-invasive, recombinant cytochrome P450 enzymes, fungicides, methionine aminopeptidase-1 (MetAP-1), herbicides, antibacterial, anti-tumor antiangiogenic, antioxidant, antiviral, and HIV-1 integrase inhibitors. Furthermore, recent literature accounts have shown that barbituric acid derivatives may also perform as immune modulators. Barbituric acid has been exploited in designing and preparing various types of carbocyclic and heterocyclic compounds. An extensive range of multicomponent reactions utilize barbituric acid as a starting material. By using the Knoevenagel condensation reaction, a wide range of barbiturate drugs, that act as central nervous system depressants can be synthesized using barbituric acid. Barbituric acid is a precursor in the laboratory production of riboflavin (vitamin B_2).

Keywords: Barbituric acid, Chromeno[2,3-*d*]pyrimidine-triones, Condensation Product, Heterocyclic Compounds, Knoevenagel Condensation, N-acylation, N-alkylation, O-methylation, Spirofuropyrimidines, Spirooxindoles, Thiobarbituric acid, Trisheterocyclic System.

* **Corresponding author Sundaram Singh:** Department of Chemistry, Indian Institute of Technology (BHU), Varanasi 221005, Uttar Pradesh, India; Tel: +919451658650; E-mail: sundaram.apc@iitbhu.ac.in

Shazia Anjum (Ed.)

INTRODUCTION

Barbituric acid (pyrimidine 2,4,6(1*H*,3*H*,5*H*)-trione, H_3BA) is an organic compound containing pyrimidine heterocyclic skeleton (Scheme **1**). It is a water-soluble and odorless compound. The chemistry of barbituric acid derivatives, commonly recognized as barbiturates (BAs), has got considerable attention owing to their importance in medicine and biology, and they have appeared in a large number of biologically active compounds (Fig. **1**) [1-11b]. Barbituric acid itself has no bioactivity of its own, but its derivatives exhibit good pharmacological activities [12 - 19]. After invention, barbiturates first came into medical treatment in 1904, modifying access to psychiatric and neurological disorders [20 - 22]. In 1912, a significant barbituric acid derivative, Phenobarbital (Fig. **2**), was discovered and utilized as an antiepileptic drug [23, 24].

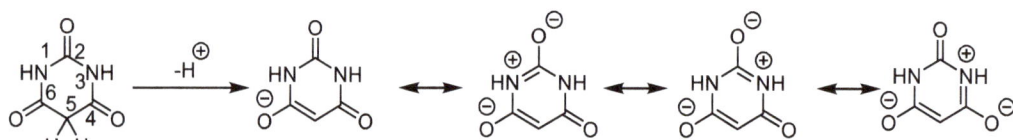

Scheme (1). Molecular structure of barbituric acid (H_3BA) and resonance structures of its deprotonated form.

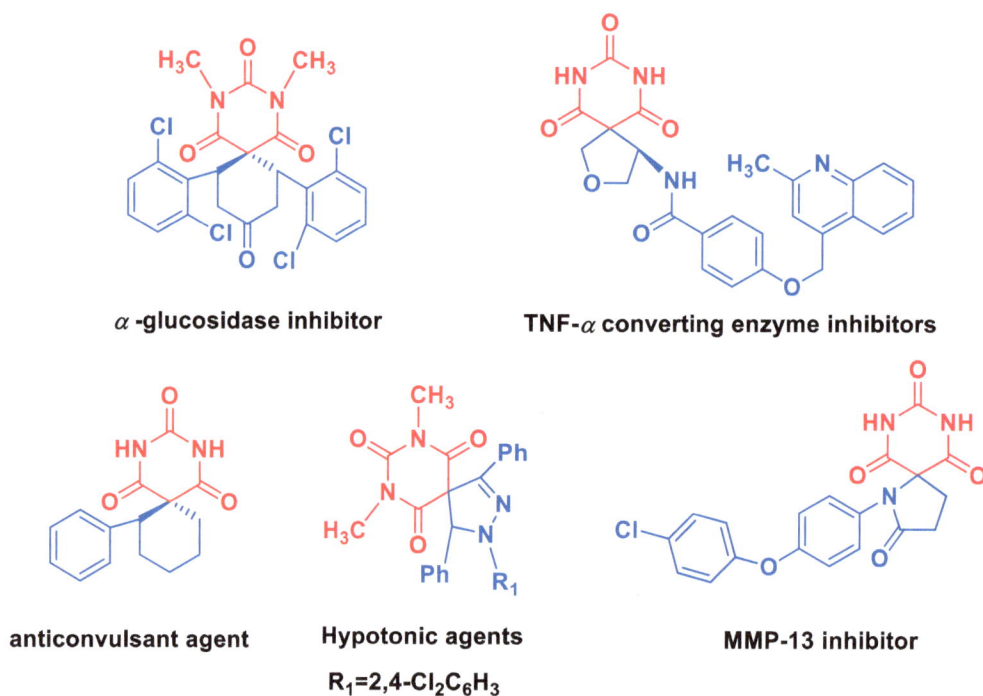

Fig. (1). Biologically active compounds containing barbituric acid moieties.

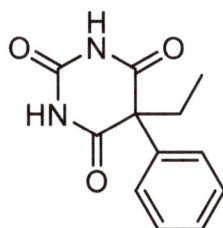

Fig. (2). Phenobarbital: an antiepileptic drug.

Barbituric acids have been generally categorized as compounds that affect the central nervous system and utilized for therapeutic uses such as sedatives [25], anticonvulsants [26, 27], and hypnotics [28, 29]. Current investigations have provided the information that barbituric acids have significant applications in anti-inflammatory [30], antifungal [31], antioxidant [32a-33], antibacterial [34-36c], antitumor [11a], antimalarial [36d], antimicrobial[36e,f], anticancer [36 g-i] as well as anti-viral [37] treatments (Fig. **3**).

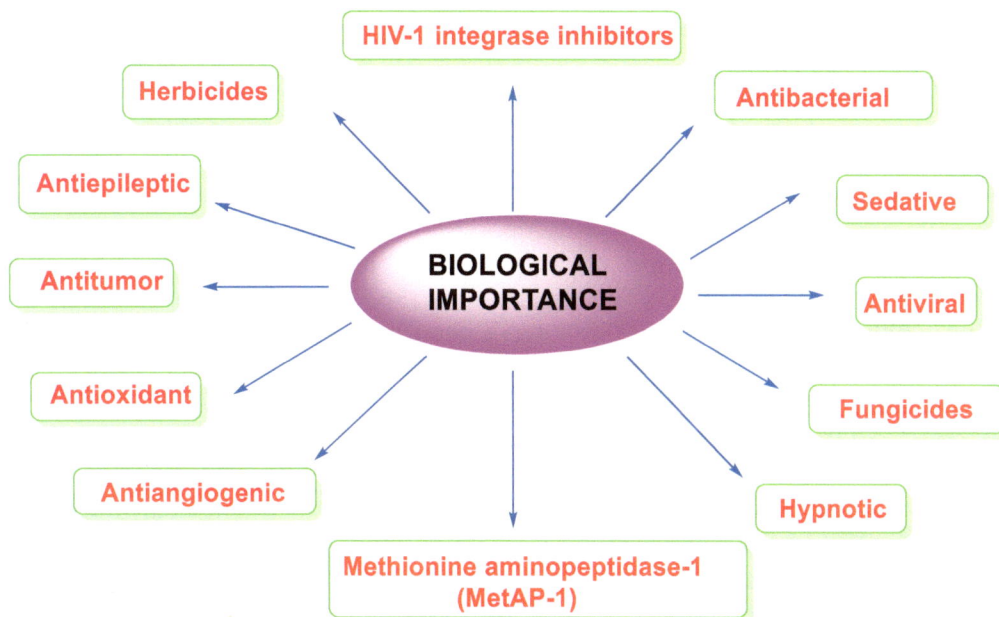

Fig. (3). Biological importance of barbituric acid derivatives.

The existence of the pyrimidine-trione ring and the type of the substituent on the C-5 position regulate the nature of the biological property of the barbiturate derivative [38]. The modifications of barbiturates produce many compounds with varying biological activities, enabling the transformation of barbiturates into a widely used pharmaceutical agents [39-43j]. 5-Aryl barbituric acid has been used to prevent the tumor necrosis factor-alpha (TNF-α) converting enzyme (TACE) [44, 45] and matrix metalloproteinases (MMPs) [46 - 50], prompting their application *in vivo* imaging [51] and cancer treatment [52 - 54].

In this chapter, we have briefly summarized various barbiturates reported in the literature and moreover our focus will be on on the use of barbituric acid as a precursor for the synthesis of biologically active organic compounds.

Barbituric acid comprises five active metal binding (donor-acceptor) sites (three *O* and two *N*, (Scheme 1), making it a resourceful polyfunctional ligand. The pyrimidine ring is further stabilized by resonance due to the capacity of the activated CH_2 group to remove one of the protons (1); however, the donor-acceptor tendency of the heteroatoms fluctuates along the molecule [55]. Consequently, the most acidic proton in barbituric acid is one of the methylenic CH_2 hydrogens with pK_a of 4.03 [56], removal of proton at this site permits the development of a planar carbanion. Additionally, the pyrimidine ring of barbituric acid can be functionalized very easily at position 5. For example, substituting a diazo moiety at position 5 leads to the formation of aryl hydrazones of BAs, which offer fascinating coordination and solvatochromic properties [57, 58].

SYNTHESIS AND PHYSICAL PROPERTIES OF BARBITURIC ACID

Synthesis of Barbituric Acid

Adolf von Baeyer first synthesized barbituric acid in 1864 by condensing diethyl malonate with urea [2, 59, 60]. The barbiturates were incorporated for clinical use in the 1900s; furthermore 2500 BAs have been prepared; among them, 50 were utilized predominantly for pharmacological purposes, showing their broad ubiquity. Various methods have been described in the literature to synthesize BAs [61 - 64], and the most significant of them are shown in (Scheme 2). For example, urea condenses with malonic esters or malonic acids, malonyl dichlorides, to give H_3BA (Scheme 2a and b). The most widely recognized method for barbituric acid synthesis is the Michael method. The condensation of suitable diethyl malonate with urea takes place in anhydrous alcohol in the presence of sodium ethoxide (Scheme 2a) [63]. This procedure has been commonly implemented for the in-dustrial synthesis of barbituric acids and signifies the most likely laboratory procedure. In the same way, H_2DEBA [diethylbarbituric acid] can be synthesized by condensation of diethyl-2,2-diethyl malonate with urea in sodium ethoxide followed by the removal of two ethanol molecules (Scheme 2c) [65]. A modification of synthesis of the BA includes a condensing agent like alkali hydroxide in liquid ammonia resulting in the dialkylbarbituric acids [63]. A mixture of urea, diethyl dialkylmalonate, and an alkali hydroxide in liquid ammonia produces the desired product in this process. Nevertheless, none of the corresponding barbituric acids were obtained when monoalkylated diethyl malonates or diethyl malonate were used. Likewise, ethyl carbonate and malonamide or its C-alkyl derivatives in the presence of alkali in liquid ammonia condense to give the corresponding derivatives of barbituric acid (Scheme **2d**) [63]. This method seems very general, resulting in the barbituric acids in good yields from C,C-dialkylmalonamides, C-alkylmalonamides, or malonamide [63]. Various examples are also reported for the synthesis of barbiturates by the reaction of 2,2-disubstituted malonic acids with 1,3-disubstituted urea. Consequently, when reaction of two moles of 1,3-disubstituted urea and 2,2-disubstituted malonic acid was carried out in THF solution, it gave a crystalline 1,3,5,5-tetrasubstitutedpyrimidine-2,4,6(1*H*,3*H*,5*H*)- trione through an exothermic pathway (Scheme **2e**) [62].

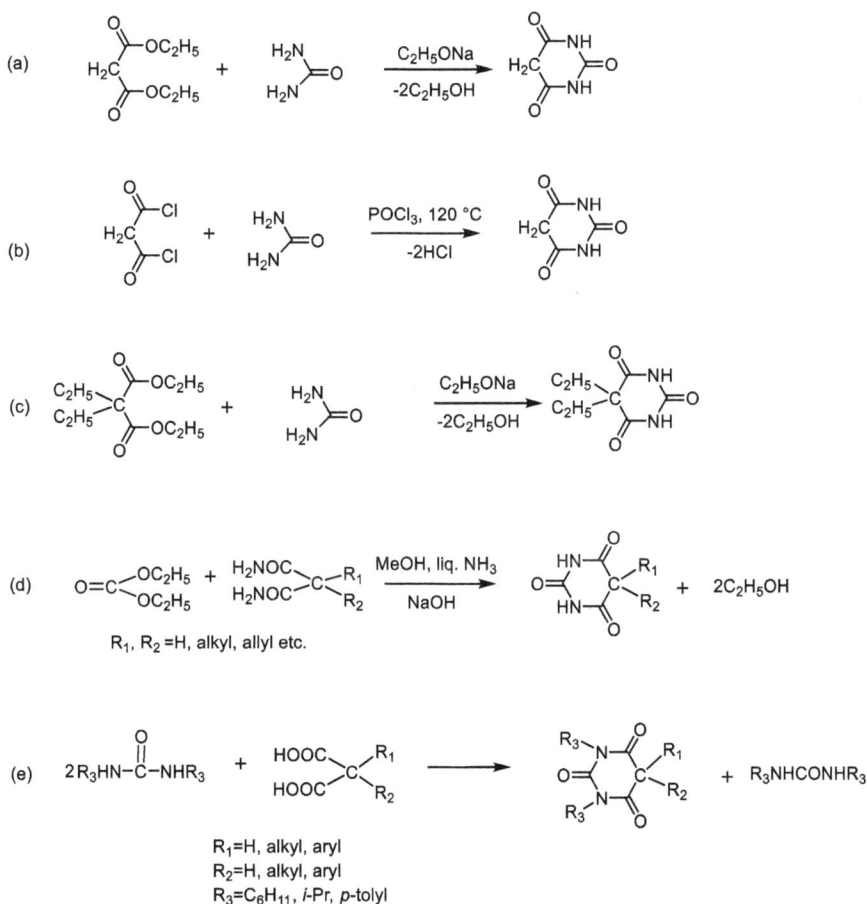

Scheme (2). Most common synthetic approaches for the preparation of barbituric acids.

There are some most common barbituric acids, viz., thiobarbituric acid (H$_3$TBA), barbituric acid (H$_3$BA), diethyl barbituric acid (H$_2$DMBA), violuric acid (H$_3$VA), *N, N'*-dimethylbarbituric acid (HNNDMBA), purpuric acid (H$_5$PURP), 5-nitrobarbituric acid (H$_3$NBA), diethylbarbituric acid (H$_2$DBA), *N,N'*- dimethyltiovioluric acid (HNNDTVA), and *N,N'*-dimethylvioluric acid (HNNDVA) shown in Scheme (**3**).

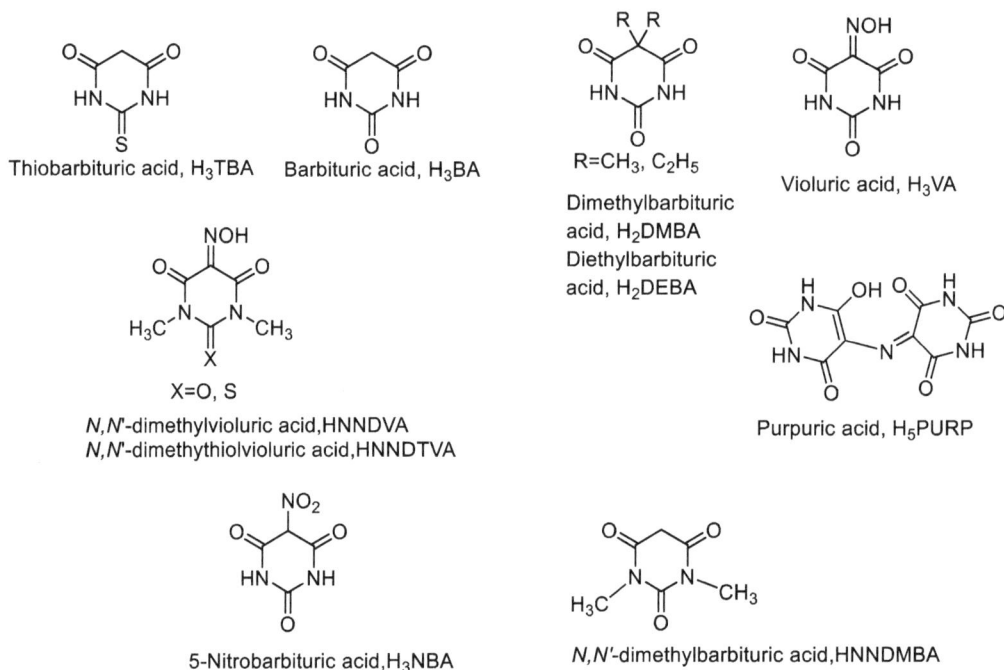

Thiobarbituric acid, H$_3$TBA   Barbituric acid, H$_3$BA

R=CH$_3$, C$_2$H$_5$

Dimethylbarbituric acid, H$_2$DMBA
Diethylbarbituric acid, H$_2$DEBA

Violuric acid, H$_3$VA

X=O, S

N,N'-dimethylvioluric acid,HNNDVA
N,N'-dimethythiolvioluric acid,HNNDTVA

Purpuric acid, H$_5$PURP

5-Nitrobarbituric acid,H$_3$NBA

N,N'-dimethylbarbituric acid,HNNDMBA

Scheme (3). Most common barbituric acids.

Physical Properties of Barbituric Acid

Tautomerization and Acid-Base Properties

Most pyrimidine derivatives prefer the keto configuration over the enol configuration [66 - 79]. However, BAs are pyrimidine-related compounds that show two types of schemetautomers by transfer of either methylene or imino hydrogen atoms to keto oxygen atoms, through a process known as lactam-lactim tautomerization, NHC=O $\rightleftharpoons$ NC–OH. Since BAs have three lactam groups, one, two, or all three groups can theoretically be converted to lactim moieties (Scheme **4**) [67, 80, 81].

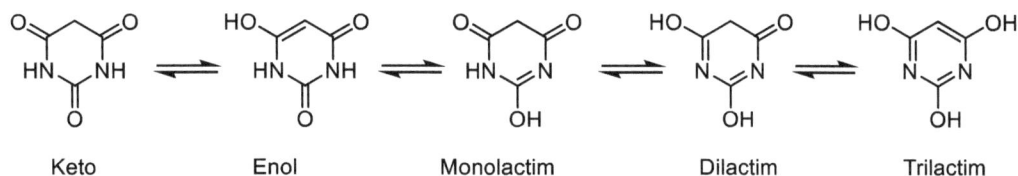

| Keto | Enol | Monolactim | Dilactim | Trilactim |

Scheme (4). Possible tautomeric structures of barbituric acid.

Generally, BA is represented in the keto form because it is the most stable form in solution [67, 76, 77], gas phase, and solid [77]. The 2-hydroxy and 4-hydroxy forms are not much commons; however, they are also stable and impact the properties of BAs and related moieties [66, 82 - 90]. The development of additional strong hydrogen bonds explains the stability of the enol form in the solid-state [87, 91]. The tautomerization of barbiturates need more investigations to explain the formation of new coordination compounds [92 - 94], molecular rotors [95], switching systems [96 - 100], solvatochromic colorants [57, 101, 102], habit modifiers for rock salt crystals [103], the construction of supramolecular architectures [58, 104], and artificial systems mimicking the biological ones [105].

Barbituric acid itself has no anticonvulsant, hypnotic, or anesthetic properties. These properties are only observed when H- atoms of the C-5 group are substituted by aryl or alkyl groups. The reactive hydrogen-containing carbon of barbituric acid is moderately acidic (pK_a = 4.03) than other diketone species (*cf.* acetylacetone with pK_a 8.95 and dimedone with pK_a 5.23) due to extra aromatic stabilization of the carbanion. It was proposed that, for a barbituric acid to have good hypnotic activity, it must be a weak acid with a lipid/water partition coefficient within certain limits [106].

The hydrophilicity of the pyrimidinetrione ring of 5,5-disubstituted barbituric acids is associated with the pK_a and the nature and number of N-substituents. Noticeably, the acidity, in this case, results from the tendency of N-atoms to lose protons and stability of the resulting conjugate base through resonance delocalization. The acidity of all 5,5-disubstituted barbituric acids is lower than that of the parent H_3BA due to the lack of an active methylene group flanked between two carbonyl groups and the inability to produce a symmetrical conjugated ring structure. Alkyl groups further diminish the acidity of alkyl-substituted barbituric acids by donating electrons [107]. Therefore, although unsubstituted barbituric acid has a pK of 4.03 [56], the values of pK_1 of 5,5-disubstituted barbituric acids vary from 7.1 to 8.1 [80]. In aqueous solutions, either the dioxo tautomeric (in alkaline medium) or the trioxo tautomeric (in acidic medium) form predominates [80]. However, the 5,5-disubstituted barbituric acids are mainly found in the trioxo tautomeric form and are weak acids; salts of these barbiturates are readily synthesized by treatment with bases. Subsequently, second ionization occurs (Scheme **5**), with pK_2 values ranging from 11.7 to 12.7 [108a].

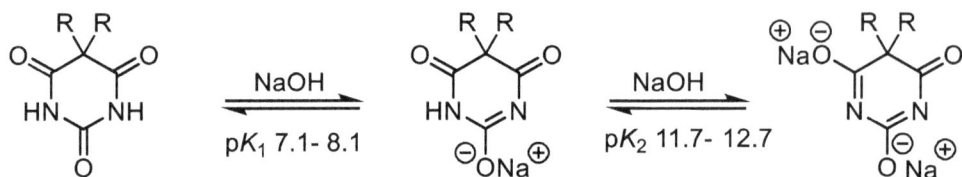

Scheme (5A). Deprotonation of 5,5 disubstituted barbituric acids.

More emphasis has been given on barbituric acid's acid-base behaviour and tautomerism. Although it is well known that keto form is the most stable neutral form in solutions, which has been established in many studies using both computational and experimental approaches, while the presence of a crystalline enol polymorph has just recently been demonstrated. The development of mono- and dianions, as well as their spectral features, has received a lot of attention. Their protonation is less investigated, and Zuccarello examines the electronic spectra of barbituric acid and demonstrates that in acidic conditions (up to 10 M HCl), there is a cation that occurs mainly in two equilibrium forms (I) and (II) with some fraction of (III). Olah's study utilizing 1H NMR spectroscopy demonstrated the development of the tri protonated form of (IV) in the medium trifluoromethanesulfonic acid, but no data on pK_b is provided [108b].

Chemical Properties

Reactions at the C-5 Position

Two active hydrogens are present at the 5-position of barbituric acid, which can be substituted by different groups to make it biologically active. Simple alkyl substituents can be easily introduced into this position, while complex substituents are introduced with the help of indole derivatives. According to Rao and Chalmers, the reaction of barbituric acid with 3-isopropylaminoethyl-2-methy-lindol in the presence of piperidine yields 5-(2'-(isopropy-lamino)ethyl) barbituric acid (Scheme **6**) [109].

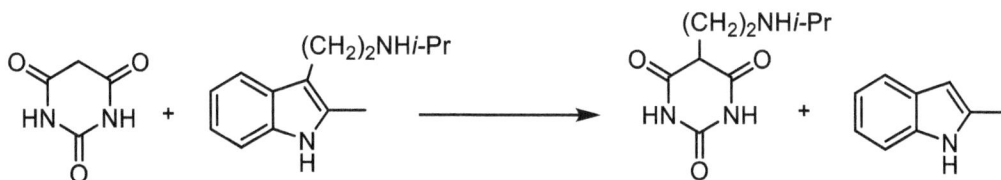

Scheme (6). Alkylation of barbituric acid at the C-5 position.

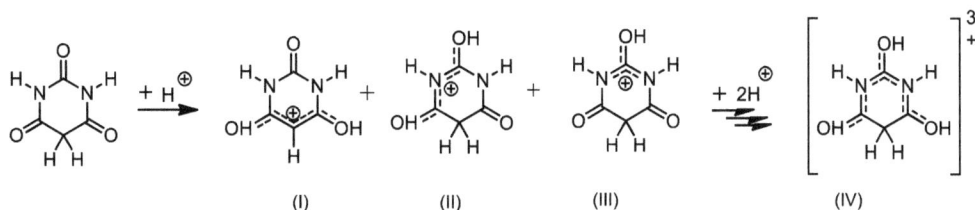

Scheme (6B). Protonation of barbituric acid.

The substitution of 2-thiobarbituric/barbituric acids with indole derivatives was also examined (Scheme 7) [110].

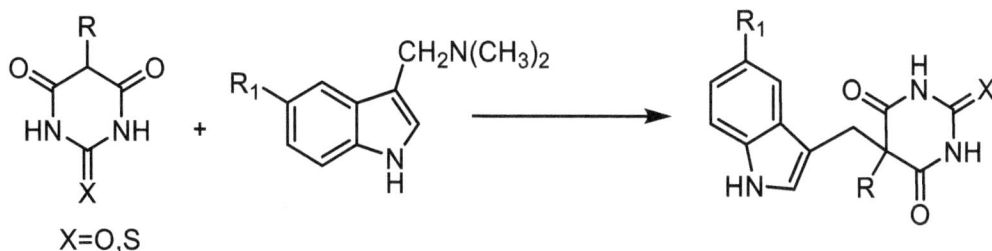

X=O,S

Scheme (7). Substitution of thiobarbituric/ barbituric acid.

Sekiya *et al.* described the synthesis of various 5-aryl or 5-alkylmethylbarbituric acids by reducing barbituric acid derivatives containing a methylidene bond at the 5-position (Scheme 8) [111 - 113].

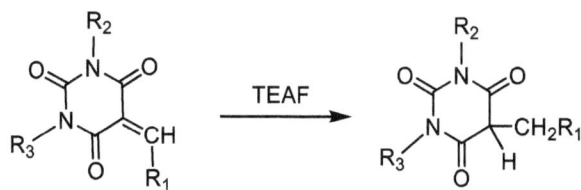

TEAF= triethylammonium formate

Scheme (8). Reduction of barbituric acid derivatives with triethylammonium formate.

Ethier and Neville demonstrated the efficient method for oxidative methylation of 5-benzylidene barbituric acid by methyl iodide and Ag$_2$O in the presence of DMF, resulting in the formation of 1,3,5,5-tetramethylbarbituric acid (Scheme **9**) [114].

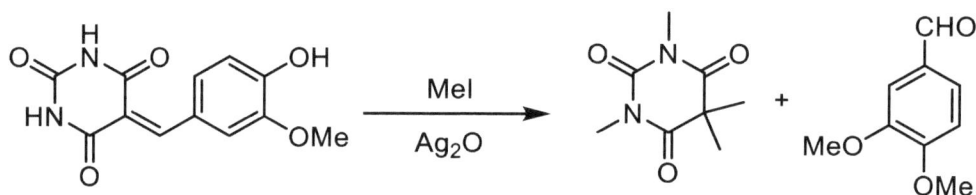

Scheme (9). Oxidative methylation of 5-benzylidene barbituric acid.

Barbituric acid gives an addition reaction with various compounds. 5-alkyl barbituric acids and barbituric or 1-substituted barbituric acids undergo an addition reaction with α, β-unsaturated ketones, and phenyl isocyanate to yield 5,5-disubstituted derivatives [110] and 5- phenyl carbamoyl barbituric acid, respectively (Schemes **10** and **11**) [115].

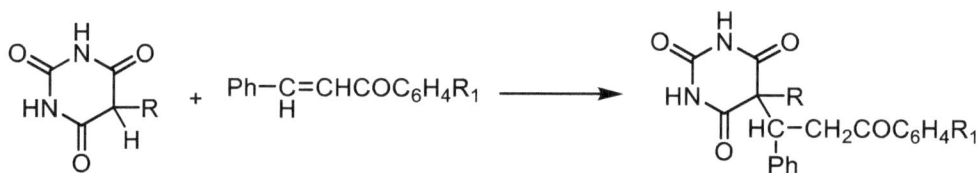

Scheme (10). Addition reaction with α, β-unsaturated ketones.

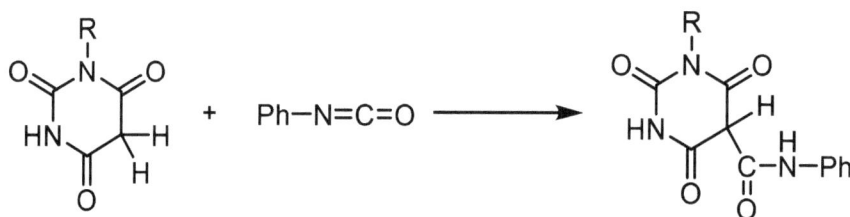

Scheme (11). Addition reaction with phenyl isocyanate.

A significant aspect of reactivity includes the capability of BA to be modified further very easily, *e.g.*, at active methylene position 5. Therefore, the reaction of barbituric acid with aromatic diazonium salts in the presence of base in ethanolic solution (Japp-Klingemann reaction, (Scheme **12**) leads to AHBAs (arylhydra-zones of barbituric acids), which can be further utilized as ligands in coordination chemistry or as intermediates in organic reactions [58, 116 - 118].

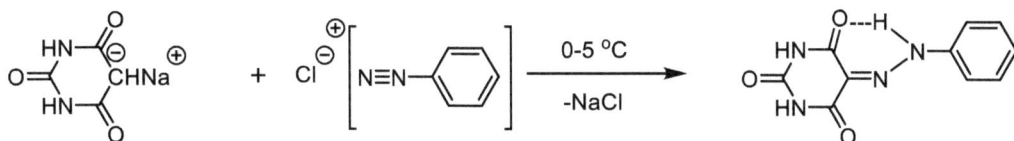

Scheme (12). Japp-Klingemann reaction of barbituric acid.

Likewise, barbituric acid reacts with nitrite in an acidic medium to result in the nitroso derivative, 5-(hydroxyimino)pyrimidine-2,4,6(1*H*,3*H*,5*H*)-trione (H$_3$VA, violuric acid, (Scheme **13**). Further oxidation of this violuric acid readily gives dilituric acid [119, 120]. These compounds' solvatochromic and other associated properties can be of great interest because of the color-dependent tautomerism (pink nitrozoenolic form and colorless oximino-ketonic form) [121].

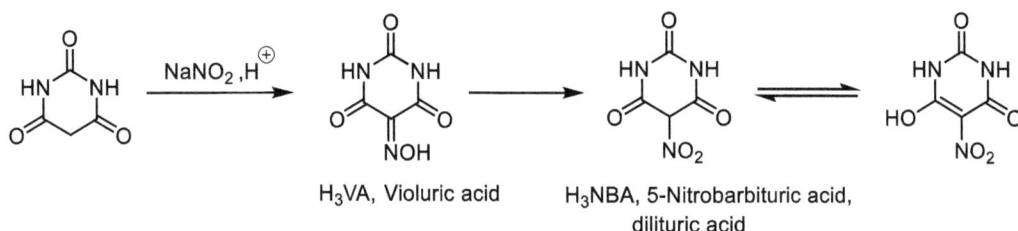

Scheme (13). Reaction of barbituric acid with sodium nitrite.

Substitution at Nitrogen

Alkylation of barbituric acids results in forming a mixture of *O*- and *N*-alkyl derivatives because these compounds contain imide hydrogen atoms, showing tautomerism. *O*- and *N*- alkylated products were obtained by the alkylation of 5- or 1-monosubstituted barbiturates. However, alkylation of 1,5,5-trisubstituted and 5,5-disubstituted barbiturates yielded *N*-alkylated product predominantly. 5,5-Disubstituted thiobarbituric acid reacts with chloromethyl methyl ether to yield *N*, *S*-bis(methoxymethyl) derivatives (Scheme **14**) which upon oxidation produces an excellent yield of *N*-methoxymethyl barbituric acid derivatives [122].

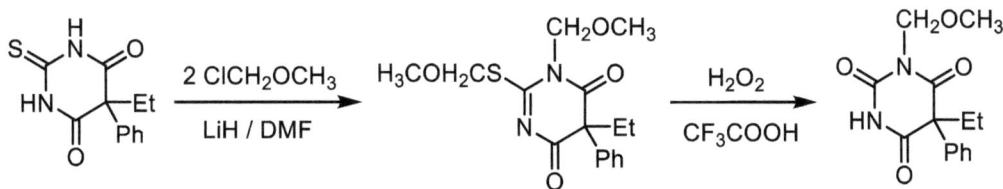

Scheme (14). *N*-alkylation of barbituric acid derivatives.

The acyl residue can also substitute the hydrogen atom of the NH group. 5-Ethy--5-phenyl- or 5,5-diethylbarbituric acids react with benzoyl chloride in two steps.

The reaction of benzoyl chloride with *N*-unsubstituted barbiturates occurs in the first step, resulting in the *N*-monosubstituted derivatives. In the second step, the self-acylation reaction occurs between these *N*-monosubstituted derivatives (Scheme **15**) [123, 124].

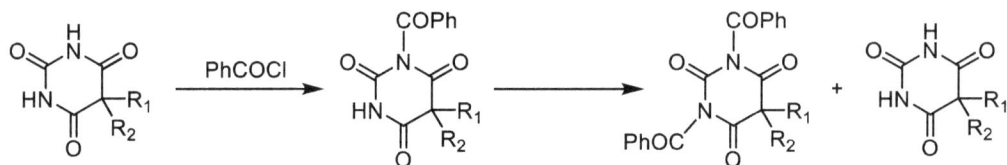

Scheme (15). *N*- acylation of barbituric acid.

Reactions of Carbonyl Groups

At the 2-position of barbituric acid, the Carbonyl group exhibits a different behaviour from the other two carbonyl groups at 4 and 6-positions in the ring. The reaction of *N*-methylated barbituric acids with $LiAlH_4$ resulted in reducing all the three carbonyl groups, whereas in the presence of AlCl3 and $LiAlH_4$, only two carbonyl groups were reduced (Scheme **16**) [125].

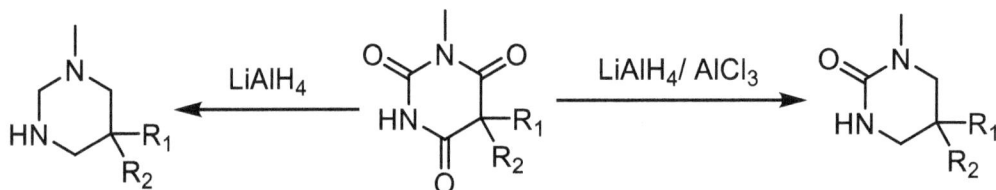

Scheme (16). Reduction of the carbonyl group of barbituric acid.

It was proposed by E.E. Smissman *et al.* and H.C. Brown *et al.* [126 a, 126 b] that borohydrides would not be able to reduce barbiturates unless they contain a phenyl ring at 5- position. It was considered that $NaBH_4$ is too insignificant reducing agent for barbiturates. Despite this, sodium borohydride was observed to react with 5-ethyl-5-phenylbarbituric acid in both organic as well as aqueous media [127 (a), 127 (b)]. Furthermore, when the reaction was performed in CH_3OH at r.t. for 1 h, 1,3-dimethyl 5,5-dibenzylbarbituric acid was also reduced [128]. Rautio examined the reduction of barbiturates with $NaBH_4$ in a variety of solvents. For each barbiturate, four major reduction products were obtained viz., di- and trihydrobarbiturates (A and B respectively) which were formed by reducing one or two carbonyl groups to a secondary hydroxyl group, primary alcohols (C), produced by the reductive cleavage of the ring, and urea derivatives (D), which were generated simultaneously with the primary alcohols from the rest of the barbituric acid ring (Scheme **17**) [129].

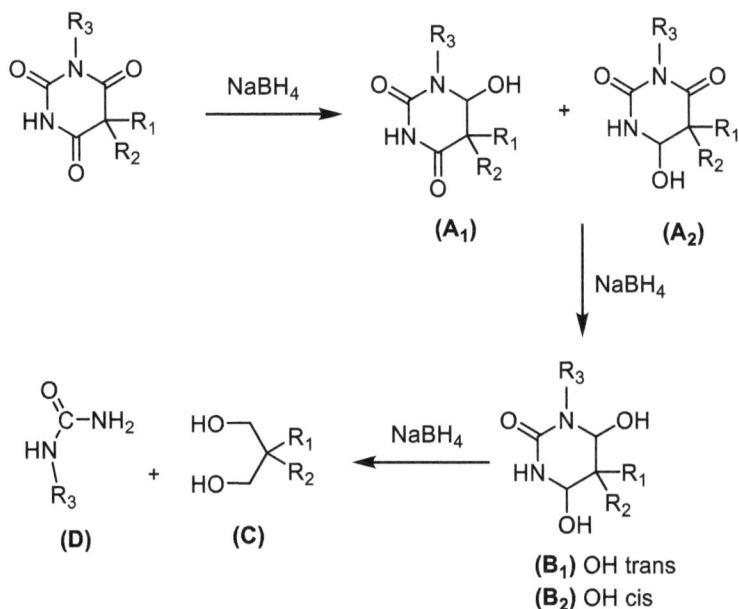

Scheme (17). Reduction of the carbonyl group of barbituric acid with NaBH$_4$.

Furthermore, *O*-methylation products have been produced by the action of HCl and methanol on *N*-phenylbarbituric and barbituric acids (Scheme **18**) [130].

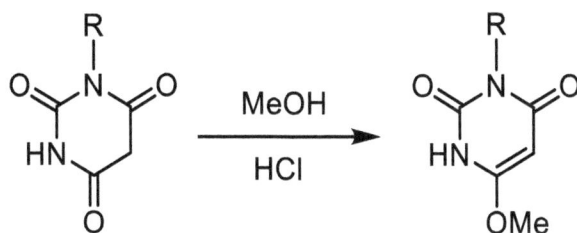

Scheme (18). *O*-methylation of barbituric acid.

Photochemical Reactions

Otsuji *et al*. have investigated that UV light (254 nm) accelerates the hydrolysis of 5,5-diethylbarbituric acid under an alkaline medium [131]. Bojarski and co-workers describe the photochemical ring-opening of monoanion of barbituric acid to yield 5,5-diethyl-3-methylhydantoin and *N*-(2-ethylbutanoyl)-*N*-methylurea (Scheme **19**) [132].

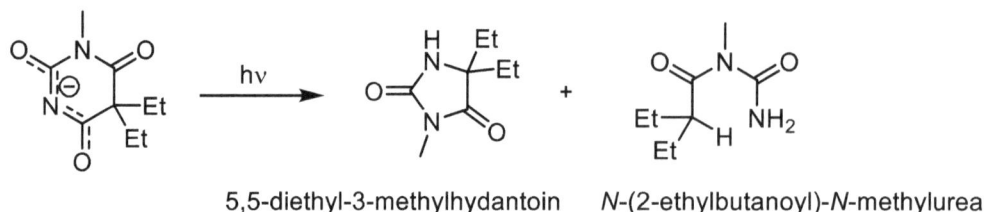

5,5-diethyl-3-methylhydantoin   *N*-(2-ethylbutanoyl)-*N*-methylurea

Scheme (19). Photochemical ring-opening reaction of barbituric acid.

Barton *et al.* reported the synthesis of *N*-(-2-ethylbutyryl)-*N'*-substituted imidodicarbonic diamides by the photochemical reaction of barbituric acid with amines (Scheme **20**) [133].

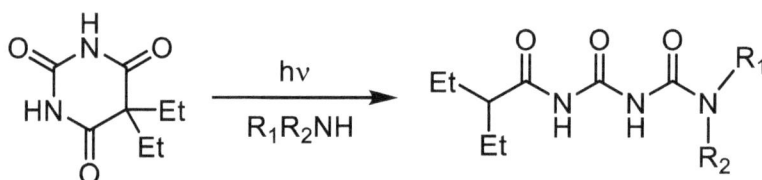

Scheme (20). Photochemical reaction of barbituric acid with amines.

Other Reactions

Bicyclo compounds were formed during the cyclization of 1,5,5-tri- and/or 5,5-disubstituted barbiturates. Therefore, pyranopyrimidine and furanopyrimidine are produced respectively due to intramolecular *O*-alkylation of 5,5-disubstituted barbiturates (Scheme **21**) [134, 135].

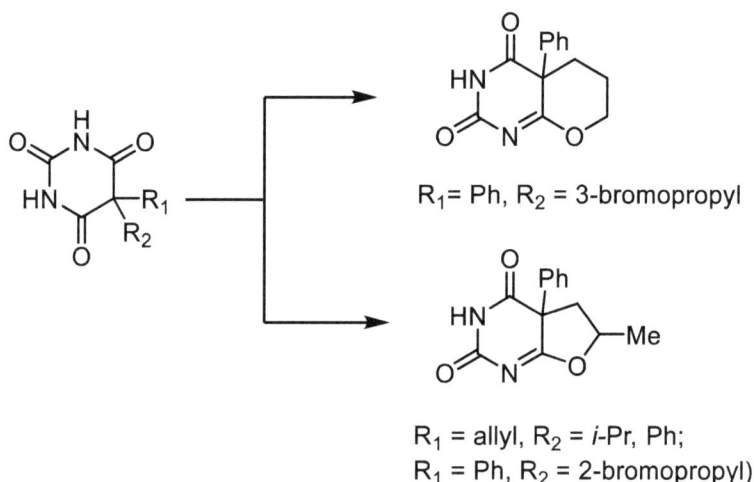

R_1= Ph, R_2 = 3-bromopropyl

R_1 = allyl, R_2 = *i*-Pr, Ph;
R_1 = Ph, R_2 = 2-bromopropyl)

Scheme (21). Intramolecular *O*-alkylation of 5,5-disubstituted barbiturates.

Barbiturates undergo ring contraction during hydrolytic degradation. *N,N*-unsymmetrically substituted barbiturates give a mixture of two 5-substituted oxazolidine-2,4-diones (Scheme **22**) [136].

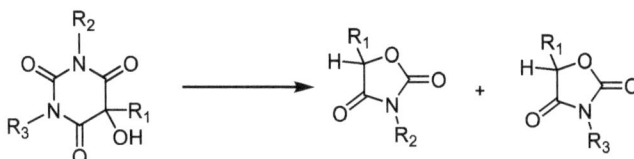

Scheme (22). Ring contraction of barbiturates.

In an acidic medium, 2-(5-oxo-4,5-dihydro-1*H*-1,2,3-triazol-3-y1) aliphatic acids are produced by the isomerization of 1-aminobarbituric acid. The reaction was observed to be started by hydrolysis of the C-6 − N-1 amide bond of 1-aminobarbituric acid under an acidic medium, followed by the cyclization of intermediate (I) to the triazole ring system (Scheme **23**) [137].

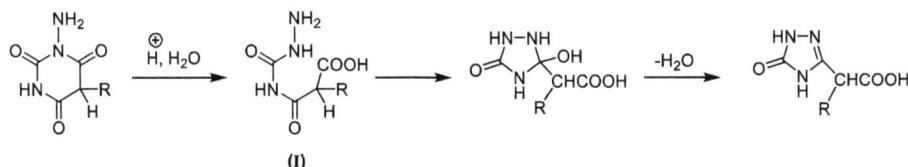

(I)

Scheme (23). Isomerization of 1-aminobarbituric acid.

APPLICATION OF BARBITURIC ACID IN ORGANIC SYNTHESIS

Synthesis of Condensation Product

1,3-Dicarbonyl substituted methylaminobenzene-sulfonamide derivative (**4**) is efficiently synthesized by one-pot, three-component reaction of barbituric acid (**1**) sulfanilamide (**2**) and triethyl[orthoformate] (**3**) in refluxing ethanol (Scheme **24**). The inhibitory action of the compounds on the properties of purified human carbonic anhydrase (hCA) I and hCA II were estimated [138].

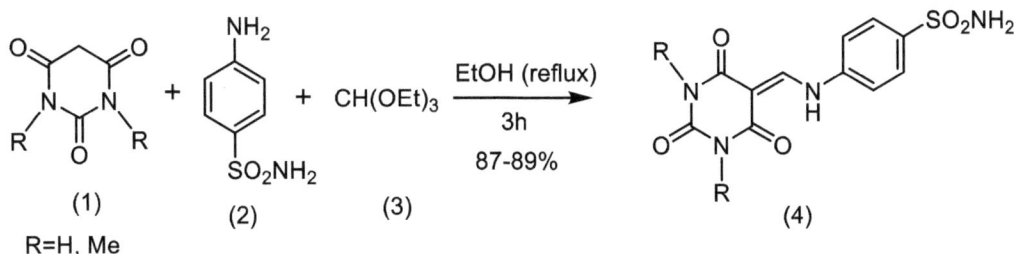

Scheme (24). Synthesis of 1,3-dicarbonyl substituted methylaminobenzene-sulfonamide derivative.

A novel class of trisheterocyclic systems, bisthiadiazolyl /bisoxadiazolyl thioxopyrimidinediones /pyrimidinetriones **(7)**, was synthesized by the condensation of 1,3,4- thiadiazole /oxadiazole **(5)**, carbon disulfide **(6)**, and barbituric acid derivatives **(1)**, under base-catalyzed medium (Scheme **25**). The antimicrobial property of the produced compounds was studied, and it was concluded that compounds having thioxopyrimidinedione along with bisthiadiazole unit displayed high activity [139].

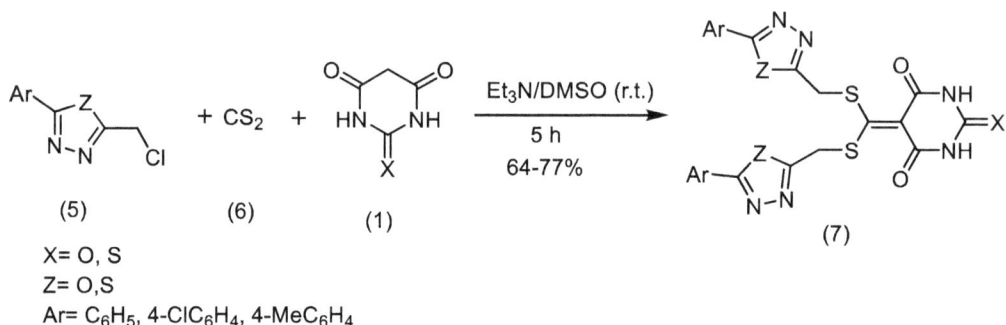

Scheme (25). Synthesis of trisheterocyclic systems.

Knoevenagel condensation reaction of barbituric acid derivatives **(1)** with ethylene glycol-based aromatic aldehyde **(8)** resulted in the synthesis of novel benzylidene barbituric acid derivatives **(9)** (Scheme **26**) [140, 141].

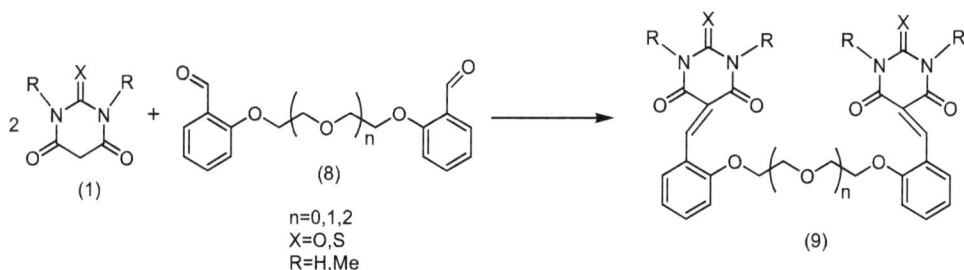

Scheme (26). Synthesis of benzylidene barbituric acid derivatives.

Synthesis of Oxygen Containing Heterocyclic Compounds

5-Membered Heterocyclic Compounds

The one-pot reaction of isatins **(10)** and barbituric acids **(1)** by the action of the bromine in ethanolic solution lead to the synthesis of substituted 2″*H*-dispiro[indole-3,5′-furo[2,3-*d*] pyrimidine] system **(11)** in 71–87% yields through cascade process (Scheme **27**) [142].

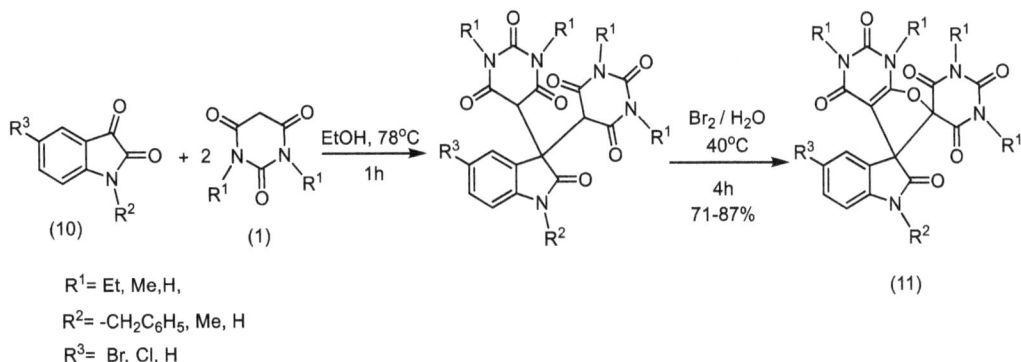

Scheme (27). Synthesis of 2″*H*-dispiro[indole-3,5′-furo[2,3-*d*] pyrimidine] system.

A series of novel furo[2,3-*d*]pyrimidine derivatives **(14)** were synthesized *via* multicomponent condensation reaction of 1,3-dimethylbarbituric acid **(1)**, isocyanides **(12)**, and aldehydes **(13)** at room temperature (Scheme **28**) [143].

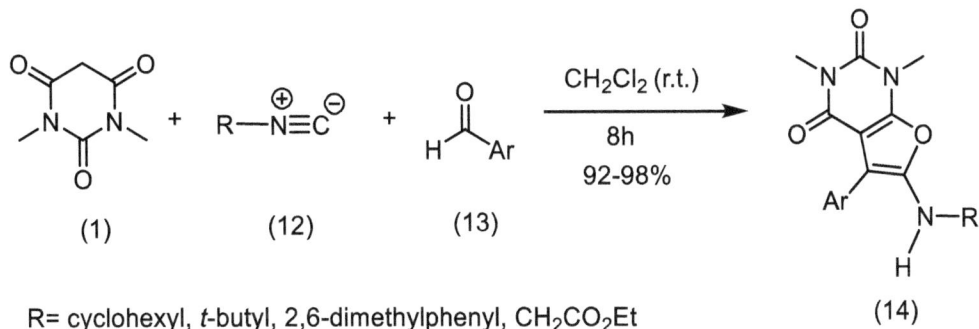

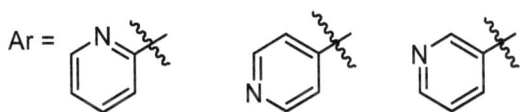

R= cyclohexyl, *t*-butyl, 2,6-dimethylphenyl, CH₂CO₂Et

Scheme (28). Synthesis of furo[2,3-*d*]pyrimidine derivatives.

Electrocatalytic assembling of *N,N′*-dialkylbarbituric acids **(1)**, and aldehydes **(13)** result in the selective synthesis of substituted spirofuropyrimidines **(15)** by complex cascade process (Scheme **29**) [144].

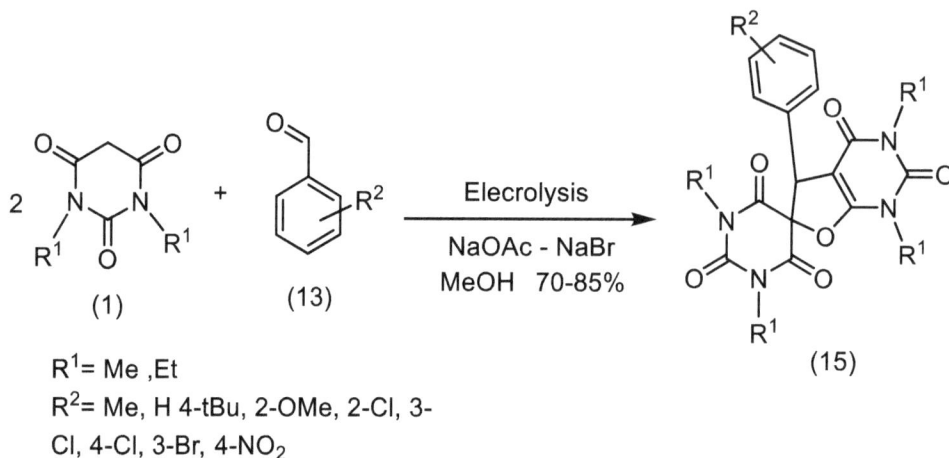

Scheme (29). Synthesis of substituted spirofuropyrimidines.

Furo(2,3-*d*)pyrimidine-2,4(1*H*,3*H*)-dione derivatives **(17)** have been prepared by the three-component condensation reaction of aryl isocyanide **(16)** aromatic aldehyde **(13)**, and barbituric acid **(1)**, in water under uncatalyzed conditions (Scheme **30**). The obtained compounds were tested for their antibacterial and antifungal properties, and the majority of the compounds displayed good results [145].

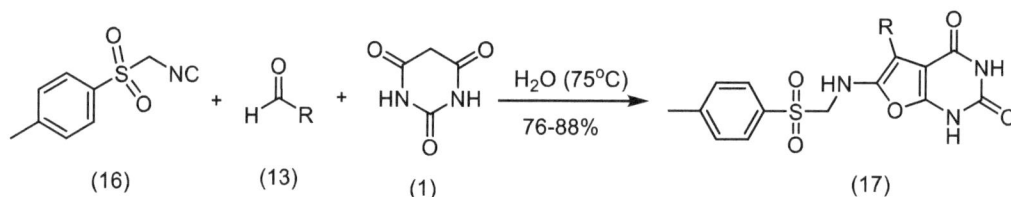

Scheme (30). Synthesis of furo(2,3-*d*)pyrimidine-2,4(1*H*,3*H*)-dione derivatives.

6-Membered Heterocyclic Compounds

Synthesis of pyrano[2,3-*d*]pyrimidine-2,4,7-triones **(19)** has been done by a three-component reaction of barbituric acid **(1)**, Meldrum's acid **(18)**, and aromatic aldehydes **(13)**, under microwave irradiation in the presence of K_2CO_3 (Scheme **31**) [146].

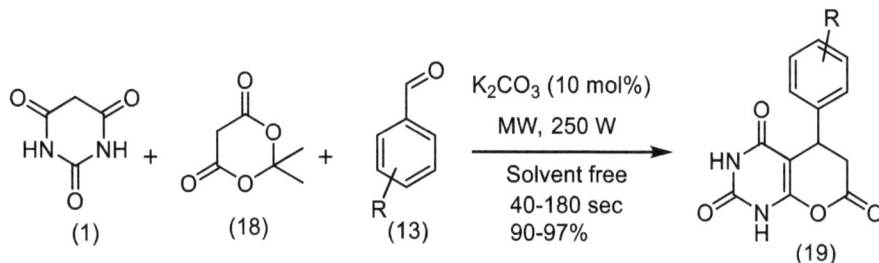

Scheme (31). Synthesis of pyrano[2,3-*d*]pyrimidine-2,4,7-triones.

A proficient one-pot three-component procedure has been reported for the preparation of novel spironaphthopyrano[2,3-*d*]pyrimidine-5,3′-indolines **(21)** *via* condensation reaction of isatins **(10)**, *β*-naphthol **(20)**, and barbituric acid derivatives **(1)**, by exploitation of [Hmim][HSO₄] as a reusable and efficient catalyst (Scheme **32**) [147].

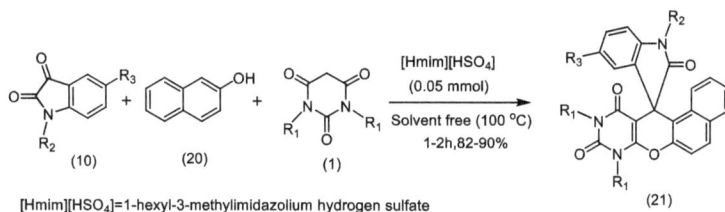

[Hmim][HSO₄]=1-hexyl-3-methylimidazolium hydrogen sulfate

Scheme (32). Synthesis of spironaphtho-pyrano [2,3-*d*]pyrimidine-5,3′-indolines.

A series of 5-(2,3,4,5-tetrahydro-1*H*-chromeno[2,3-*d*]pyrimidin-5-yl)pyrimidione derivatives **(22)** have been prepared from 2-thiobarbituric acid or barbituric acid **(1)** and substituted salicylaldehydes **(13)** (Scheme **33**). These derivatives displayed significant antioxidant and *in vitro* antibacterial properties [148 - 150].

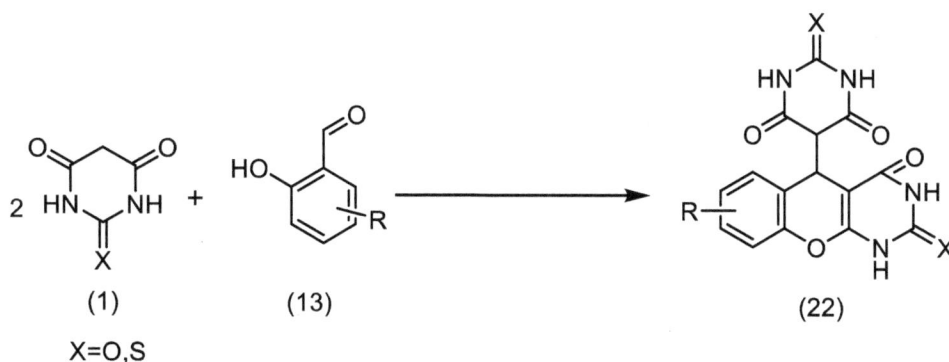

X=O,S

Scheme (33). Synthesis of 5-(2,3,4,5-tetrahydro-1*H*-chromeno[2,3-*d*]pyrimidin-5-yl)pyrimidione derivatives.

An efficient, one-pot, three-component method for synthesis of novel chromeno[3′,4′:5,6]pyrano[2,3-*d*]pyrimidines **(24-25)** has been reported by utilizing tetrabutylammonium fluoride (TBAF) as a supporting electrolyte and acetonitrile as a base *via* condensation of 4-hydroxycoumarin **(23)**, isatin **(10)** or an aromatic aldehyde **(13)**, and barbituric acid **(1)**. (Scheme **34**) [151].

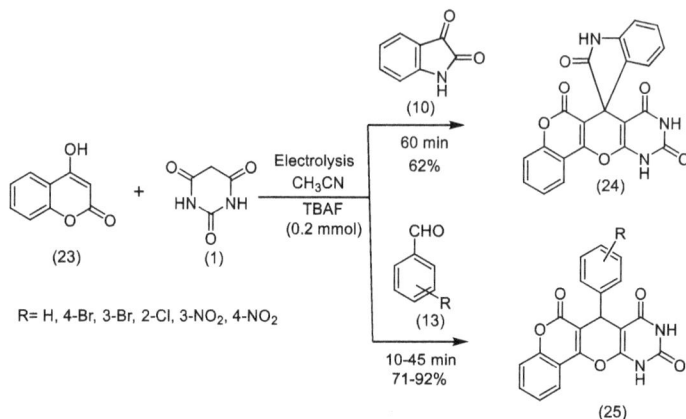

Scheme (34). Synthesis of novel chromeno[3′,4′:5,6]pyrano[2,3-*d*]pyrimidines.

One-pot three-component reaction of barbituric acid **(1)**, 5- (1*H*-imidazol-1-y-)-3-methyl-1-phenyl-1*H*-pyrazole-4-carbaldehyde **(26)**, and malononitrile/ethyl-cyanoacetate **(27)** was effectively performed with piperidine as a primary catalyst (Scheme **35**). All the products have been examined for their anti-malarial, anti-tuberculosis, and antibacterial activities [152].

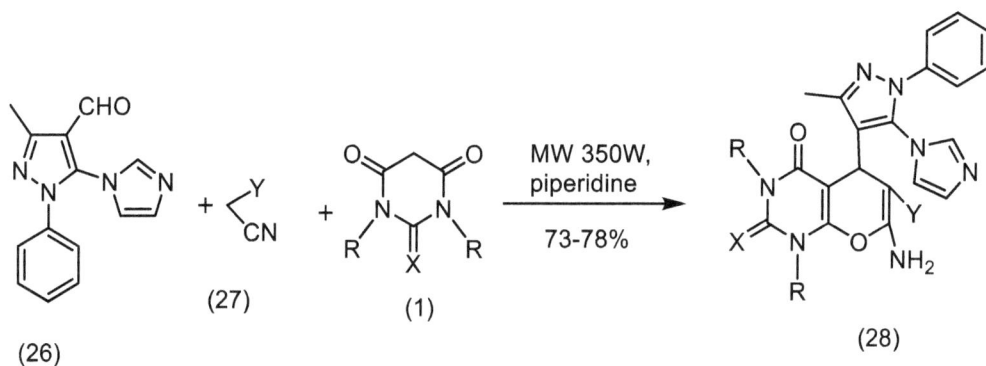

Scheme (35). Synthesis of pyranopyrimidines.

A significant, simple, and environmentally benign method for the preparation of chromeno[2,3-*d*]pyrimidine-trione derivatives **(30)** was reported *via* the one-pot, the three-component reaction of aromatic aldehydes **(13)**, dimedone/cyclohexane-

1,3-dione **(29)** and barbituric acid **(1)**, exploiting $Sc(OTf)_3$ as a catalyst (Scheme **36**) [153].

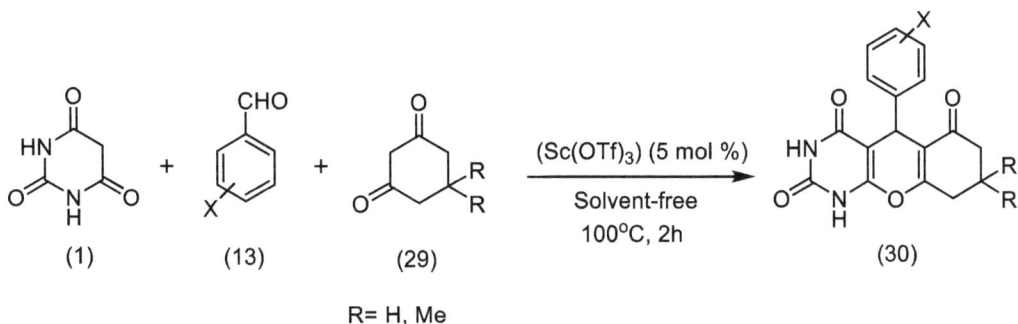

R= H, Me

Scheme (36). Synthesis of chromeno[2,3-*d*]pyrimidine-trione derivatives.

Synthesis of Nitrogen-Containing Heterocyclic Compounds

5-Membered Heterocyclic Compounds

The synthesis of spirooxindoles spiroannulated with pyrazolopyrimido-phthalazines **(33)** has been done by a one-pot four-component domino reaction of phthalic anhydride **(31)**, hydrazine hydrate **(32)**, isatins **(10)**, and barbituric acid **(1)** in a deep eutectic solvent (choline chloride: urea: 1: 2) (Scheme **37**) [154].

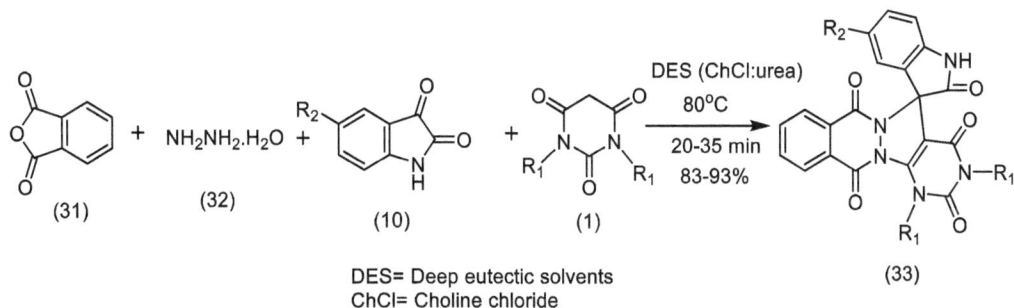

DES= Deep eutectic solvents
ChCl= Choline chloride

Scheme (37). Synthesis of spirooxindoles.

6-Membered Heterocyclic Compounds

A one- pot four component condensation reaction of aldehydes **(13)**, amines **(34-35)**, dimedone **(29)** and barbituric acid **(1)** has been reported for the preparation of novel 8,9- dihydro-8,8-dimethyl-5,10-diphenylpyrimido[4,5-*b*]quinoline-2,4,6 (1*H*, 3*H*,5*H*,7*H*,10*H*)-trione derivatives **(36-37)** in the presence of tungstopho-sphoric acid ($H_3PW_{12}O_{40}$) as a catalyst (Scheme **38**) [155].

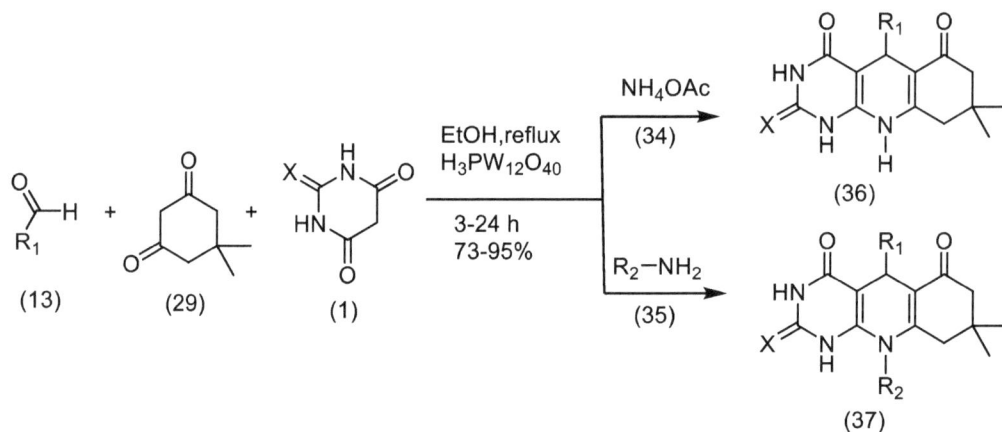

Scheme (38). Synthesis of 8,9- dihydro-8,8-dimethyl-5,10-diphenylpyrimido[4,5-*b*]quinoline-2,4,6 (1*H*,3*H*,5*H*,7*H*,10*H*)-trione derivatives.

An efficient and catalyst-free synthesis of spiro[dihydropyridine-oxindole] moiety **(39)** has been reported by a one-pot three-component reaction of barbituric acids **(1)** 2-naphthylamine **(38)** and isatins **(10)** in poly(ethylene glycol) 400/H$_2$O (Scheme **39**) [156].

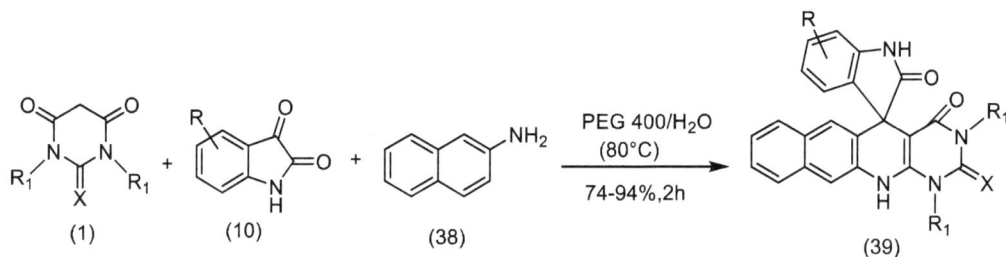

Scheme (39). Synthesis of spiro[dihydropyridine-oxindole].

Rahmati and Khalesi reported the *p*-toluenesulfonic acid catalysed synthesis of spiro[indolin-isoxazolo[40,30:5,6]pyrido[2,3-*d*]pyrimidine]triones **(41)** *via* one-pot, three-component condensation reaction of 5-amino-3-methyl isoxazole **(40)**, isatins **(10)** and barbituric acids **(1)** in water (Scheme **40**) [157].

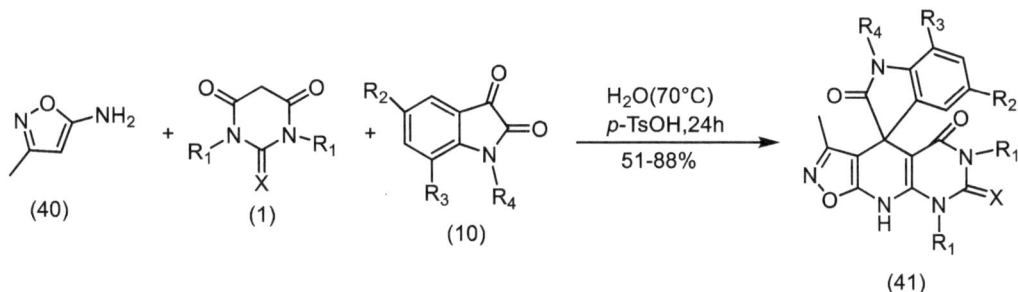

Scheme (40). Synthesis of spiro[indolin-isoxazolo[40,30:5,6]pyrido[2,3-*d*]pyrimidine]triones.

The synthesis of pyrimidine-fused nucleoside analogues **(43)** has been done by a pseudo four-component coupling reaction of barbituric acid derivatives **(1)**, aldehydes **(13)**, and nucleosides **(42)** (Scheme **41**) [158, 159].

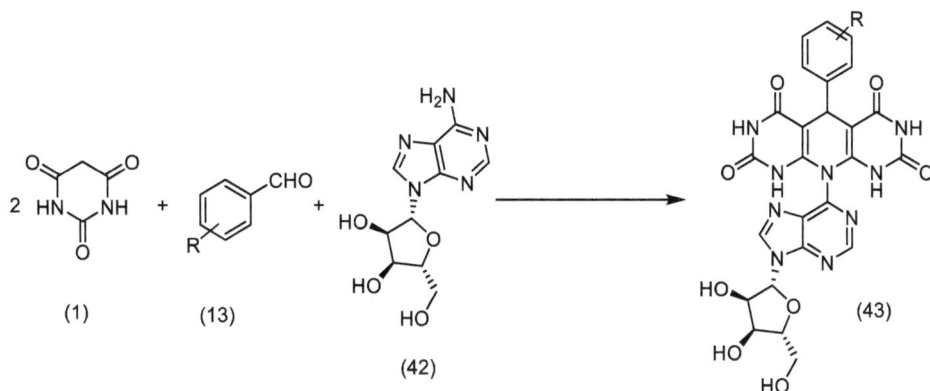

Scheme (41). Synthesis of pyrimidine-fused nucleoside analogues.

An efficient and green synthesis of 5-aryl-pyrimido[4,5-*b*]quinoline derivatives **(44)** was carried by a one-pot three-component reaction barbituric acids **(1)**, aldehydes **(13)**, and anilines **(35)** using L-proline as a catalyst in an aqueous medium (Scheme **42**) [160].

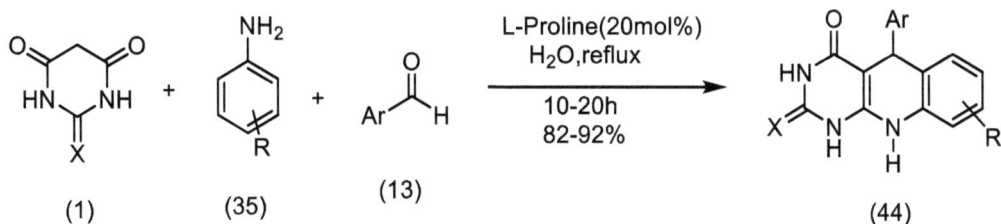

Scheme (42). Synthesis of 5-aryl-pyrimido[4,5-*b*]quinoline derivatives.

Recent Application of Barbituric Acid in Organic Synthesis

Trifluoromethylated dispirobarbituric acid derivatives have been obtained by asymmetric Michael/Mannich [3+2] cycloaddition reaction between barbiturate-based olefins and N-2,2,2-trifluoroethyl isatin ketimines [161].

N,N'-disubstituted barbituric acids have been synthesized by F. Portier *et al.* [162].

G. Bogdanov *et al.* crystallized and studied two compounds, 1,3-diethyl-5-{(2E,4E)-6-[(E)-1,3,3-trimethylindolin-2-ylidene]hexa-2,4-dien-1-ylidene} pyrimidine-2,4,6(1H,3H,5H)-trione and 1,3-diethyl-2-sulfanylidene-5-[2-(1,3,3-tri-methylindolin-2-ylidene)ethylidene]dihydropyrimidine-4,6(1H,5H)-dione [163].

Pyrano[2,3-*d*]pyrimidine derivatives and 5-arylidene barbituric acids have been prepared by condensation reaction between malononitrile, barbituric acid and aromatic aldehydes, or barbituric acid and aromatic aldehydes [164].

By reaction of *tert*-butyl hydroperoxide and substituted barbituric acids, *α-tert*-butylperoxybarbiturates were synthesized by O. V Bityukov *et al.* [165].

Piperazine-1,4-diium dihydrogen phosphate catalyzed synthesis of pyrano[2,3-d]pyrimidinone (thione) and 5-arylidenepyrimidine-2,4,6(1H,3H,5H)-trione derivatives have been done by S. Darvishzad *et al.* [166].

1,4-diionic compounds were synthesized by reaction between barbituric/thiobarbituric acid, dialkyl acetylenedicarboxylates and N-heterocycles [167].

5-(furan-3-yl) thiobarbiturate and 5-(furan-3-yl)barbiturate derivatives have been prepared by reaction of thiobarbituric acid or barbituric acid, arylglyoxals, and acetylacetone [168].

By thermal polymerization of barbituric acid (BA), photocatalytic material is prepared by N. Keshavarzi *et al.* [169].

Selective oxidation of aromatic alcohols was performed in the presence of C_3N_4 photocatalysts derived from the polycondensation of barbituric acid, cyanuric acid and melamine [170].

Multicomponent synthesis of diphenyl -1,3-thiazole-barbituric acid have been done by A. Mahata *et al.* [171].

Arylidene barbituric acid derivatives and dibarbiturates of oxindole have been prepared *via* condensation of barbituric acid with isatin/aryl aldehyde [172].

CONCLUDING REMARKS

Barbituric acid is an important precursor of many biologically active heterocyclic compounds. In this chapter, we have summarised the synthesis, physical and chemical properties of barbituric acid. There is an extensive range of organic reactions that comprise barbituric acid in preparing several organic moieties. The main objective of this chapter is to represent the overview of barbituric acid. It is expected that this chapter will be beneficial in the further development of barbituric acid chemistry.

REFERENCES

[1]   Shafiq, N.; Arshad, U.; Zarren, G.; Parveen, S.; Javed, I.; Ashraf, A. A comprehensive review: Bio-potential of barbituric acid and its analogues. *Curr. Org. Chem.,* **2020**, *24*(2), 129-161.
[http://dx.doi.org/10.2174/1385272824666200110094457]

[2]   Guillén Sans, R.; Guzmán Chozas, M. Historical aspects and applications of barbituric acid derivatives. A review. *Pharmazie,* **1988**, *43*(12), 827-829.
[PMID: 3073393]

[3]   Shaker, R.M.; Ishak, E.A. Barbituric acid utility in multi-component reactions. *Z. Naturforsch. B. J. Chem. Sci.,* **2011**, *66*(12), 1189-1201.
[http://dx.doi.org/10.1515/znb-2011-1201]

[4]   Moussier, N.; Bruche, L.; viani, F.; Zanda, M. Fluorinated barbituric acid derivatives: synthesis and bio-activity. *Curr. Org. Chem.,* **2003**, *7*(11), 1071-1080.
[http://dx.doi.org/10.2174/1385272033486567]

[5]   Mulinos, M.G Iso-amyl-ethyl-barbituric acid (amytal) as a laboratory anesthetic for cats. *J. Pharmacol. Exp. Ther.,* **1928**, *34*(4), 425-435.

[6]   Gallagher, B.; Freer, L. Barbituric acid derivatives. In: *Antiepileptic Drugs*; Springer, **1985**; pp. 421-447.
[http://dx.doi.org/10.1007/978-3-642-69518-6_14]

[7]   C. M.; ROBERTS, S. J., IV. The Effect of Sodium Phenobarbital and Some Other Barbituric Acid Derivatives upon the Coronary Circulation. *J. Pharmacol. Exp. Ther.,* **1926**, *27*(4), 327-334.

[8]   E. E.; PAGE, I. H., The comparative anesthetic efficacy of isoamylethyl barbituric acid and diethyl barbituric acid. *J. Pharmacol. Exp. Ther.,* **1927**, *31*(1), 1-9.

[9]   Ziarani, G.M.; Aleali, F.; Lashgari, N. Recent applications of barbituric acid in multicomponent reactions. *RSC Advances,* **2016**, *6*(56), 50895-50922.
[http://dx.doi.org/10.1039/C6RA09874F]

[10]   Mahmudov, K.T.; Kopylovich, M.N.; Maharramov, A.M.; Kurbanova, M.M.; Gurbanov, A.V.; Pombeiro, A.J. Barbituric acids as a useful tool for the construction of coordination and supramolecular compounds. *Coord. Chem. Rev.,* **2014**, *265*, 1-37.
[http://dx.doi.org/10.1016/j.ccr.2014.01.002]

[11]   Maquoi, E.; Sounni, N.E.; Devy, L.; Olivier, F.; Frankenne, F.; Krell, H.W.; Grams, F.; Foidart, J.M.; Noël, A. Anti-invasive, antitumoral, and antiangiogenic efficacy of a pyrimidine-2,4,6-trione derivative, an orally active and selective matrix metalloproteinases inhibitor. *Clin. Cancer Res.,* **2004**, *10*(12 Pt 1), 4038-4047.

[http://dx.doi.org/10.1158/1078-0432.CCR-04-0125] [PMID: 15217936]

[11] Liao, Y-J.; Hsu, S-M.; Chien, C-Y.; Wang, Y-H.; Hsu, M-H.; Suk, F-M. (b). Liao, Y.-J.; Hsu, S.-M.; Chien, C.-Y.; Wang, Y.-H.; Hsu, M.-H.; Suk, F.-M., Treatment with a new Barbituric acid derivative exerts Antiproliferative and Antimigratory effects against Sorafenib resistance in hepatocellular carcinoma. *Molecules,* **2020**, *25*(12), 2856.
[http://dx.doi.org/10.3390/molecules25122856]

[12] F. A.; FIELD, J., The relationship between chemical structure and inhibitory action of barbituric acid derivatives on rat brain respiration *in vitro. J. Pharmacol. Exp. Ther.,* **1943**, *77*(4), 392-400.

[13] Anderson, H.H. CHEN, M.-Y.; LEAKE, C. D., The Effects of Barbituric Acid Hypnotics on Basal Metabolism in Humans. *J. Pharmacol. Exp. Ther.,* **1930**, *40*(2), 215-228.

[14] Barakat, A.; Soliman, S.M.; Ali, M.; Elmarghany, A.; Al-Majid, A.M.; Yousuf, S.; Ul-Haq, Z.; Choudhary, M.I.; El-Faham, A. Synthesis, crystal structure, evaluation of urease inhibition potential and the docking studies of cobalt (III) complex based on barbituric acid Schiff base ligand. *Inorg. Chim. Acta,* **2020**, *503*, 119405.
[http://dx.doi.org/10.1016/j.ica.2019.119405]

[15] Perlstein, M. Gemonal®(5, 5-diethyl 1-methyl barbituric acid): new drug for convulsive and related disorders. *Pediatrics,* **1950**, *5*(3), 448-451.
[http://dx.doi.org/10.1542/peds.5.3.448]

[16] Halbeisen, W.A.; Gruber, C.M., Jr; Gruber, C.M. A toxicologic and pharmacologic investigation of cycloheptenylethyl barbituric acid. *J. Pharmacol. Exp. Ther.,* **1948**, *93*(1), 101-108.
[PMID: 18865191]

[17] Bush, M.T.; Butler, T.C.; Dickison, H.L. The metabolic fate of 5-(1-cyclohexen-1-yl)-1, 5-dimethyl barbituric acid (hexobarbital, evipal) and of 5-(1-cyclohexen-1-yl)-5-methyl barbituric acid ("nor-evipal"). *J. Pharmacol. Exp. Ther.,* **1953**, *108*(1), 104-111.
[PMID: 13053428]

[18] Badige, N.P.; Shetty, N.S.; Lamani, R.S.; Khazi, I.A.M. Synthesis and antimicrobial activitiy of novel barbituric acid and thiohydantoin derivatives of imidazo [2, 1-b][1, 3, 4] thiadiazoles. *Heterocycl. Commun.,* **2009**, *15*(6), 433-442.
[http://dx.doi.org/10.1515/HC.2009.15.6.433]

[19] Gibson, W.R.; Doran, W.J.; Wood, W.C.; Swanson, E.E. Pharmacology of stereoisomers of 1-methy--5-(1-methyl-2-pentynyl)-5-allyl-barbituric acid. *J. Pharmacol. Exp. Ther.,* **1959**, *125*(1), 23-27.
[PMID: 13621387]

[20] Meerloo, A. On the action of barbituric acid compounds: A contribution to the prolonged narcosis treatment of mental symptoms. *J. Ment. Sci.,* **1933**, *79*(325), 336-367.
[http://dx.doi.org/10.1192/bjp.79.325.336]

[21] Smith, M.C.; Riskin, B.J. The clinical use of barbiturates in neurological disorders. *Drugs,* **1991**, *42*(3), 365-378.
[http://dx.doi.org/10.2165/00003495-199142030-00003] [PMID: 1720379]

[22] Robinson, G., Sr Barbituric acid and mental health. *J. Nerv. Ment. Dis.,* **1942**, *96*(6), 716.
[http://dx.doi.org/10.1097/00005053-194212000-00030]

[23] Yasiry, Z.; Shorvon, S.D. How phenobarbital revolutionized epilepsy therapy: the story of phenobarbital therapy in epilepsy in the last 100 years. *Epilepsia,* **2012**, *53* Suppl. 8, 26-39.
[http://dx.doi.org/10.1111/epi.12026] [PMID: 23205960]

[24] Brodie, M.J.; Kwan, P. Current position of phenobarbital in epilepsy and its future. *Epilepsia,* **2012**, *53* Suppl. 8, 40-46.
[http://dx.doi.org/10.1111/epi.12027] [PMID: 23205961]

[25] Kliethermes, C.L.; Metten, P.; Belknap, J.K.; Buck, K.J.; Crabbe, J.C. Selection for pentobarbital withdrawal severity: correlated differences in withdrawal from other sedative drugs. *Brain Res.,* **2004**,

1009(1-2), 17-25.
[http://dx.doi.org/10.1016/j.brainres.2004.02.040] [PMID: 15120579]

[26]   Andrews, P.R.; Jones, G.P.; Lodge, D. Convulsant, anticonvulsant and anaesthetic barbiturates. 5-Ethyl-5-(3'-methyl-but-2'-enyl)-barbituric acid and related compounds. *Eur. J. Pharmacol.,* **1979**, *55*(2), 115-120.
[http://dx.doi.org/10.1016/0014-2999(79)90382-0] [PMID: 456410]

[27]   Archana, ; Srivastava, V.K.; Kumar, A. Synthesis of some newer derivatives of substituted quinazolinonyl-2-oxo/thiobarbituric acid as potent anticonvulsant agents. *Bioorg. Med. Chem.,* **2004**, *12*(5), 1257-1264.
[http://dx.doi.org/10.1016/j.bmc.2003.08.035] [PMID: 14980637]

[28]   Shonle, H.; Moment, A. Some new hypnotics of the barbituric acid series. *J. Am. Chem. Soc.,* **1923**, *45*(1), 243-249.
[http://dx.doi.org/10.1021/ja01654a033]

[29]   Nielsen, C. HIGGINS, J. A.; SPRUTH, H. C., A comparative study on hypnotics of the barbituric acid series. *J. Pharmacol. Exp. Ther.,* **1925**, *26*(5), 371-383.

[30]   Xu, C.; Wyman, A.R.; Alaamery, M.A.; Argueta, S.A.; Ivey, F.D.; Meyers, J.A.; Lerner, A.; Burdo, T.H.; Connolly, T.; Hoffman, C.S.; Chiles, T.C. Anti-inflammatory effects of novel barbituric acid derivatives in T lymphocytes. *Int. Immunopharmacol.,* **2016**, *38*, 223-232.
[http://dx.doi.org/10.1016/j.intimp.2016.06.004] [PMID: 27302770]

[31]   Rathee, P.; Tonk, R.; Dalal, A.; Ruhil, M.; Kumar, A. Synthesis and application of thiobarbituric acid derivatives as antifungal Agents. *Cell. Mol. Biol,* **2016**, *62*, 4172.

[32]   Khan, K.M.; Ali, M.; Wadood, A.; Zaheer-ul-Haq, ; Khan, M.; Lodhi, M.A.; Perveen, S.; Choudhary, M.I.; Voelter, W. (a). Khan, K. M.; Ali, M.; Wadood, A.; Khan, M.; Lodhi, M. A.; Perveen, S.; Choudhary, M. I.; Voelter, W., Molecular modeling-based antioxidant arylidene barbiturates as urease inhibitors. *J. Mol. Graph. Model.,* **2011**, *30*, 153-156.
[http://dx.doi.org/10.1016/j.jmgm.2011.07.001]

[32]   Altowyan, M.S.; Barakat, A.; Soliman, S.M.; Al-Majid, A.M.; Ali, M.; Elshaier, Y.A.M.M.; Ghabbour, H.A. (b). Altowyan, M. S.; Barakat, A.; Soliman, S. M.; Al-Majid, A. M.; Ali, M.; Elshaier, Y. A.; Ghabbour, H. A., A new barbituric acid derivatives as reactive oxygen scavenger: Experimental and theoretical investigations. *J. Mol. Struct.,* **2019**, *1175*, 524-535.
[http://dx.doi.org/10.1016/j.molstruc.2018.07.105]

[33]   Moon, K.M.; Lee, B.; Jeong, J.W.; Kim, D.H.; Park, Y.J.; Kim, H.R.; Park, J.Y.; Kim, M.J.; An, H.J.; Lee, E.K.; Ha, Y.M.; Im, E.; Chun, P.; Ma, J.Y.; Cho, W.K.; Moon, H.R.; Chung, H.Y. Thio-barbiturate-derived compounds are novel antioxidants to prevent LPS-induced inflammation in the liver. *Oncotarget,* **2017**, *8*(53), 91662-91673.
[http://dx.doi.org/10.18632/oncotarget.21714] [PMID: 29207675]

[34]   Sokmen, B.B.; Ugras, S.; Sarikaya, H.Y.; Ugras, H.I.; Yanardag, R. Antibacterial, antiurease, and antioxidant activities of some arylidene barbiturates. *Appl. Biochem. Biotechnol.,* **2013**, *171*(8), 2030-2039.
[http://dx.doi.org/10.1007/s12010-013-0486-6] [PMID: 24018846]

[35]   Yan, Q.; Cao, R.; Yi, W.; Chen, Z.; Wen, H.; Ma, L.; Song, H. Inhibitory effects of 5-benzylidene barbiturate derivatives on mushroom tyrosinase and their antibacterial activities. *Eur. J. Med. Chem.,* **2009**, *44*(10), 4235-4243.
[http://dx.doi.org/10.1016/j.ejmech.2009.05.023] [PMID: 19552984]

[36]   (a) Sweidan, K.; Abu Rayyan, W.; Abu Zarga, M.M.; El-Abadelah, M.A.Y.; Mohammad, H. Synthesis and antibacterial evaluation of model fluoroquinolone-benzylidene barbiturate hybrids. *Lett. Org. Chem.,* **2014**, *11*(6), 422-425.
[http://dx.doi.org/10.2174/1570178611666140401220850]
(b) Wu, K.; Li, J.; Chen, X.; Yao, J.; Shao, Z. Synthesis of novel multi-hydroxyl *N* -halamine

precursors based on barbituric acid and their applications in antibacterial poly(ethylene terephthalate) (PET) materials. *J. Mater. Chem. B Mater. Biol. Med.,* **2020**, *8*(37), 8695-8701.
[http://dx.doi.org/10.1039/D0TB01497D] [PMID: 32857090]
(c) Shukla, S.; Bishnoi, A.; Devi, P.; Kumar, S.; Srivastava, A.; Srivastava, K.; Fatma, S. Synthesis, characterization, and *in vitro* antibacterial evaluation of barbituric acid derivatives. *Russ. J. Org. Chem.,* **2019**, *55*(6), 860-865.
[http://dx.doi.org/10.1134/S1070428019060174]
(d) Chaudhari, M.B.; Mohanta, N.; Pandey, A.M.; Vandana, M.; Karmodiya, K.; Gnanaprakasam, B. Peroxidation of 2-oxindole and barbituric acid derivatives under batch and continuous flow using an eco-friendly ethyl acetate solvent. *React. Chem. Eng.,* **2019**, *4*(7), 1277-1283.
[http://dx.doi.org/10.1039/C9RE00068B]
(e) Fahad, M.M.; Zimam, E.H.; Mohamad, M.J. Synthesis and Antimicrobial Activity of Some New Barbituric Acid Derivatives Containing Thiazole Moiety from Sulfadiazine. *Nano Biomed. Eng.,* **2019**, *11*(2), 124-137.
[http://dx.doi.org/10.5101/nbe.v11i2.p124-137]
(f) Fahad, M.M.; Zimam, E.H.; Mohamad, M.J. A Series of barbituric acid derivatives from sulfa drug: synthesis and antimicrobial activity. *Nano Biomed. Eng.,* **2019**, *11*(1), 67-83.
[http://dx.doi.org/10.5101/nbe.v11i1.p67-83]
(g) Liu, Y.; Li, P.X.; Mu, W.W.; Sun, Y.L.; Liu, R.M.; Yang, J.; Liu, G.Y. Design, synthesis, and anticancer activity of cinnamoylated barbituric acid derivatives. *Chem. Biodivers.,* **2022**, *19*(2), e202100809.
[http://dx.doi.org/10.1002/cbdv.202100809] [PMID: 34931450]
(h) Ghadami, S.A.; Hosseinzadeh, L.; Eskandari, E.; Yarmohammadi, N.; Adibi, H. *In vitro* evaluation of the anticancer activity of barbituric/thiobarbituric acid-based chromene derivatives. *Mol. Biol. Rep.,* **2021**, *48*(12), 7637-7646.
[http://dx.doi.org/10.1007/s11033-021-06738-7] [PMID: 34741706]
(i) Liu, H-J.; Huang, X.; Shen, Q-K.; Deng, H.; Li, Z.; Quan, Z-S. Design, Synthesis, and Anticancer Activity Evaluation of Hybrids of Azoles and Barbituric Acids. *Iran. J. Pharm. Res.,* **2021**, *20*(2), 144-155.
[PMID: 34567152]

[37]   Naguib, F.N.; Levesque, D.L.; Wang, E-C.; Panzica, R.P.; el Kouni, M.H. 5-Benzylbarbituric acid derivatives, potent and specific inhibitors of uridine phosphorylase. *Biochem. Pharmacol.,* **1993**, *46*(7), 1273-1283.
[http://dx.doi.org/10.1016/0006-2952(93)90477-E] [PMID: 8216379]

[38]   Patrick, G.L. *An introduction to medicinal chemistry,* 4th ed; Oxford University Press: Oxford, **2009**, p. 752.

[39]   Kesharwani, S.; Sahu, N.K.; Kohli, D. Synthesis and biological evaluation of some new spiro derivatives of barbituric acid. *Pharm. Chem. J.,* **2009**, *43*(6), 315-319.
[http://dx.doi.org/10.1007/s11094-009-0298-8]

[40]   Srivastava, P.K.; Tiwari, U.K. Synthesis of S-[(arylamino) formimidoyl]-2-thiobarbituric acid hydrochloride antituberculous agents. *J. Chem. Eng. Data,* **1985**, *30*(1), 133-134.
[http://dx.doi.org/10.1021/je00039a038]

[41]   Barakat, A.; Al-Majid, A.M.; Lotfy, G.; Arshad, F.; Yousuf, S.; Choudhary, M.I.; Ashraf, S.; Ul-Haq, Z. Synthesis and dynamics studies of barbituric acid derivatives as urease inhibitors. *Chem. Cent. J.,* **2015**, *9*(1), 63.
[http://dx.doi.org/10.1186/s13065-015-0140-1] [PMID: 26583043]

[42]   Ma, L.; Li, S.; Zheng, H.; Chen, J.; Lin, L.; Ye, X.; Chen, Z.; Xu, Q.; Chen, T.; Yang, J.; Qiu, N.; Wang, G.; Peng, A.; Ding, Y.; Wei, Y.; Chen, L. Synthesis and biological activity of novel barbituric and thiobarbituric acid derivatives against non-alcoholic fatty liver disease. *Eur. J. Med. Chem.,* **2011**, *46*(6), 2003-2010.
[http://dx.doi.org/10.1016/j.ejmech.2011.02.033] [PMID: 21429633]

[43]   (a) Laxmi, S.V.; Rajitha, G.; Rajitha, B.; Rao, A.J. Photochemical synthesis and anticancer activity of

barbituric acid, thiobarbituric acid, thiosemicarbazide, and isoniazid linked to 2-phenyl indole derivatives. *J. Chem. Biol.,* **2016**, *9*(2), 57-63.
[http://dx.doi.org/10.1007/s12154-015-0148-y] [PMID: 27118996]
(b) O'Brien, A.G.; Liu, Y.C.; Hughes, M.J.; Lim, J.J.; Hodnett, N.S.; Falco, N. Investigation of a Weak Temperature–Rate Relationship in the Carbamoylation of a Barbituric Acid Pharmaceutical Intermediate. *J. Org. Chem.,* **2019**, *84*(8), 4948-4952.
[http://dx.doi.org/10.1021/acs.joc.9b00411] [PMID: 30840462]
(c) Segovia, C.; Lebrêne, A.; Levacher, V.; Oudeyer, S.; Brière, J.F. Enantioselective catalytic transformations of barbituric acid derivatives. *Catalysts,* **2019**, *9*(2), 131.
[http://dx.doi.org/10.3390/catal9020131]
(d) Marecki, J.C.; Aarattuthodiyil, S.; Byrd, A.K.; Penthala, N.R.; Crooks, P.A.; Raney, K.D. N-Naphthoyl-substituted indole thio-barbituric acid analogs inhibit the helicase activity of the hepatitis C virus NS3. *Bioorg. Med. Chem. Lett.,* **2019**, *29*(3), 430-434.
[http://dx.doi.org/10.1016/j.bmcl.2018.12.026] [PMID: 30578035]
(e) Al-Romaizan, A. N. **2019**.
(f) Sedaghati, S.; Azizian, H.; Montazer, M.N.; Mohammadi-Khanaposhtani, M.; Asadi, M.; Moradkhani, F.; Ardestani, M.S.; Asgari, M.S.; Yahya-Meymandi, A.; Biglar, M.; Larijani, B.; Sadat-Ebrahimi, S.E.; Foroumadi, A.; Amanlou, M.; Mahdavi, M. Novel (thio)barbituric-phenoxy-N-phenylacetamide derivatives as potent urease inhibitors: synthesis, *in vitro* urease inhibition, and in silico evaluations. *Struct. Chem.,* **2021**, *32*(1), 37-48.
[http://dx.doi.org/10.1007/s11224-020-01617-6]
(g) Wang, C.; Liu, H.; Zhao, W.; Li, P.; Ji, L.; Liu, R.; Lei, K.; Xu, X. Synthesis and Herbicidal Activity of 5-(1-Amino-2-phenoxyethylidene)barbituric Acid Derivatives. *Youji Huaxue,* **2021**, *41*(5), 2063-2073.
[http://dx.doi.org/10.6023/cjoc202010042]
(h) Wang, Y.H.; Suk, F.M.; Liu, C.L.; Chen, T.L.; Twu, Y.C.; Hsu, M.H.; Liao, Y.J. Antifibrotic effects of a barbituric acid derivative on liver fibrosis by blocking the NF-κB signaling pathway in hepatic stellate cells. *Front. Pharmacol.,* **2020**, *11*, 388.
[http://dx.doi.org/10.3389/fphar.2020.00388] [PMID: 32296336]
(i) Ghadami, S.A.; Shevidi, S.; Hosseinzadeh, L.; Adibi, H. Synthesis and *in vitro* quantification of amyloid fibrils by barbituric and thiobarbituric acid-based chromene derivatives. *Biophys. Chem.,* **2021**, *269*, 106522.
[http://dx.doi.org/10.1016/j.bpc.2020.106522] [PMID: 33352334]
(j) Gao, H.; Bao, P.; Dai, S.; Liu, R.; Ji, S.; Zeng, S.; Shen, J.; Liu, Q.; Ding, D. Far-red/near-infrared emissive (1, 3-dimethyl) barbituric acid-based AIEgens for high-contrast detection of metastatic Tumors in the lung. *Chem. Asian J.,* **2019**, *14*(6), 871-876.
[http://dx.doi.org/10.1002/asia.201801660] [PMID: 30548916]

[44]   Duan, J.J-W.; Lu, Z.; Wasserman, Z.R.; Liu, R-Q.; Covington, M.B.; Decicco, C.P. Non-hydroxamate 5-phenylpyrimidine-2,4,6-trione derivatives as selective inhibitors of tumor necrosis factor-α converting enzyme. *Bioorg. Med. Chem. Lett.,* **2005**, *15*(12), 2970-2973.
[http://dx.doi.org/10.1016/j.bmcl.2005.04.039] [PMID: 15908214]

[45]   DasGupta, S.; Murumkar, P.R.; Giridhar, R.; Yadav, M.R. Current perspective of TACE inhibitors: a review. *Bioorg. Med. Chem.,* **2009**, *17*(2), 444-459.
[http://dx.doi.org/10.1016/j.bmc.2008.11.067] [PMID: 19095454]

[46]   Kim, S-H.; Pudzianowski, A.T.; Leavitt, K.J.; Barbosa, J.; McDonnell, P.A.; Metzler, W.J.; Rankin, B.M.; Liu, R.; Vaccaro, W.; Pitts, W. Structure-based design of potent and selective inhibitors of collagenase-3 (MMP-13). *Bioorg. Med. Chem. Lett.,* **2005**, *15*(4), 1101-1106.
[http://dx.doi.org/10.1016/j.bmcl.2004.12.016] [PMID: 15686921]

[47]   Grams, F.; Brandstetter, H.; D'Alò, S.; Geppert, D.; Krell, H.W.; Leinert, H.; Livi, V.; Menta, E.; Oliva, A.; Zimmermann, G.; Gram, F.; Brandstetter, H.; D'Alò, S.; Geppert, D.; Krell, H.W.; Leinert, H.; Livi VMenta, E.; Oliva, A.; Zimmermann, G. Pyrimidine-2,4,6-Triones: a new effective and selective class of matrix metalloproteinase inhibitors. *Biol. Chem.,* **2001**, *382*(8), 1277-1285.
[http://dx.doi.org/10.1515/BC.2001.159] [PMID: 11592410]

[48]   Foley, L.H.; Palermo, R.; Dunten, P.; Wang, P. Novel 5,5-disubstitutedpyrimidine-2,4,6-triones as selective MMP inhibitors. *Bioorg. Med. Chem. Lett.,* **2001**, *11*(8), 969-972.
[http://dx.doi.org/10.1016/S0960-894X(01)00104-4] [PMID: 11327602]

[49]   Wang, J.; O'Sullivan, S.; Harmon, S.; Keaveny, R.; Radomski, M.W.; Medina, C.; Gilmer, J.F. Design of barbiturate-nitrate hybrids that inhibit MMP-9 activity and secretion. *J. Med. Chem.,* **2012**, *55*(5), 2154-2162.
[http://dx.doi.org/10.1021/jm201352k] [PMID: 22248361]

[50]   Wang, J.; Radomski, M.W.; Medina, C.; Gilmer, J.F. MMP inhibition by barbiturate homodimers. *Bioorg. Med. Chem. Lett.,* **2013**, *23*(2), 444-447.
[http://dx.doi.org/10.1016/j.bmcl.2012.11.063] [PMID: 23246356]

[51]   Schrigten, D.; Breyholz, H.J.; Wagner, S.; Hermann, S.; Schober, O.; Schäfers, M.; Haufe, G.; Kopka, K. A new generation of radiofluorinated pyrimidine-2,4,6-triones as MMP-targeted radiotracers for positron emission tomography. *J. Med. Chem.,* **2012**, *55*(1), 223-232.
[http://dx.doi.org/10.1021/jm201142w] [PMID: 22118188]

[52]   Singh, P.; Kaur, M.; Verma, P. Design, synthesis and anticancer activities of hybrids of indole and barbituric acids--identification of highly promising leads. *Bioorg. Med. Chem. Lett.,* **2009**, *19*(11), 3054-3058.
[http://dx.doi.org/10.1016/j.bmcl.2009.04.014] [PMID: 19398334]

[53]   Lee, S.Y.; Slagle-Webb, B.; Sharma, A.K.; Connor, J.R. Characterization of a novel barbituric acid and two thiobarbituric acid compounds for lung cancer treatment. *Anticancer Res.,* **2020**, *40*(11), 6039-6049.
[http://dx.doi.org/10.21873/anticanres.14625] [PMID: 33109542]

[54]   Penthala, N.R.; Ketkar, A.; Sekhar, K.R.; Freeman, M.L.; Eoff, R.L.; Balusu, R.; Crooks, P.A. 1-Benzyl-2-methyl-3-indolylmethylene barbituric acid derivatives: Anti-cancer agents that target nucleophosmin 1 (NPM1). *Bioorg. Med. Chem.,* **2015**, *23*(22), 7226-7233.
[http://dx.doi.org/10.1016/j.bmc.2015.10.019] [PMID: 26602084]

[55]   Muthiah, P.T.; Hemamalini, M.; Bocelli, G.; Cantoni, A. Hydrogen-bonded supramolecular motifs in crystal structures of trimethoprim barbiturate monohydrate (I) and 2-amino-4, 6-dimethylpyrimidinium barbiturate trihydrate (II). *Struct. Chem.,* **2007**, *18*(2), 171-180.
[http://dx.doi.org/10.1007/s11224-006-9083-4]

[56]   Braga, D.; Cadoni, M.; Grepioni, F.; Maini, L.; Rubini, K. Gas–solid reactions between the different polymorphic modifications of barbituric acid and amines. *CrystEngComm,* **2006**, *8*(10), 756-763.
[http://dx.doi.org/10.1039/B603910C]

[57]   Karcı, F.; Karcı, F. The synthesis and solvatochromic properties of some novel heterocyclic disazo dyes derived from barbituric acid. *Dyes Pigments,* **2008**, *77*(2), 451-456.
[http://dx.doi.org/10.1016/j.dyepig.2007.07.009]

[58]   Mahmudov, K.T.; da Silva, M.F.C.G.; Glucini, M.; Renzi, M.; Gabriel, K.C.; Kopylovich, M.N.; Sutradhar, M.; Marchetti, F.; Pettinari, C.; Zamponi, S. Water-soluble heterometallic copper (II)-sodium complex comprising arylhydrazone of barbituric acid as a ligand. *Inorg. Chem. Commun.,* **2012**, *22*, 187-189.
[http://dx.doi.org/10.1016/j.inoche.2012.06.008]

[59]   Baeyer, A. Experiments on uric acid group. *Ann,* **1864**, *131*, 291-302.

[60]   Baeyer, A. Mittheilungen aus dem organischen Laboratorium des Gewerbeinstitutes in Berlin: Untersuchungen über die Harnsäuregruppe. *Justus Liebigs Ann. Chem.,* **1864**, *130*(2), 129-175.
[http://dx.doi.org/10.1002/jlac.18641300202]

[61]   Foye, W.O. *Foye's principles of medicinal chemistry,* 6th ed; Lippincott williams & wilkins, **2008**, p. 1377.

[62]   Bose, A.K.; Garratt, S. A new synthesis of barbituric acids. *J. Am. Chem. Soc.,* **1962**, *84*(7), 1310-

1311.
[http://dx.doi.org/10.1021/ja00866a054]

[63] Shimo, K.; Wakamatsu, S. K.; WAKAMATSU, S., A New Barbituric Acid Synthesis in Liquid Ammonia-Alkali Hydroxide. II. 1 Condensation of Malonamide Derivatives with Ethyl Carbonate. *J. Org. Chem.,* **1959**, *24*(1), 19-21.
[http://dx.doi.org/10.1021/jo01083a006]

[64] Kar, A. *Medicinal chemistry,* 4th; New Age International (P) Limited, Publishers: New Delhi, **2006**, pp. 171-194.

[65] Kar, A. *Medicinal chemistry*; New Age International, **2005**.

[66] Ralhan, S.; Ray, N. Density functional study of barbituric acid and its tautomers. *J. Mol. Struct. THEOCHEM,* **2003**, *634*(1-3), 83-88.
[http://dx.doi.org/10.1016/S0166-1280(03)00260-4]

[67] Senthilkumar, K.; Kolandaivel, P. Quantum chemical studies on tautomerism of barbituric acid in gas phase and in solution. *J. Comput. Aided Mol. Des.,* **2002**, *16*(4), 263-272.
[http://dx.doi.org/10.1023/A:1020273219651] [PMID: 12400856]

[68] Noroozi Pesyan, N.; Khalafy, J.; Dilmaghani, A.; Rastgar, S.; Malekpoor, Z.; Mohammadzadeh, M. Tautomeric behaviors of 5-arylazobarbituric acids in different concentrations. *JJ. Iran. Chem. Res.,* **2009**, *2*(2), 133-144.

[69] Rubin, Y.V.; Morozov, Y.; Venkateswarlu, D.; Leszczynski, J. Prototropic equilibria in 4-thiouracil: A combined spectroscopic and ab initio SCF-MO investigation. *J. Phys. Chem. A,* **1998**, *102*(12), 2194-2200.
[http://dx.doi.org/10.1021/jp9726798]

[70] Luyten, I.; Pankiewicz, K.; Watanabe, K.; Chattopadhyaya, J. Determination of the tautomeric equilibrium of Ψ-uridine in the basic solution. *J. Org. Chem.,* **1998**, *63*(4), 1033-1040.
[http://dx.doi.org/10.1021/jo971348o]

[71] Wu, D-H.; Ho, J-J. Ab initio study of intramolecular proton transfer in formohydroxamic acid. *J. Phys. Chem. A,* **1998**, *102*(20), 3582-3586.
[http://dx.doi.org/10.1021/jp980242+]

[72] Gorb, L.; Leszczynski, J. Intramolecular proton transfer in mono-and dihydrated tautomers of guanine: an ab initio post Hartree- Fock study. *J. Am. Chem. Soc.,* **1998**, *120*(20), 5024-5032.
[http://dx.doi.org/10.1021/ja972017w]

[73] Leszczynski, J. The potential energy surface of guanine is not flat: an ab initio study with large basis sets and higher order electron correlation contributions. *J. Phys. Chem. A,* **1998**, *102*(13), 2357-2362.
[http://dx.doi.org/10.1021/jp972950l]

[74] Zhang, X-M.; Malick, D.; Petersson, G.A. Enolization enthalpies for aliphatic carbonyl and thiocarbonyl compounds. *J. Org. Chem.,* **1998**, *63*(16), 5314-5317.
[http://dx.doi.org/10.1021/jo972157o]

[75] Broo, A. A theoretical investigation of the physical reason for the very different luminescence properties of the two isomers adenine and 2-aminopurine. *J. Phys. Chem. A,* **1998**, *102*(3), 526-531.
[http://dx.doi.org/10.1021/jp9713625]

[76] Kakkar, R.; Katoch, V. In *A semiempirical MO study of tautomerism and the electronic structure of barbituric acid Proceedings of the Indian Academy of Sciences-Chemical Sciences,* **1998**, , p. 535.
[http://dx.doi.org/10.1007/BF02872580]

[77] Jeffrey, G.; Ghose, S.; Warwicker, J. The crystal structure of barbituric acid dihydrate. *Acta Crystallogr.,* **1961**, *14*(8), 881-887.
[http://dx.doi.org/10.1107/S0365110X61002539]

[78] Braga, D.; Grepioni, F.; Maini, L. The growing world of crystal forms. *Chem. Commun. (Camb.),*

2010, *46*(34), 6232-6242.
[http://dx.doi.org/10.1039/c0cc01195a] [PMID: 20623084]

[79] Lubczak, J.; Mendyk, E. Stable enol form of barbituric acid. *Heterocycl. Commun.,* **2008**, *14*(3), 149-154.
[http://dx.doi.org/10.1515/HC.2008.14.3.149]

[80] Foye, W.O. *Foye's principles of medicinal chemistry*; Lippincott Williams & Wilkins, **2008**.

[81] Delchev, V.B.; Ivanova, I.P. Theoretical study of the excited-state reaction paths of the OH and NH dissociation processes in barbituric acid. *Monatsh. Chem.,* **2012**, *143*(8), 1141-1150.
[http://dx.doi.org/10.1007/s00706-012-0766-9]

[82] Chierotti, M.R.; Gobetto, R.; Pellegrino, L.; Milone, L.; Venturello, P. Mechanically induced phase change in barbituric acid. *Cryst. Growth Des.,* **2008**, *8*(5), 1454-1457.
[http://dx.doi.org/10.1021/cg701214k]

[83] Chierotti, M.R.; Ferrero, L.; Garino, N.; Gobetto, R.; Pellegrino, L.; Braga, D.; Grepioni, F.; Maini, L. The richest collection of tautomeric polymorphs: the case of 2-thiobarbituric acid. *Chemistry,* **2010**, *16*(14), 4347-4358.
[http://dx.doi.org/10.1002/chem.200902485] [PMID: 20183832]

[84] Cruz-Cabeza, A.J.; Groom, C.R. Identification, classification and relative stability of tautomers in the cambridge structural database. *CrystEngComm,* **2011**, *13*(1), 93-98.
[http://dx.doi.org/10.1039/C0CE00123F]

[85] Lewis, T.C.; Tocher, D.A.; Price, S.L. An experimental and theoretical search for polymorphs of barbituric acid: the challenges of even limited conformational flexibility. *Cryst. Growth Des.,* **2004**, *4*(5), 979-987.
[http://dx.doi.org/10.1021/cg034209a]

[86] Bolton, W. The crystal structure of anhydrous barbituric acid. *Acta Crystallogr.,* **1963**, *16*(3), 166-173.
[http://dx.doi.org/10.1107/S0365110X63000438]

[87] Zencirci, N.; Gelbrich, T.; Apperley, D.C.; Harris, R.K.; Kahlenberg, V.; Griesser, U.J. Structural features, phase relationships and transformation behavior of the polymorphs I- VI of phenobarbital. *Cryst. Growth Des.,* **2010**, *10*(1), 302-313.
[http://dx.doi.org/10.1021/cg901062n]

[88] Zencirci, N.; Gstrein, E.; Langes, C.; Griesser, U.J. Temperature-and moisture-dependent phase changes in crystal forms of barbituric acid. *Thermochim. Acta,* **2009**, *485*(1-2), 33-42.
[http://dx.doi.org/10.1016/j.tca.2008.12.001]

[89] Roux, M.V.; Temprado, M.; Notario, R.; Foces-Foces, C.; Emel'yanenko, V.N.; Verevkin, S.P. Structure-energy relationship in barbituric acid: a calorimetric, computational, and crystallographic study. *J. Phys. Chem. A,* **2008**, *112*(32), 7455-7465.
[http://dx.doi.org/10.1021/jp803370u] [PMID: 18646743]

[90] Nichol, G.S.; Clegg, W. A variable-temperature study of a phase transition in barbituric acid dihydrate. *Acta Crystallogr. B,* **2005**, *61*(Pt 4), 464-472.
[http://dx.doi.org/10.1107/S0108768105017258] [PMID: 16041097]

[91] Schmidt, M.U.; Brüning, J.; Glinnemann, J.; Hützler, M.W.; Mörschel, P.; Ivashevskaya, S.N.; van de Streek, J.; Braga, D.; Maini, L.; Chierotti, M.R.; Gobetto, R. The thermodynamically stable form of solid barbituric acid: the enol tautomer. *Angew. Chem. Int. Ed. Engl.,* **2011**, *50*(34), 7924-7926.
[http://dx.doi.org/10.1002/anie.201101040] [PMID: 21744444]

[92] Tamaki, K.; Okabe, N. Diaquabis (5-isonitrosobarbiturato) cobalt (II) Dihydrate. *Acta Crystallogr. C,* **1996**, *52*(5), 1124-1125.
[http://dx.doi.org/10.1107/S0108270195015903]

[93] Sinn, E.; Flynn, C.M., Jr; Martin, R.B. Crystal and molecular structure of bis [ethylenediamine (barbiturato) palladium (II)]-4-water. *J. Am. Chem. Soc.,* **1978**, *100*(2), 489-492.

[http://dx.doi.org/10.1021/ja00470a021]

[94]   Colacio, E.; Dominguez-Vera, J.; Kivekäs, R.; Moreno, J.; Ruiz, J. Reactions of the polynuclear chain complex [Cu (L)(H2O)] n (H2L is 5-[(2'-carboxyphenyl) azo]-1, 3-dimethylbarbituric acid) with imidazole. Crystal structures of the dinuclear and mononuclear derivatives and EPR study of the dimer-monomer equilibrium in solution. *Inorg. Chim. Acta,* **1994**, *219*(1-2), 127-133.
       [http://dx.doi.org/10.1016/0020-1693(94)03838-4]

[95]   Michl, J.; Sykes, E.C.H. Molecular rotors and motors: recent advances and future challenges. *ACS Nano,* **2009**, *3*(5), 1042-1048.
       [http://dx.doi.org/10.1021/nn900411n] [PMID: 19845364]

[96]   Gilli, P.; Bertolasi, V.; Pretto, L.; Lyčka, A.; Gilli, G. The Nature of Solid-State N−H···O/O−H···N Tautomeric Competition in Resonant Systems. Intramolecular Proton Transfer in Low-Barrier Hydrogen Bonds Formed by the ···OC−CN−NH··· □ ···HO−CC−NN··· Ketohydrazone−Azoenol System. A Variable-Temperature X-ray Crystallographic and DFT Computational Study. *J. Am. Chem. Soc.,* **2002**, *124*(45), 13554-13567.
       [http://dx.doi.org/10.1021/ja020589x] [PMID: 12418911]

[97]   Gilli, P.; Bertolasi, V.; Pretto, L.; Gilli, G. Outline of a transition-state hydrogen-bond theory. *J. Mol. Struct.,* **2006**, *790*(1-3), 40-49.
       [http://dx.doi.org/10.1016/j.molstruc.2006.01.024]

[98]   Kottas, G.S.; Clarke, L.I.; Horinek, D.; Michl, J. Artificial molecular rotors. *Chem. Rev.,* **2005**, *105*(4), 1281-1376.
       [http://dx.doi.org/10.1021/cr0300993] [PMID: 15826014]

[99]   Landge, S.M.; Tkatchouk, E.; Benítez, D.; Lanfranchi, D.A.; Elhabiri, M.; Goddard, W.A., III; Aprahamian, I. Isomerization mechanism in hydrazone-based rotary switches: lateral shift, rotation, or tautomerization? *J. Am. Chem. Soc.,* **2011**, *133*(25), 9812-9823.
       [http://dx.doi.org/10.1021/ja200699v] [PMID: 21585197]

[100] Su, X.; Aprahamian, I. Switching around two axles: controlling the configuration and conformation of a hydrazone-based switch. *Org. Lett.,* **2011**, *13*(1), 30-33.
       [http://dx.doi.org/10.1021/ol102422h] [PMID: 21133395]

[101] Mahmudov, K.T.; Maharramov, A.M.; Aliyeva, R.A.; Aliyev, I.A.; Askerov, R.K.; Batmaz, R.; Kopylovich, M.N.; Pombeiro, A.J. 3-(para-Substituted phenylhydrazo) pentane-2, 4-diones: Physicochemical and solvatochromic properties. *J. Photochem. Photobiol. Chem.,* **2011**, *219*(1), 159-165.
       [http://dx.doi.org/10.1016/j.jphotochem.2011.02.006]

[102] Reichardt, C. Solvatochromic dyes as solvent polarity indicators. *Chem. Rev.,* **1994**, *94*(8), 2319-2358.
       [http://dx.doi.org/10.1021/cr00032a005]

[103] Sen, A.; Ganguly, B. Is dual morphology of rock-salt crystals possible with a single additive? The answer is yes, with barbituric acid. *Angew. Chem. Int. Ed. Engl.,* **2012**, *51*(45), 11279-11283.
       [http://dx.doi.org/10.1002/anie.201206170] [PMID: 23038042]

[104] Fyfe, M.C.; Stoddart, J.F. Synthetic supramolecular chemistry. *Acc. Chem. Res.,* **1997**, *30*(10), 393-401.
       [http://dx.doi.org/10.1021/ar950199y]

[105] Wiester, M.J.; Ulmann, P.A.; Mirkin, C.A. Enzymnachbildungen auf der Basis supramolekularer Koordinationschemie. *Angew. Chem.,* **2011**, *123*(1), 118-142.
       [http://dx.doi.org/10.1002/ange.201000380]

[106] Sandberg, F. Anaesthetic properties of some new N-substituted and N,N'.-disubstituted derivatives of 5,5-diallylbarbituric acid. *Acta Physiol. Scand.,* **1951**, *24*(1), 7-26.
       [http://dx.doi.org/10.1111/j.1748-1716.1951.tb00823.x] [PMID: 14877580]

[107] Gryff-Keller, A.; Kraska-Dziadecka, A. Acid–base equilibrium in aqueous solutions of 1, 3-

dimethylbarbituric acid as studied by 13C NMR spectroscopy. *J. Mol. Struct.,* **2011**, *1006*(1-3), 665-671.
[http://dx.doi.org/10.1016/j.molstruc.2011.10.029]

[108] Butler, T.C.; Ruth, J.M.; Tucker, G.F., Jr (a). Butler, T. C.; Ruth, J. M.; Tucker Jr, G. F., The Second Ionization of 5, 5-Disubstituted Derivatives of Barbituric Acid1. *J. Am. Chem. Soc.,* **1955**, *77*(6), 1486-1488.
[http://dx.doi.org/10.1021/ja01611a024]

[108] Vu, T.Q.; Yudin, N.V.; Kushtaev, A.A.; Nguyen, T.X.; Maltsev, S.A. Spectroscopic Study of the Basicity of 4,6-Dihydroxypyrimidine Derivatives. *ACS Omega,* **2021**, *6*(22), 14154-14163.
[http://dx.doi.org/10.1021/acsomega.1c00671] [PMID: 34124438]

[109] Rao, R.; Chalmers, A. Indole derivatives. i. preparation and reactions of some alkylaminoethylindoles and related compounds. *Indian J. Chem.,* **1968**, *6*(6), 336.

[110] Bojarski, J.T.; Mokrosz, J.L.; Bartoń, H.J.; Paluchowska, M.H. Recent progress in barbituric acid chemistry. *Adv. Heterocycl. Chem.,* **1985**, *38*, 229-297.
[http://dx.doi.org/10.1016/S0065-2725(08)60921-6]

[111] Sekiya, M.; Yanaihara, C. Barbituric acid derivatives. Reduction of 5-arylmethylenebarbituric acid and some analogous barbituric acid derivatives. *Chem. Pharm. Bull. (Tokyo),* **1969**, *17*, 747-751.
[http://dx.doi.org/10.1248/cpb.17.747]

[112] Sekiya, M.; Yanaihara, C.; Suzuki, J. M.; YANAIHARA, C.; SUZUKI, J., Formic Acid Reduction. V. Barbituric Acid Derivatives. Reaction of 5-(3-Indolylmethylene) barbituric Acid with the Formate. *Chem. Pharm. Bull. (Tokyo),* **1969**, *17*(4), 752-755.
[http://dx.doi.org/10.1248/cpb.17.752]

[113] Sekiya, M.; Yanaihara, C. M.; YANAIHARA, C., Formic Acid Reduction. VI. Barbituric Acid Derivatives. Reduction of 5-Arylaminomethylene-and 5-Alkylaminomethylene-substituted Barbituric Acids converting to 5-Methylbarbituric Acids. *Chem. Pharm. Bull. (Tokyo),* **1969**, *17*(4), 810-814.
[http://dx.doi.org/10.1248/cpb.17.810]

[114] Ethier, J.; Neville, G. Oxidative methylation of 5-benzylidenebarbituric acids with argentic oxide. *Tetrahedron Lett.,* **1972**, *13*(52), 5297-5300.
[http://dx.doi.org/10.1016/S0040-4039(01)85234-5]

[115] Ukita, T.; Kato, Y.; Hori, M.; Nishizawa, H. On the Anti-tumor Activity of Nitrogenous Cyclic β-Diketones. *Chem. Pharm. Bull. (Tokyo),* **1960**, *8*(11), 1021-1028.
[http://dx.doi.org/10.1248/cpb.8.1021]

[116] Mahmudov, K.T.; Kopylovich, M.N.; Pombeiro, A.J. Coordination chemistry of arylhydrazones of methylene active compounds. *Coord. Chem. Rev.,* **2013**, *257*(7-8), 1244-1281.
[http://dx.doi.org/10.1016/j.ccr.2012.12.016]

[117] Colacio, E.; Dominguez-Vera, J.M.; Costes, J.P.; Kivekas, R.; Laurent, J.P.; Ruiz, J.; Sundberg, M. Structural and magnetic studies of a syn-anti carboxylate-bridged helix-like chain copper (II) complex. *Inorg. Chem.,* **1992**, *31*(5), 774-778.
[http://dx.doi.org/10.1021/ic00031a016]

[118] Shchegol, E., kov, YV Burgart, OG Khudina, VI Saloutin, ON Chupakhin. Russ. Chem. Rev 2010, 79, 31.

[119] Johnson, T.B.; Hahn, D.A. Pyrimidines: Their Amino and Aminoöxy Derivatives. *Chem. Rev.,* **1933**, *13*(2), 193-303.
[http://dx.doi.org/10.1021/cr60045a002]

[120] Aydın, A.; Ercan, O.; Taşcıoğlu, S. A novel method for the spectrophotometric determination of nitrite in water. *Talanta,* **2005**, *66*(5), 1181-1186.
[http://dx.doi.org/10.1016/j.talanta.2005.01.024] [PMID: 18970107]

[121] Dutt, S.; Dass, I.N.D. Colour in relation to chemical constitution of the organic and inorganic salts of

isonitroso-pyrazolones and isooxazolones *Proceedings of the Indian Academy of Sciences-Section A Springer,* **1939**, 55-64.
[http://dx.doi.org/10.1007/BF03170541]

[122] Vida, J.A. An Unequivocal Synthesis of Mono-N-Substituted Barbiturate Derivatives. *Synth. Commun.,* **1973**, *3*(2), 105-109.
[http://dx.doi.org/10.1080/00397917308062017]

[123] Kulev, L.; Stepnova, G.; Stolyarchuk, V.; Nechaeva, O. Atsilproizvodnye barbiturovoi kisloty. 1. n-benzoillyuminal. *ZHURNAL OBSHCHEI KHIMII,* **1960**, *30*(4), 1385-1387.

[124] Sladowska, H. Synthesis and properties of 1-phenyl-3-methoxycarbonyl-5-ethyl-5-(gamma-hydroxybutyl) barbituric acid and allied compounds. *Farmaco, Sci.,* **1980**, *35*, 60-64.

[125] Knabe, J.; Geismar, W.; Urbahn, C. Über chirale Reduktionsprodukte optisch aktiver Barbiturate. 7. Mitt. Barbitursäurederivate. *Arch. Pharm. (Weinheim),* **1969**, *302*(6), 468-474.
[http://dx.doi.org/10.1002/ardp.19693020606] [PMID: 883875]

[126] Smissman, E.E.; Matuszak, A.J.; Corder, C.N. (a). Smissman, E.E.; Matuszak, A.J.; Corder, C.N., Reduction of barbiturates under hydroboration conditions. *J. Pharm. Sci.,* **1964**, *53*(12), 1541-1542.
[http://dx.doi.org/10.1002/jps.2600531229]

[126] Brown, H.C.; Rao, B.C.S. (b). Brown, H.C.; Rao, B.S., Hydroboration. II. A Remarkably Fast Addition of Diborane to Olefins-Scope and Stoichiometry of the Reaction. *J. Am. Chem. Soc.,* **1959**, *81*(24), 6428-6434.
[http://dx.doi.org/10.1021/ja01533a024]

[127] Kondo, Y.; Witkop, B. Reductive ring openings of glutarimides and barbiturates with sodium borohydride. *J. Org. Chem.,* **1968**, *33*(1), 206-212.
[http://dx.doi.org/10.1021/jo01265a039] [PMID: 5634894]

[127a] Chafetz, L.; Chen, T-M.; Greenough, R.C. (b). Chafetz, L.; CHEN, T. M.; RC, G., Ureide ring scission of phenobarbital by sodium borohydride. *J. Pharm. Sci.,* **1973**, *62*(3), 512.
[http://dx.doi.org/10.1002/jps.2600620341]

[128] Dudley, K.H.; Davis, I.J.; Kim, D.K.; Ross, F.T. Reductions of a model barbiturate system with sodium borohydride. *J. Org. Chem.,* **1970**, *35*(1), 147-153.
[http://dx.doi.org/10.1021/jo00826a033] [PMID: 5409733]

[129] Rautio, M. Reduction of Some 5, 5-disubstituted and 1, 5, 5-trisubstituted Barbituric Acids with Sodium Borohydride. *Annales Academiae Scientiarum Fennicae, Ser. A2,* **1976**, *178*, 1-34.

[130] Stankiewicz, K. O-methylation of barbituric acids. *Farmaco, Sci.,* **1978**, *33*(10), 740-742.

[131] Otsuji, Y.; Kuroda, T.; Imoto, E. Photochemical hydrolysis of barbitals. *Bull. Chem. Soc. Jpn.,* **1968**, *41*(11), 2713-2718.
[http://dx.doi.org/10.1246/bcsj.41.2713] [PMID: 5717080]

[132] Barton, H.; Bojarski, J.; Mokrosz, J. Photochemical ring opening of barbital. *Tetrahedron Lett.,* **1982**, *23*(20), 2133-2134.
[http://dx.doi.org/10.1016/S0040-4039(00)87280-9]

[133] Barton, H.; Paluchowska, M.H.; Mokrosz, J.L.; Szneler, E. Photolytic synthesis of N-(2-ethylbutyry-)-N'-substituted imidodicarbonic diamides. *Synthesis,* **1987**, *1987*(2), 156-158.
[http://dx.doi.org/10.1055/s-1987-27869]

[134] Smissman, E.E.; Robinson, R.A. Prevost reaction with 5-substituted 5-allylbarbituric acids. *J. Org. Chem.,* **1970**, *35*(10), 3532-3534.
[http://dx.doi.org/10.1021/jo00835a076]

[135] Smissman, E.E.; Robinson, R.A.; Carr, J.B.; Matuszak, A.J. The synthesis and rearrangement of 5-phenyl-7-methoxy-2,4,9-triketo-1,3-diazabicyclo[3.3.1]nonane. *J. Org. Chem.,* **1970**, *35*(11), 3821-3822.

[http://dx.doi.org/10.1021/jo00836a053] [PMID: 5474328]

[136] Takechi, H.; Machida, M. Photoreactions of thiobarbiturates. Intermolecular cycloaddition with alkenes and ring contraction reaction of trithiobarbiturate. *Chem. Pharm. Bull. (Tokyo),* **1997**, *45*(1), 1-7.
[http://dx.doi.org/10.1248/cpb.45.1]

[137] Jacobsen, N.W.; McCarthy, B.L.; Smith, S. N-Aminopyrimidines: The isomerization of 1-aminobarbituric acids to 2-(5-oxo-4, 5-dihydro-1H-1, 2, 4-triazol-3-yl) aliphatic acids. *Aust. J. Chem.,* **1979**, *32*(1), 161-165.
[http://dx.doi.org/10.1071/CH9790161]

[138] Demirci, T.; Arslan, M.; Bilen, Ç.; Demir, D.; Gençer, N.; Arslan, O. Synthesis and carbonic anhydrase inhibitory properties of 1,3-dicarbonyl derivatives of methylaminobenzene-sulfonamide. *J. Enzyme Inhib. Med. Chem.,* **2014**, *29*(1), 132-136.
[http://dx.doi.org/10.3109/14756366.2012.757603] [PMID: 23356427]

[139] Padmavathi, V.; Reddy, G.D.; Venkatesh, B.C.; Padmaja, A. One-pot synthesis of 5-[[Bis(azolylmethylthio)]methylene]pyrimidine-2,4,6-(1H,3H,5H)-triones. *Arch. Pharm. (Weinheim),* **2011**, *344*(3), 165-169.
[http://dx.doi.org/10.1002/ardp.201000085] [PMID: 21384415]

[140] Faryabi, M.; Sheikhhosseini, E. Efficient synthesis of novel benzylidene barbituric and thiobarbituric acid derivatives containing ethyleneglycol spacers. *J. Indian Chem. Soc.,* **2015**, *12*(3), 427-432.
[http://dx.doi.org/10.1007/s13738-014-0499-2]

[141] Yahyazadehfar, M.; Ahmadi, S.A.; Sheikhhosseini, E.; Ghazanfari, D. Synthesis of arylidene (thio) barbituric acid derivatives using bentonite as a natural and reusable catalyst in green media. *J. Appl. Chem. Res.,* **2020**, *14*(2), 36-47.
[http://dx.doi.org/10.1001.1.20083815.2020.14.2.3.7]

[142] Elinson, M.N.; Merkulova, V.M.; Ilovaisky, A.I.; Nikishin, G.I. Cascade Assembling of Isatins and Barbituric Acids: Facile and Efficient Way to 2″ H-Dispiro [indole-3, 5′-furo [2, 3-d] pyrimidine-6′, 5″-pyrimidine]-2, 2′, 2″, 4′, 4″, 6″-(1H, 1′ H, 1″ H, 3′ H, 3″ H)-hexone Scaffold. *J. Heterocycl. Chem.,* **2013**, *50*(5), 1236-1241.
[http://dx.doi.org/10.1002/jhet.1699]

[143] Maghsoodlou, M.T.; Marandi, G.; Hazeri, N.; Habibi-Khorassani, S.M.; Mirzaei, A.A. Synthesis of 5-aryl-1,3-dimethyl-6-(alkyl- or aryl-amino) furo [2,3-d]pyrimidine derivatives by reaction between isocyanides and pyridinecarbaldehydes in the presence of 1,3-dimethylbarbituric acid. *Mol. Divers.,* **2011**, *15*(1), 227-231.
[http://dx.doi.org/10.1007/s11030-010-9257-2] [PMID: 20623369]

[144] Vereshchagin, A.N.; Elinson, M.N.; Dorofeeva, E.O.; Zaimovskaya, T.A.; Stepanov, N.O.; Gorbunov, S.V.; Belyakov, P.A.; Nikishin, G.I. Electrocatalytic and chemical assembling of N, N′-dialkylbarbituric acids and aldehydes: efficient cascade approach to the spiro-[furo [2, 3-d] pyrimidine-6, 5′-pyrimidine]-2, 2′, 4, 4′, 6′-(1′ H, 3H, 3′ H)-pentone framework. *Tetrahedron,* **2012**, *68*(4), 1198-1206.
[http://dx.doi.org/10.1016/j.tet.2011.11.057]

[145] Sambavekar, P.; Aitawade, M.; Kolekar, G.; Deshmukh, M.; Anbhule, P. Uncatalyzed synthesis of furo (2, 3-d) pyrimidine-2, 4 (1H, 3H)-diones in water and their antimicrobial activity. *Indian J. Chem.,* **2014**, *53*, 1454-1461.

[146] Azzam, S.H.S.; Pasha, M. Microwave-assisted, mild, facile, and rapid one-pot three-component synthesis of some novel pyrano [2, 3-d] pyrimidine-2, 4, 7-triones. *Tetrahedron Lett.,* **2012**, *53*(52), 7056-7059.
[http://dx.doi.org/10.1016/j.tetlet.2012.10.056]

[147] Yang, X.; Yang, L.; Wu, L. [Hmim][HSO 4]: An Efficient and Reusable Catalyst for the Synthesis of Spiro [dibenzo [a, i]-xanthene-14, 3′-indoline]-2′, 8, 13-triones and Spironaphthopyran [2, 3-d]

pyrimidine-5, 3′-indolines. *Bull. Korean Chem. Soc.,* **2012**, *33*(2), 714-716.
[http://dx.doi.org/10.5012/bkcs.2012.33.2.714]

[148] Cheng, Q.; Wang, Q.; Tan, T.; Wang, M.; Chen, N. Synthesis and *in vitro* Antibacterial Activities of 5-(2, 3, 4, 5-Tetrahydro-1H-chromeno [2, 3-d] pyrimidin-5-yl) pyrimidione Derivatives. *Chin. J. Chem.,* **2012**, *30*(2), 386-390.
[http://dx.doi.org/10.1002/cjoc.201100112]

[149] Naidu Kalla, R.M.; Karunakaran, R.S.; Balaji, M.; Kim, I. Catalyst-Free Synthesis of Xanthene and Pyrimidine-Fused Heterocyclic Derivatives at Water-Ethanol Medium and Their Antioxidant Properties. *ChemistrySelect,* **2019**, *4*(2), 644-649.
[http://dx.doi.org/10.1002/slct.201803449]

[150] Lohar, T.; Mane, A.; Kamat, S.; Kumbhar, A.; Salunkhe, R. Trifluoroethanol and liquid-assisted grinding method: a green catalytic access for multicomponent synthesis. *Res. Chem. Intermed.,* **2018**, *44*(3), 1919-1933.
[http://dx.doi.org/10.1007/s11164-017-3206-y]

[151] Kazemi-Rad, R.; Azizian, J.; Kefayati, H. Electrogenerated acetonitrile anions/tetrabutylammonium cations: an effective catalytic system for the synthesis of novel chromeno [3′, 4′: 5, 6] pyrano [2, 3-d] pyrimidines. *Tetrahedron Lett.,* **2014**, *55*(50), 6887-6890.
[http://dx.doi.org/10.1016/j.tetlet.2014.10.099]

[152] Kalaria, P.N.; Satasia, S.P.; Raval, D.K. Synthesis, characterization and biological screening of novel 5-imidazopyrazole incorporated fused pyran motifs under microwave irradiation. *New J. Chem.,* **2014**, *38*(4), 1512-1521.
[http://dx.doi.org/10.1039/c3nj01327h]

[153] Kumari, S.; Kumar, D.; Gajaganti, S.; Srivastava, V.; Singh, S. Sc (OTf) 3 catalysed multicomponent synthesis of chromeno [2, 3-d] pyrimidinetriones under solvent-free condition. *Synth. Commun.,* **2019**, *49*(3), 431-443.
[http://dx.doi.org/10.1080/00397911.2018.1560471]

[154] Rajawat, A.; Khandelwal, S.; Kumar, M. Deep eutectic solvent promoted efficient and environmentally benign four-component domino protocol for synthesis of spirooxindoles. *RSC Advances,* **2014**, *4*(10), 5105-5112.
[http://dx.doi.org/10.1039/c3ra44600j]

[155] Khalafi-Nezhad, A.; Panahi, F. Synthesis of new dihydropyrimido [4, 5-b] quinolinetrione derivatives using a four-component coupling reaction. *Synthesis,* **2011**, *2011*(06), 984-992.
[http://dx.doi.org/10.1055/s-0030-1258446]

[156] Lu, G.; Cai, C. An efficient, one-pot synthesis of spiro [dihydropyridine-oxindole] compounds under catalyst-free conditions. *J. Chem. Res.,* **2011**, *35*(9), 547-551.
[http://dx.doi.org/10.3184/174751911X13157531279974]

[157] Rahmati, A.; Khalesi, Z. A one-pot, three-component synthesis of spiro [indoline-isoxazolo [4′, 3′: 5, 6] pyrido [2, 3-d] pyrimidine] triones in water. *Tetrahedron,* **2012**, *68*(40), 8472-8479.
[http://dx.doi.org/10.1016/j.tet.2012.07.073]

[158] Khalafi-Nezhad, A.; Divar, M.; Panahi, F. Nucleosides as reagents in multicomponent reactions: one-pot synthesis of heterocyclic nucleoside analogues incorporating pyrimidine-fused rings. *Tetrahedron Lett.,* **2013**, *54*(3), 220-222.
[http://dx.doi.org/10.1016/j.tetlet.2012.11.003]

[159] Kordnezhadian, R.; Shekouhy, M.; Karimian, S.; Khalafi-Nezhad, A. DBU-functionalized MCM-4-coated nanosized hematite (DBU-F-MCM-41-CNSH): A new magnetically separable basic nanocatalyst for the synthesis of some nucleoside-containing heterocycles. *J. Catal.,* **2019**, *380*, 91-107.
[http://dx.doi.org/10.1016/j.jcat.2019.10.020]

[160] Khalafi-Nezhad, A.; Sarikhani, S.; Shahidzadeh, E.S.; Panahi, F. l-Proline-promoted three-component

reaction of anilines, aldehydes and barbituric acids/malononitrile: regioselective synthesis of 5-arylpyrimido [4, 5-b] quinoline-diones and 2-amino-4-arylquinoline-3-carbonitriles in water. *Green Chem.,* **2012**, *14*(10), 2876-2884.
[http://dx.doi.org/10.1039/c2gc35765h]

[161] An, T.L.; Du, D.M. Chiral Squaramide Catalyzed Asymmetric [3+ 2] Cycloaddition Reaction for Synthesis of Trifluoromethylated Barbituric Acid Derivatives. *ChemistrySelect,* **2019**, *4*(38), 11302-11306.
[http://dx.doi.org/10.1002/slct.201903146]

[162] Portier, F.; Solier, J.; Halila, S. N, N′-Disubstituted Barbituric Acid: A Versatile and Modular Multifunctional Platform for Obtaining β-C-Glycoconjugates from Unprotected Carbohydrates in Water. *Eur. J. Org. Chem.,* **2019**, *2019*(36), 6158-6162.
[http://dx.doi.org/10.1002/ejoc.201901251]

[163] Bogdanov, G.; Tillotson, J.P.; Bustos, J.; Timofeeva, T.V. Synthesis and structure of push-pull merocyanines based on barbituric and thio-barbituric acid. *Acta Crystallogr. E Crystallogr. Commun.,* **2019**, *75*(Pt 9), 1306-1310.
[http://dx.doi.org/10.1107/S2056989019011071] [PMID: 31523455]

[164] Karami, S.; Momeni, A.R.; Albadi, J. Preparation and application of triphenyl (propyl-3-hydrogen sulfate) phosphonium bromide as new efficient ionic liquid catalyst for synthesis of 5-arylidene barbituric acids and pyrano [2, 3-d] pyrimidine derivatives. *Res. Chem. Intermed.,* **2019**, *45*(6), 3395-3408.
[http://dx.doi.org/10.1007/s11164-019-03798-0]

[165] Bityukov, O.V.; Vil, V.A.; Sazonov, G.K.; Kirillov, A.S.; Lukashin, N.V.; Nikishin, G.I.; Terent'ev, A.O. Kharasch reaction: Cu-catalyzed and non-Kharasch metal-free peroxidation of barbituric acids. *Tetrahedron Lett.,* **2019**, *60*(13), 920-924.
[http://dx.doi.org/10.1016/j.tetlet.2019.02.042]

[166] Darvishzad, S.; Daneshvar, N.; Shirini, F.; Tajik, H. Introduction of piperazine-1, 4-diium dihydrogen phosphate as a new and highly efficient dicationic brönsted acidic ionic salt for the synthesis of (thio) barbituric acid derivatives in water. *J. Mol. Struct.,* **2019**, *1178*, 420-427.
[http://dx.doi.org/10.1016/j.molstruc.2018.10.053]

[167] Anary-Abbasinejad, M.; Nejad-Shahrokhabadi, F. Formation of zwitterionic salts *via* three-component reactions of barbituric/thiobarbituric acid, N-heterocyclic compounds and dialkyl acetylenedicarboxylates. *ARKIVOC,* **2019**, *2019*(part vi), 149-157.
[http://dx.doi.org/10.24820/ark.5550190.p011.018]

[168] Dehghanzadeh, F.; Shahrokhabadi, F.; Anary-Abbasinejad, M. A simple route for synthesis of 5-(furan-3-yl) barbiturate/thiobarbiturate derivatives *via* a multi-component reaction between arylglyoxals, acetylacetone and barbituric/thiobarbituric acid. *ARKIVOC,* **2019**, *2019*(part v), 133-141.
[http://dx.doi.org/10.24820/ark.5550190.p010.837]

[169] Keshavarzi, N.; Cao, S.; Antonietti, M. A new conducting polymer with exceptional visible-light photocatalytic activity derived from barbituric acid polycondensation. *Adv. Mater.,* **2020**, *32*(16), e1907702.
[http://dx.doi.org/10.1002/adma.201907702] [PMID: 32129563]

[170] García-López, E.I.; Abbasi, Z.; Di Franco, F.; Santamaria, M.; Marcì, G.; Palmisano, L. Selective oxidation of aromatic alcohols in the presence of C3N4 photocatalysts derived from the polycondensation of melamine, cyanuric and barbituric acids. *Res. Chem. Intermed.,* **2021**, *47*(1), 131-156.
[http://dx.doi.org/10.1007/s11164-020-04330-5]

[171] Mahata, A.; Bhaumick, P.; Panday, A.K.; Yadav, R.; Parvin, T.; Choudhury, L.H. Multicomponent synthesis of diphenyl-1, 3-thiazole-barbituric acid hybrids and their fluorescence property studies. *New J. Chem.,* **2020**, *44*(12), 4798-4811.
[http://dx.doi.org/10.1039/D0NJ00406E]

[172] Kumari, S.; Kumar Maury, S.; Kumar Singh, H.; Kamal, A.; Kumar, D.; Singh, S.; Srivastava, V. Visible Light Mediated, Photocatalyst-Free Condensation of Barbituric Acid with Carbonyl Compounds. *ChemistrySelect,* **2021**, *6*(12), 2980-2987.
[http://dx.doi.org/10.1002/slct.202100051]

CHAPTER 4

Ionic Liquids as Solvents and/or Catalysts for Organic Synthesis

Vaishali Khokhar[1], Shruti Trivedi[2], Shreya Juneja[1], Komal[3], Siddharth Pandey[1,*], Gyandshwar K. Rao[3,*], Kamalakanta Behera[4, *] and Kamal Nayan Sharma[3,*]

[1] *Department of Chemistry, Indian Institute of Technology Delhi, New Delhi - 110016, India*

[2] *Department of Chemistry, Institute of Science, Banaras Hindu University Varanasi - 221 005 (U.P.), India*

[3] *Department of Chemistry, Biochemistry and Forensic Science, Amity School of Applied Sciences, Amity University Haryana, Gurgaon 122413, India*

[4] *Department of Chemistry, Faculty of Science, University of Allahabad, Prayagraj - 211002, India*

Abstract: Ionic liquids (ILs) are receiving increased enticement from synthetic organic chemists; world-wide due to their extraordinary physicochemical properties. The wide-ranging applications of ionic liquids as solvents and catalysts in organic synthesis are mainly due to their non-volatile nature which arises from very low vapor pressures. Since the past few decades, researchers have explored the efficacy of these designer solvents as green substitutes of toxic and volatile organic solvents for a variety of value added synthetic organic reactions. Furthermore, the tremendous potential of ILs as catalysts is also worth mentioning. Unlike organic solvents of comparable polarity, they often act as catalysts in various organic reactions. Thus, the present chapter aims at observing and exploring the application of ionic liquids as solvents and catalysts in various synthetic organic reactions. The green chemistry aspects of the solvent as well as the catalytic use of ionic liquids in order to develop environmentally benign organic synthesis is also the focus of discussion in this chapter.

Keywords: Catalyst, Diels-Alder reactions, Epoxidation, Friedel-Crafts Reaction, Green Solvent, Hydrogenation, Isomerization Reactions, Oxidation.

---

* **Corresponding authors Kamal Nayan Sharma, Kamalakanta Behera, Gyandshwar K. Rao, and Siddharth Pandey:** Department of Chemistry, Biochemistry and Forensic Science, Amity School of Applied Sciences, Amity University Haryana, Gurgaon 122413, India; Department of Chemistry, Indian Institute of Technology Delhi, New Delhi - 110016, India and Department of Chemistry, Faculty of Science, University of Allahabad, Prayagraj - 211002, India;
E-mails: kamalnayaniitd@gmail.com; kamala.iitd@gmail.com; gyan23iitd@gmail.com and sipandey@chemistry.iitd.ac.in

1. INTRODUCTION

Ionic liquid is a topic of immense importance due to their unusual physicochemical properties and wide-spread application potential. In the past two decades, ionic liquids (ILs) have shown burgeoning interest in various fields of science and technology due to their several striking properties, such as, low melting point, negligible vapour pressure, wide electrochemical window, high thermal stability and high ionic conductivity. More importantly, these unique materials are termed as designer solvents as their physicochemical properties can be effectively tuned just by tuning the structure of the cation and/or anion. Owing to several benefits of ILs over conventional molecular solvents, these find numerous applications in the field of organic synthesis. They have become attractive and alternative reaction media for various useful and important organic reactions. A series of reactions that can be performed in ILs has grown tremendously in the mid-1990s with the advancement of water stable neutral ILs. However, the reaction chemistry in ILs was initially restricted to the use of chloroaluminate(III) ionic liquids. Currently, ILs have shown tremendous application potential as designer solvents for variety of organic reactions [1 - 8]. The distinct characteristics and unique features of ILs have prompted researchers to use them as a reaction media in a variety of chemical transformations.

Further, ILs have shown vast applications as solvents or co-solvent, for the catalysis of various organic and inorganic reactions along with their self-catalysis activity. Catalysis is integral in the greater part of both the industrial sector and academic research [1 - 8]. A catalyst accelerates a chemical reaction and is not consumed during the reaction, thus allowing the catalysts to be employed in multiple cycles. However, the recyclability of a catalyst depends on the whole reaction system including the solvent, co-solvents, reactants and the reaction conditions as well. It is a primary requisite to design system with not only efficient catalyst but good recyclability and reusability. The use of catalysts also helps in minimalizing the side products, cutting the energy requirements, and increasing the selectivity. Consequently, catalysts render the various chemical reaction sustainable and environment-friendly. However, the use of certain organic solvents lessens the credibility of the process. In this regard, ILs have come up as a fascinating alternative solvent for biphasic catalysis and in IL thin film catalysis. From early 2000, there was a striking boost in publications for catalysis in ionic liquids. Ionic liquids are good solvents for a wide range of organic and inorganic materials, and their temperature-dependent miscibility with water, non-volatility and thermal stability brand them as favorable media for numerous chemical processes [1 - 8]. The liquid-liquid biphasic catalysis using ILs has found application in a variety of catalysis processes; organic reactions, such as Friedel-craft alkylation, Diels-alder reaction, and transition metal-catalyzed reactions, and many more [9 - 12]. The phase separation enables the easy removal of the products and catalyst for re-use. Apart from the biphasic system, the Supported Ionic Liquid Phase Catalysis (SILPC) and Solid Catalyst with Ionic Liquid Layer (SCILL) have also garnered much-deserved attention [13 - 15]. There are two ways to immobilize IL on support, (a) IL is fixed onto the support *via* physisorption (b) the IL phase is impregnated onto the support by chemical bonds. In SILPC, a thin film of ILs containing dissolved catalysts is dispersed on the support, *i.e.*, a porous material surface (alumina, silica, or active carbon) characterizing supported homogeneous catalysis. Since the dissolved catalysts are near the reaction interface and the diffusion rate is reduced, these systems often result in a high reaction rate. The SLIPC material is synthesized using the following possible methods: (a) the physical confinement/or encapsulation method; (b) the immersion method; (c) the covalent anchoring method. SCILL material is characterized by a solid support catalytic surface, or a catalyst deposited on a solid support, coated with a thin IL film, resulting in a modified heterogeneous catalyst system.

These different possible approaches of incorporating ILs in catalysis has opened a wide door of opportunities for ILs as solvents or as catalyst/co-catalyst in a varied array of reactions. Several task-specific ILs (TSILs) of mostly imidazolium salts have been designed and synthesized by incorporating acidic or basic functional groups into cation/anion of IL which can serve as both the reaction media and catalyst [16, 17]. Acidic ILs have been generated in which carboxylate groups or alkane sulfonic have been covalently bonded to the cation of different ILs such as imidazolium, ammonium, and pyridinium or by the addition of Bronsted acid to the halide based ILs. In basic type ILs, usually organic amine bases are tethered to the IL cations. A class of modified ILs, polyoxometalates-based acidic ILs (POM-ILs) were reported to behave like a co-catalyst and reactant activator in coupling reactions and biomass conversion [18, 19]. Thermoregulated ILs have shown great potential in catalysis. The conversion of a homogeneous to a heterogeneous system with a change in temperature allows easy separation of the reaction system. The assimilation of these ILs and catalysis have found numerous applications. This review presents a brief discussion on the diverse classes of chemical reactions including transition metal-catalyzed hydrogen transfer reactions, coupling reactions, ring-closing reactions, oxidation/reduction reactions, and various organic reactions. The present book chapter highlights the importance of ILs as a solvent and/or as a catalyst in organic synthesis. In Chart **1**, structure of some commonly used cations in combination of different anions are depicted.

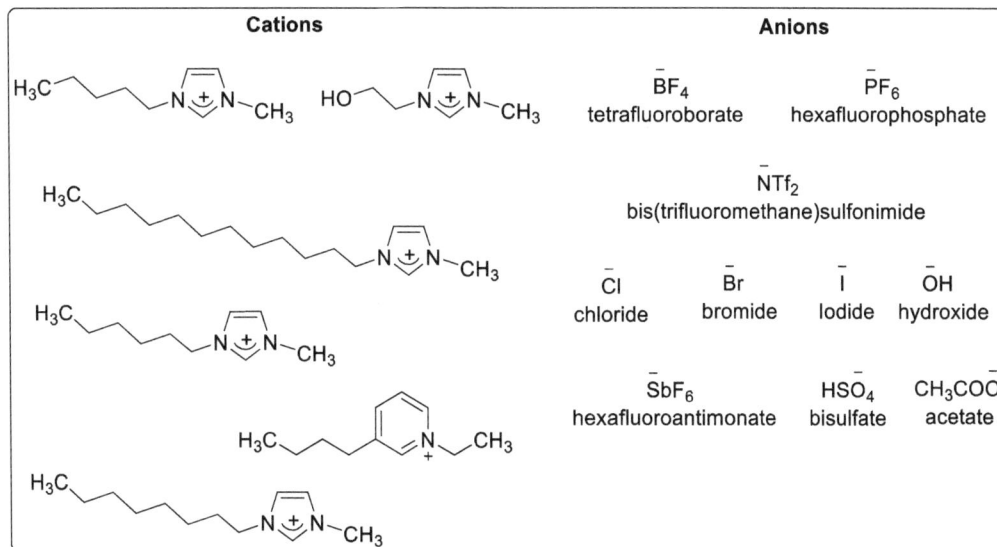

Chart (1). Some Common Ionic Liquids.

2. IONIC LIQUIDS AS SOLVENTS FOR ORGANIC SYNTHESIS

Applications of ILs as solvents for various reactions is straight forward and appropriate as compared with analogous reactions in conventional solvents. Several organic reactions have been synthesized using ILs as solvents as follows.

2.1. Demethylation and Cyclization Reactions

Several uses of molten pyridinium chloride in chemical synthesis have been reported so far. The specific property of pyridinium chloride as an acid and as a nucleophile allows its use in the dealkylation reactions of aromatic ethers. For instance, the demethylation of 2-methoxynaphthalene to 2-naphthol with pyridinium chloride (Fig. **1**) is shown below [20, 21].

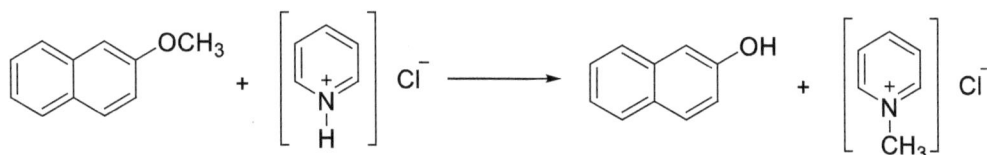

Fig. (1). Demethylation of 2-methoxynaphthalene catalyzed with pyridinium chloride.

Further, molten pyridinium chloride has also been used in various cyclization reactions of aryl ethers (Fig. **2**) [20, 21].

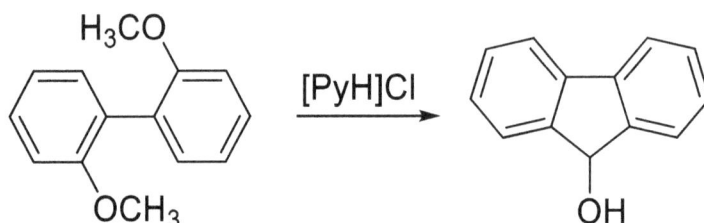

Fig. (2). Cyclization reactions of aryl ethers catalyzed with pyridinium chloride.

2.2. Scholl Reaction

ILs belong to a unique class of molten salts and presumably the first reaction which was carried out in a molten salt was the Scholl reaction. Scholl reaction involves the coupling of two aromatic rings which occur through the oxidation of the attached aromatic rings. However, an oxidizing agent is not required when the reaction is performed in NaCl.AlCl$_3$. It has been proposed that the atmospheric oxygen plays the role of oxidant in this case which further needs validation. Carlin and co-workers have shown the cyclization of 1-phenylpyrene to indeno[1,2,3-cd]pyrene (Fig. **3**) in 1990 [22].

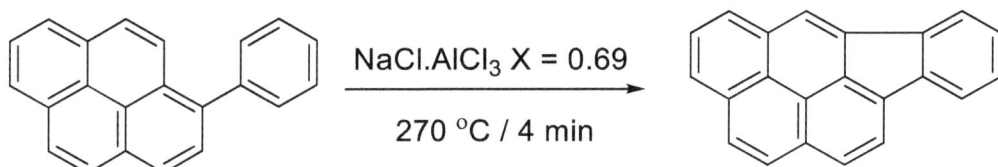

Fig. (3). Scholl reaction of 1-phenylpyrene in NaCl.AlCl$_3$.

In another study, Buchanan *et al.* observed the performance of several aromatic compounds in the presence of Sb^{3+} molten salts [23] The molten salts of antimony behave as mild Lewis acid as well as base which promote the redox reaction. The acidic and basic melts are formed by introducing chloride donor (KCl) and chloride acceptors (AlCl$_3$), respectively. The antimony(III) chloride molten salt used is in the cyclization of 1,2- bis-(1-naphthyl)-ethane Fig. (**4**) [23, 24]. Pagni has thoroughly assessed and published the chemistry of antimony(III) chloride molten salts in 1987 [25].

Green *et al.* investigated the reactions of benzoyl chloride with ethers to produce alkyl benzoates in chloroaluminate(III) ionic liquids which proceeds *via* acylative cleavage of ethers [26]. The esterification of benzoyl chloride with cyclic and acyclic ethers has been studied in the presence of ionic liquid 1-ethyl-3-butylpyridinium iodide-aluminium chloride mixture ([emim]I/AlCl$_3$) [26].

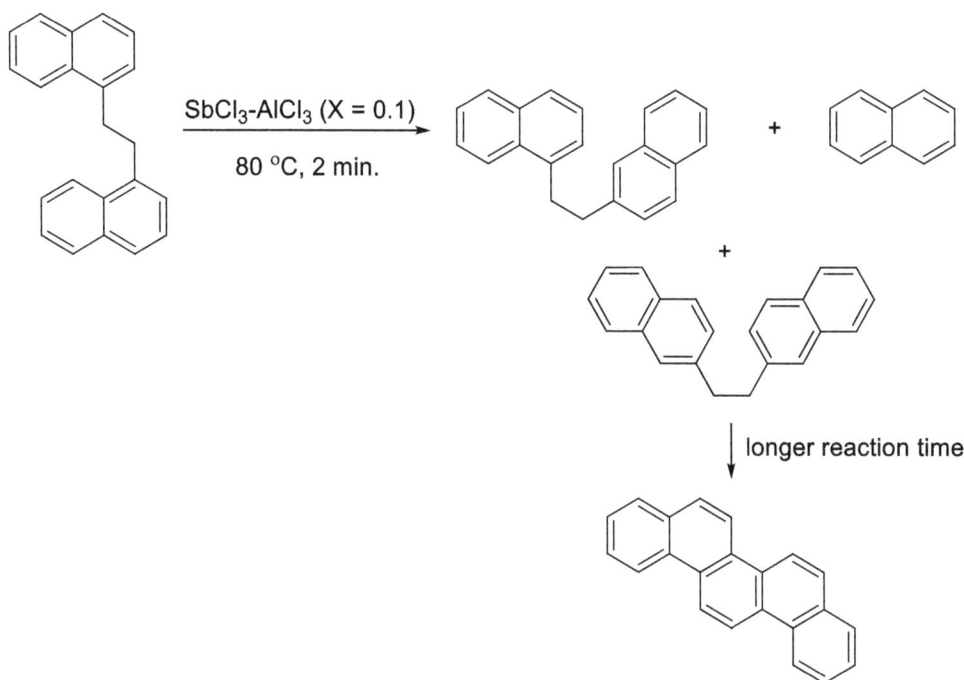

Fig. (4). SbCl$_3$.AlCl$_3$ molten salt used for the cyclization of 1,2- bis-(1-naphthyl)-ethane.

2.3. Diels-Alder Reaction

Owing to the moisture issue and separation complications, chloroaluminate(III) ionic liquids could not continue in organic synthesis and the research direction has been shifted towards the investigation of ionic liquids which are more stable to water. This approach has benefits with regard to the product separation and handling. In fact, several ionic liquids are available now-a-days that are immiscible with water, however, readily dissolve several organic substances. It has been observed that neutral ionic liquids play a significant role in Diels-Alder reaction. Jaeger and Trucker have reported the first Diels-Alder reaction in an ionic liquid [EtNH$_3$][NO$_3$] [27]. It is interesting to note that a substantial improvement in rate was observed when methyl acrylate was reacted with cyclopentadiene in [EtNH$_3$][NO$_3$] (Fig. 5) [27]. Earle *et al.* reported the Diels-Alder reactions in different neutral ionic liquids with enhanced rate and selectivity like to those of reactions achieved in lithium perchlorate-diethyl ether mixtures [28]. Various ionic liquids, such as 1-butyl-3-methylimidazolium trifluoromethanesulfonate ([bmim][Tf$_2$N]), 1-butyl-3-methylimidazolium hexafluoropho-sphate ([bmim][PF$_6$]), 1-butyl-3-methylimidazolium tetrafluoroborate ([bmim] [BF$_4$) and 1-butyl-3-methyl-imidazolium lactate ([bmim]lactate, have been explored for their ability towards Diels-Alder reaction. The outcomes of employing the aforementioned ionic liquids in Diels-Alder reaction opened the novel pathways from industrial point of view.

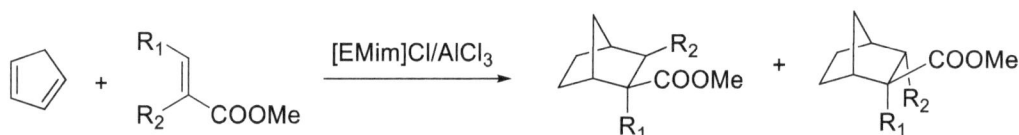

Fig. (5). Diels-Alder reaction catalyzed with [emim]Cl/AlCl$_3$.

In another report, Ludley *et al.* have shown the potential of phosphonium tosylates as a very effective solvents for Diels-Alder reactions with high regioselectivity and regiospecificity. These solvents have high thermal stability and are air and moisture insensitive. It has been found that the phosphonium tosylates can be used for the Diels-Alder reactions of isoprene with methylacrylate with high selectivity [29].

Davies and co-workers observed the higher yields for Diels-Alder reactions in the presence of water insensitive quaternary ammonium-based zinc- or tin-containing ionic liquids. The advantage of using such water insensitive ionic liquids (choline chloride: MCl$_2$, 1:2; M = Zinc or Tin) is their reusability without no significant reduction in activity [30]. Chiappe and coworkers have published a review exhibiting the role of ionic liquids as solvent for Diels-Alder reaction *via* correlation studies using empirical parameters and theoretical calculations. They have also mentioned the rate and selectivity of the reaction depends on the Lewis acidity of ILs [31].

2.4. Friedel-Crafts Alkylation

Friedel-Crafts reaction is one of the most important methods for carbon-carbon bond formation which occurs in the presence of strong Brønsted or Lewis acids (Fig. **6**). In a typical Friedel-Crafts alkylation reaction, an alkylating agent (alkyl halide, alkene, *etc.*) reacts with an aromatic moiety to generate an alkylated aromatic compound.

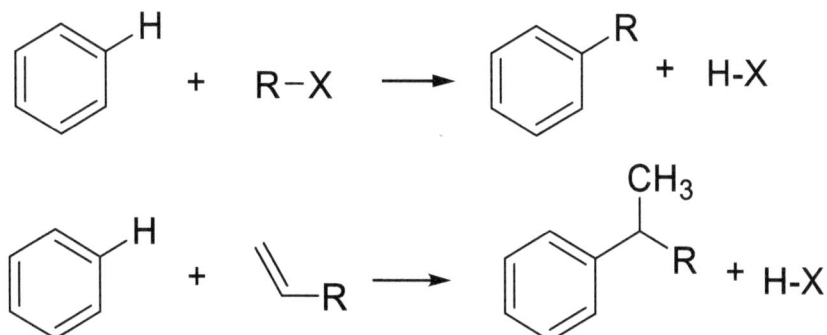

Fig. (6). Friedel-Crafts reaction.

However, Friedel-Crafts alkylation reactions are not always straightforward and are associated with a few complications, such as polyalkylation, isomerization, and rearrangement, leading to the formation of a number of undesired products [32, 33].

The very first report on Friedel-Crafts alkylation reaction in molten salts was reported by Baddeley in 1952 [34, 35]. They studied several intra-molecular cyclization, including alkenes in the NaCl/AlCl$_3$ molten salt (Fig. 7).

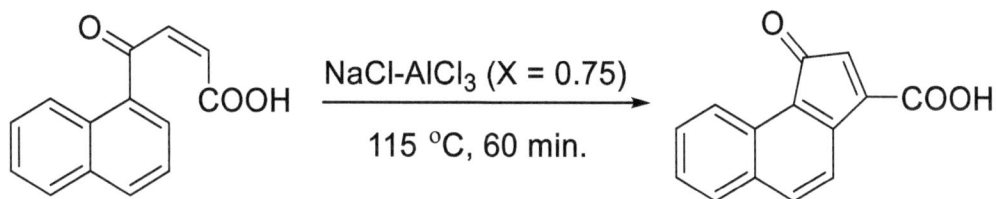

Fig. (7). Intra-molecular Friedel-Crafts reaction of cyclization of alkenes.

Wilkes and coworkers have shown the alkylation reactions of benzene and toluene in [EMIM]Cl/AlCl$_3$ [36]. The alkylation of benzene with various chloropentanes in [emim]Cl/AlCl$_3$ [X(AlCl$_3$) = 0.55] has been reported. It was found that a mixture of products was obtained when 1-chloropentane was reacted with benzene in [emim]Cl/AlCl$_3$. Only 1% yield of the unisomerized *n*-pentylbenzene as a major product as the substantial portion of the product was isomerized (Fig. **8**) [37].

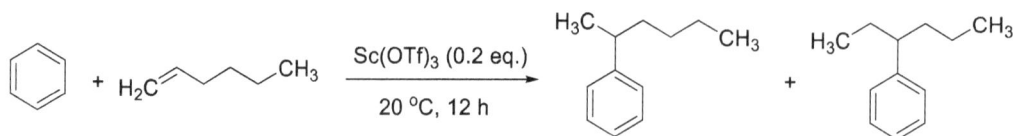

Fig. (8). Reaction of benzene with 1-penetene.

Wasserscheid and co-workers have reported several alkylation reactions of aromatic compounds in ionic liquids [bmim][HSO$_4$] without any Lewis acids [38]. The study exemplifies that esterification reactions can be performed in these ionic liquid-acid based systems. Song *et al.* reported scandium(III) triflate in [bmim][PF$_6$] for the coupling of benzene with hexane [39]. It is noteworthy to mention that when ionic liquid, such as [bmim]Cl/AlCl$_3$ supported on silica, alumina, and zirconia have been employed for the alkylation of benzene, monoalkylated dodecylbenzenes were produced (Fig. **9**) [40, 41].

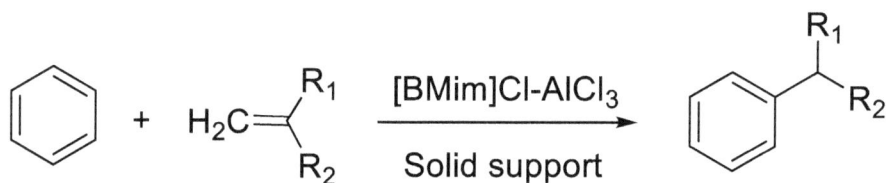

$R_1 = CH_3, CH_2CH_3, (CH_2)_2CH_3, (CH_2)_3CH_3, (CH_2)_4CH_3$

$R_2 = (CH_2)_9CH_3, (CH_2)_8CH_3, (CH_2)_7CH_3, (CH_2)_6CH_3, (CH_2)_5CH_3$

Fig. (9). Supported [bmim]Cl/AlCl$_3$ for Friedel-Crafts reaction.

2.5. Friedel-Crafts Acylation Reactions

The Friedel-Crafts acylation reaction is an electrophilic substitution reaction in which aromatic compounds react with acyl chlorides or anhydrides in the presence of a strong Lewis acid catalyst to form a carbon-carbon bond [32, 33]. Unlike alkylation reactions, the product of acylation reaction is less reactive as its starting material and thus the formation of monoacylated products takes place. The Friedel-Craft acylation reaction of phthalic anhydride and hydroquinone in a molten salt [NaCl/AlCl$_3$ (X(AlCl$_3$) = 0.69)] has been found to be very first. After this report, several other acylation reactions have been carried out by Scholl's and Bruce *et al* in similar molten salts at different temperature [42, 43]. One of the disadvantages associated with the use of NaCl/AlCl$_3$ molten salt is their high melting point. Consequently, the higher reaction temperatures would end up with the formation of unwanted side products and the decomposition of the products. Hence, the scientific community started looking at acylation reactions which could be performed under milder conditions. Several reactions have been performed using room-temperature ionic liquids under milder conditions. For example, Wilkes group investigated Friedel-Crafts acylation reaction in [emim]Cl/AlCl$_3$ (X(AlCl3) = 0.67) [36, 44]. In another report, 5-acetyl-1,1,2-6-tetramethyl-3-isopropylindane has been synthesized by Friedel-Craft acylation reaction in high yield using [emim]Cl/AlCl$_3$ (X(AlCl3) = 0.67) ionic liquid under milder condition [45]. Davey *et al.* reported Friedel-Craft acylation reaction in FeCl$_3$-based ionic liquids [46]. It has been revealed that the [emim]Cl/FeCl$_3$ based ionic liquids are noble acylating agents, and the resulting product can be extracted easily when X$_{(FeCl3)}$ varies between 0.51–0.55 [47]. The benzoylation and acetylation reactions have also been carried out using hydrophilic ionic liquid 1-butyl-3-methylimidazolium tetrafluoroborate in the presence of [Cu(OTf)$_2$] (OTf = trifluoromethanesulfonate). A good conversion rate of benzoyl chloride and anisole to methoxybenzophenone has been observed in a short duration with a ratio of 4:96 (ortho:para product). The authors have also performed this reaction using conventional molecular solvents (acetonitrile and dichloroethane). It has been revealed that the Friedel-Crafts acylation reactions were slower, and regioselectivity declined when the reaction was carried out in organic solvents [45].

Friedel-Crafts reactions at ambient temperature have been carried out in molten salts. The Friedel-Crafts alkylation and acylation of benzene with methyl chloride and acetyl chloride as the alkylating and acylating agents, respectively, have been carried out in 1-methyl-3-ethylimidazolium-aluminum trichloride ionic liquid ([emim]Cl–AlCl$_3$) (Fig. **10**) [36, 44]. It has been observed that the rate of the reaction is highly influenced by the Lewis acidity of the ionic liquid.

Earle *et al.* investigated the alkylation and acylation of several organic compounds in [emim]Cl/AlCl$_3$. Furthermore, the outcomes of such reactions were compared with the conventional solvents [28, 48]. Friedel-Crafts reactions have also been reported in room-temperature ionic liquids [45, 49]. In a recent study, Rogers and coworkers demonstrated that the Lewis acidity and the activity of metal-based ionic liquids in Friedel-Crafts acylation and alkylation reactions were enhanced due to dynamic anionic speciation [50]. It has been found that mixed-metal ionic liquid {[HN$_{222}$][xAlCl$_3$+(2– x)GaCl$_3$]Cl} displayed superior catalytic activity in comparison to AlCl$_3$, GaCl$_3$, [HN$_{222}$][Al$_2$Cl$_7$], or [HN$_{222}$][Ga$_2$Cl$_7$]. NMR, MALDI-TOF and FTIR studies revealed the presence of mixed-metal-comprising anionic species {[Cl$_2$GaOClAlClOGaCl$_2$]$^-$} responsible for superior catalytic activity [51].

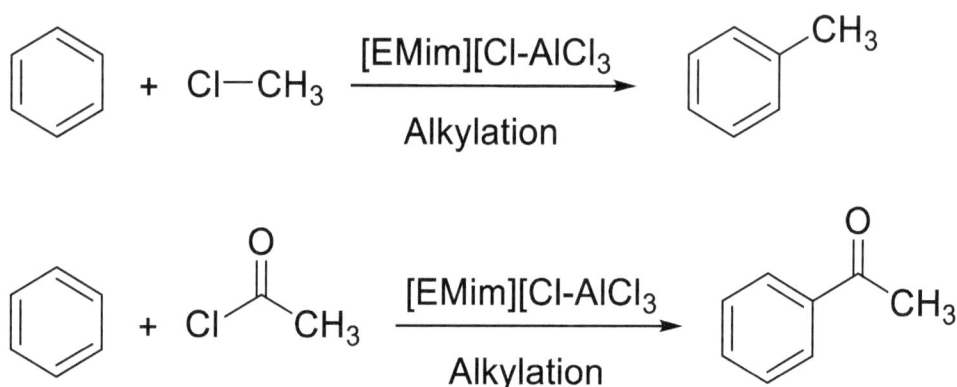

Fig. (10). Friedel-Crafts alkylation and acylation of benzene.

Gmouh *et al.* reported superior Friedel-Craft's acylating system for the acylation of aromatic compounds [51]. The catalytic activity of Bismuth(III) derivatives increased drastically when mixed with ionic liquids [emim][NTf$_2$] and [bmim][NTf$_2$]. Furthermore, these catalytic systems can be easily recycled in comparison to conventional solvents.

2.6. Friedel-Craft's Sulfonylation

Salunkhe *et al.* reported the Friedel-Crafts sulfonylation reaction of benzene and substituted benzene derivatives with 4-methyl-benzene-sulfonyl chloride for the first time in 1-butyl-3-methylimidazolium chloroaluminate ionic liquids as Lewis acid catalyst. The [bmim] chloroaluminate ionic liquids have been chosen as a solvent due to the fact that they remain liquid at room temperature over a substantial composition range of apparent mole fraction of AlCl$_3$, (N = ca. 0.30-0.67). The products, diaryl sulfones, have been isolated in high yield under ambient conditions in [bmim]Cl-AlCl$_3$ (N = 0.67) (Fig. **11**) [52].

The systematic investigations on the extent of conversion as a function of time for different substituted benzene with 4-methyl benzenesulfonyl chloride have been studied (Fig. **12**). It has been found that the conversion rates were high in the presence of chloroaluminate ionic liquids under milder conditions. The ^{27}Al NMR studies revealed the presence of [Al$_2$Cl$_7$]$^-$ species in [bmim]Cl-AlCl$_3$ (N = 0.67) in the presence of 4-methyl benzenesulfonyl chloride and [Al$_2$Cl$_7$]$^-$ has been found to be transformed into [AlCl$_4$]$^-$ with the introduction of benzene or its derivative. This transformation suggests the interaction of the Lewis acidic species [Al$_2$Cl$_7$]$^-$ with the ionic liquid resulting in the formation of HCl during the sulfonylation reaction [52].

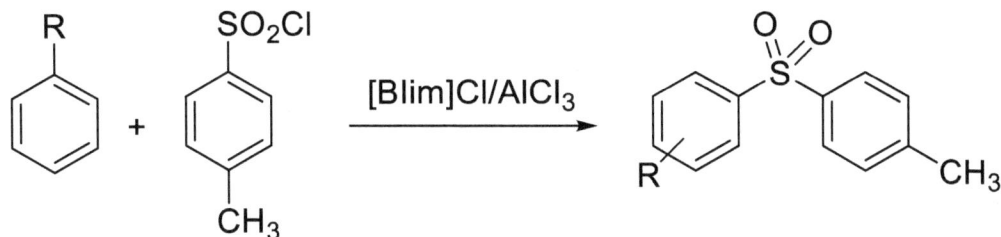

Fig. (11). Friedel-Crafts sulfonylation of substituted benzene with 4-methyl benzenesulfonyl chloride in ionic liquid [Bmim]Cl/AlCl₃ $[0.67 \geq N > 0.50]$.

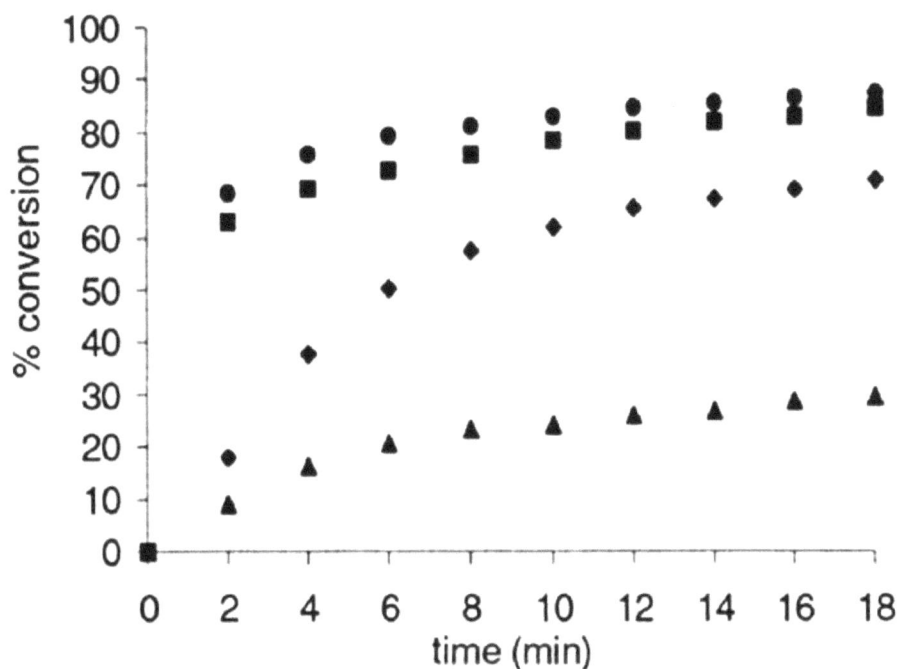

Fig. (12). Friedel-Craft sulfonylation of benzene (▲); chlorobenzene; (♦), methylbenzene (●), and 1,4-dimethylbenzene (●) with 4-methyl benzenesulfonyl chloride at room temperature in [bmim]Cl-AlCl₃ $N = 0.67$. ♦ [Reprinted with permission from [52] (Copyright 2021, American Chemical Society).

2.7. Isomerization and Cracking in Ionic Liquids

The isomerization of alkanes has attracted researchers worldwide, which provides a safer method to improve gasoline numbers [53 - 56]. The isomerization reactions can be performed readily in chloroaluminate ionic liquids. Catalytic cracking of polyethylene to a mixture of gaseous alkanes in ionic liquids has been reported by Seddon and coworkers [49]. Perumal and coworkers have also reported the isomerization reactions of n-pentane, and n-hexane to isoalkanes using chloroaluminate-based ionic liquids [emim][OTf] and [emim]Cl immobilized on SBA-15 (0.5% platinum) supported on mesoporous material as heterogeneous catalyst. They have systematically shown the influence of several key factors, such as temperature, pressure, flow rate of hydrogen, *etc.*, on isomerization reaction in ILs.

2.8. Knoevenagel Condensation Reaction

Knoevenagel condensation reaction of several aldehydes/ketones (cyclic / acyclic) with compounds having active methylene groups have been studied in the presence of various 1,4-diazabicyclo[2.2.2]octane (DABCO) based ionic liquids (Fig. **13**) [57]. The [(CH$_3$CH$_2$CH$_2$CH$_2$)DABCO]BF$_4^-$ has been found to be the best among all ionic liquids tested in terms of yield of the resulting compound with E selectivity. The catalyst has been recycled seven times without a decrease in yield, and purification of products is not required.

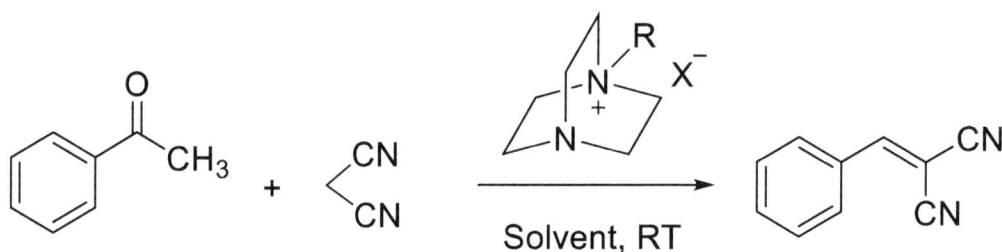

Fig. (13). Knoevenagel condensation reaction of benzaldehyde with malonitrile.

Task-specific ionic liquids (TSILs) having a neutral basic unit N-(i-Pr)$_2$ have been explored for the Knoevengeal reaction (Fig. **14**) [58]. The neutral basic unit has been found to deprotonate the acidic hydrogen of active methylene compound to produce a nucleophilic center. These TSILs act as a catalyst as well as a solvent for this reaction.

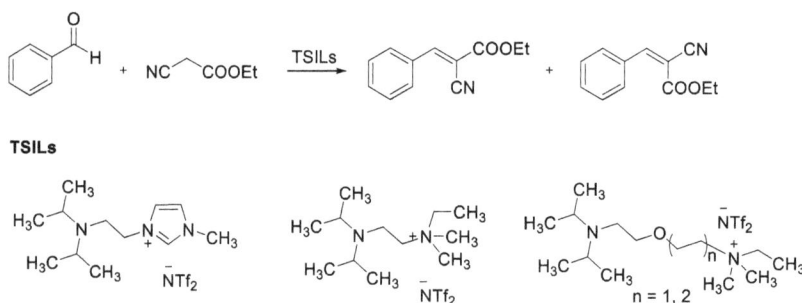

Fig. (14). Knoevenagel reaction catalyzed with TSILs.

2.9. Synthesis of Urea Derivative

IL [bmim]OH has been explored for its potential for the synthesis of urea derivatives using various amines and carbon dioxide (Fig. **15**) [59]. A variety of such cyclic/acyclic/aromatic/aliphatic has been successfully concerted to corresponding urea derivatives. The advantage of using this ionic liquid is that the reaction was completed very smoothly without the formation of side products, easy recyclability of [bmim]OH and nonrequirement of a dehydrating agent.

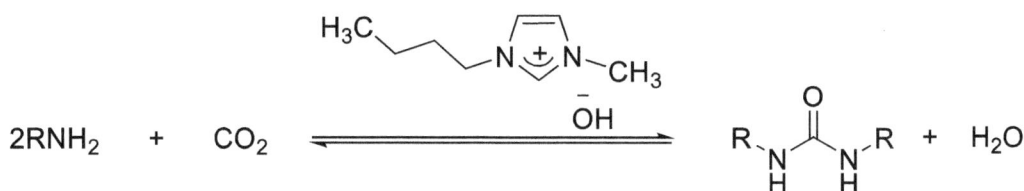

Fig. (15). Synthesis of urea derivative.

2.10. Thiol-Ene Reaction

A thiol-ene reaction has been reported to be catalyzed with [hmim]Br to produce an anti-Markovnikov product (Fig. **16**). The role of secondary T-shape л⁺----л interaction seems to be the driving force of the reaction [60]. This interaction results in cyclic six-member (H---Br---H---S---terminal alkene double bond) bond formation followed by the release of the HBr molecule, which catalyzes the reaction. This reaction is regioselective in conventional solvents, whereas in the present case, it is regiospecific. Also, a catalyst is required when this reaction is carried out in conventional solvents.

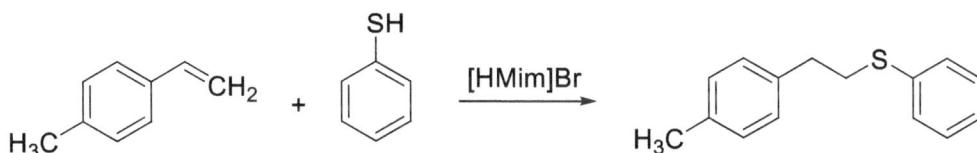

Fig. (16). The thiol–ene reaction.

2.11. Michael Addition Reaction

Michael addition reaction is one of the most important tools for the synthesis of C−C bond formation. Michael addition of N-heterocycles to β-unsaturated compounds has been carried out using [bmim]OH at room temperature (Fig. **17**) [61]. The reactions were completed in very short reaction times (10-20) minutes. Furthermore, cyclic/acyclic/aliphatic/aromatic amines have been successfully coupled with various alkenes. The additional advantage of using this ionic liquid is that it could be used several times without loss of its catalytic activity.

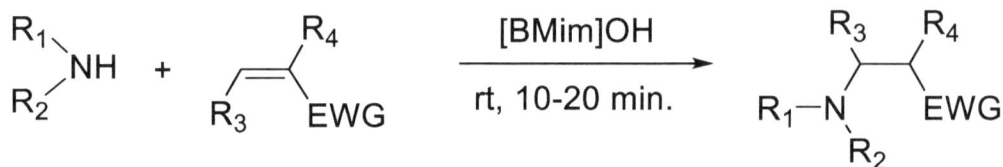

EWG: CN, COCH$_3$, COOCH$_3$

Fig. (17). Michael addition of N-heterocycles to α,β-unsaturated compounds.

2.12. Fischer Indole Synthesis Using Ketones

Fischer Indole synthesis is an important method for producing substituted Indole derivatives in the presence of some Brønsted or Lewis acids. 1-Butylpyridiniu--AlCl$_3$ [n-BPC-AlCl$_3$] has been explored for their potential for Fisher Indole synthesis of various substituted aldehydes/ketones (Fig. **18**) [62]. Several aldehydes/ketones have been successfully coupled to indole derivative at very low reaction time and lower catalyst loading.

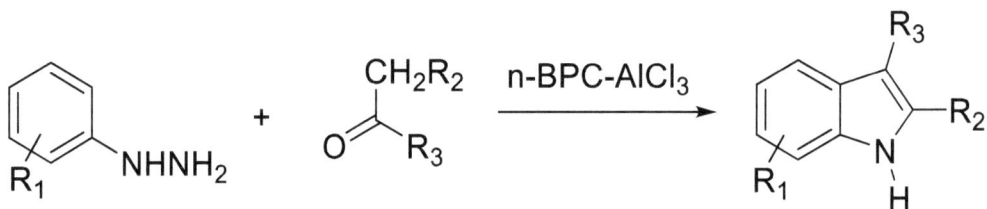

Fig. (18). Fischer-Indole synthesis using ketones.

2.13. Henry Reactions

Various substituted aldehydes with nitromethane have been carried out in imidazolium and pyridinium substituted ionic liquids (Fig. **19**) [63]. It has been found that in the presence of chloroaluminate, the reactions have been accelerated. Both, aromatic and aliphatic aldehydes have been successfully coupled in the product in high yield. The ILs have been recycled five times without loss of any catalytic efficiency.

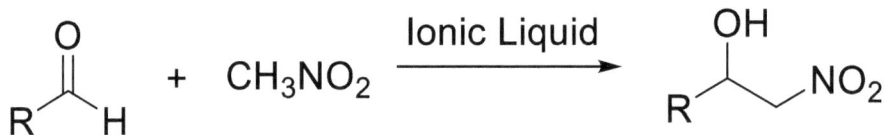

Fig. (19). Henry reactions in various.

2.14. Unimolecular, Bimolecular Substitution and Nucleophilic Addition Reactions

Unimolecular reaction of bromodiphenylmethane with 3-chloropyridine has been studied by Harper and co-workers (Fig. **20**) [64]. It has been found that the reactions were faster in ionic liquids except [mtoa][NTf$_2$], as compared to traditional solvents. The reaction in [mtoa][NTf$_2$] is slower, probably due to the presence of bulky ammonium cation, which causes steric hindrance. It has also been found that the smaller and spherical anions on IL have a higher reaction rate.

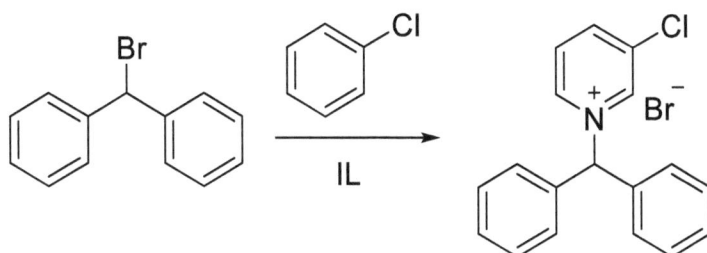

IL = [BMPyrr][NTf$_2$]; [MOta][NTf$_2$]; [BMim][BF$_4$]; [BMim][PF$_6$]

Fig. (20). Unimolecular reaction of bromodiphenylmethane with 3-chloropyridine.

Harper and coworkers reported that the bimolecular reaction of triphenyl phosphine with benzylic electrophiles can be carried out in ionic liquid [bmim][NTf$_2$] (Fig. **21**) [65]. They also studied the effect of different substituent's present at the para position of the CH$_2$X group. It has been observed that the order of the rate of the reactions having different substituents was faster or slower depending on the benzylic electrophiles as compared to traditional solvents such as acetonitrile. This effect arises due to the fact that the interactions of different benzylic groups with ionic liquid in the transition state were different.

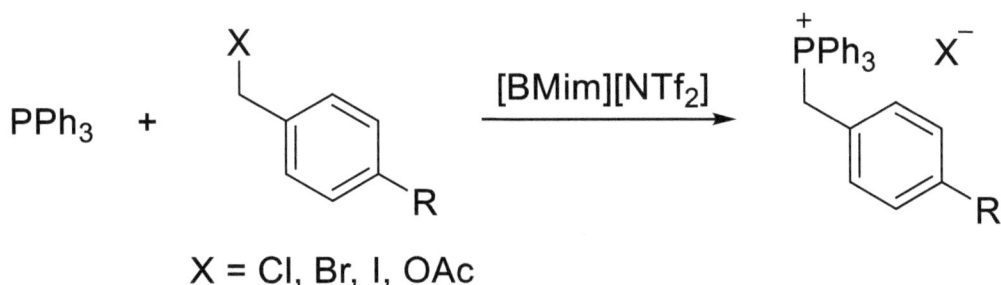

X = Cl, Br, I, OAc

Fig. (21). Bimolecular reaction of triphenyl phosphine with benzylic electrophiles.

Nucleophilic addition reactions of substituted aldehydes with 1-hexaneamine have been carried out in [bmim][NTf₂] (Fig. **22**) [66]. Once again, it has been found that the reaction was faster than when carried out in IL compared to traditional solvents such as acetonitrile. A kinetic study showed that the rate in each step was higher as compared to acetonitrile.

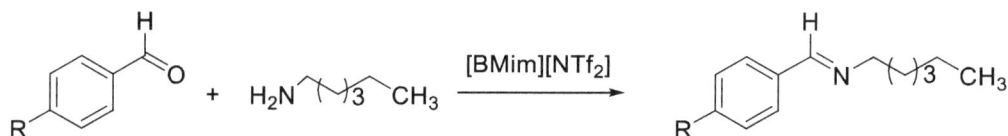

Fig. (22). Nucleophilic addition reactions of substituted aldehydes with 1-hexane amine.

3. IONIC LIQUIDS FOR CATALYSIS AS SOLVENT/CATALYST

3.1. Oxidation Reactions

3.1.1. Desulfurization

Desulfurization is an imperative chemical process, specifically in the petroleum industry, as the sulfur-containing materials in fuel are hazardous to the environment. Reddy and Verkade were the first to investigate the formation of sulfones by the oxidation of organic sulfides using H_2O_2 as oxidant and $Ti_4[(OCH_2)_3CMe]_2(i\text{-}OPr)$ as a catalyst in MeOH and three ILs: 1-butyl-3-methylimidazolium hexafluorophosphate ([bmim].PF_6), 1-butyl-3-methylimidazolium tetrafluoroborate ([bmim].BF_4), and 1-ethyl-3-methylimidazolium tetrafluoroborate ([emim].BF_4) as solvents [67]. Their study revealed that the catalyst showed ~30% acceleration in ILs as compared to methanol for selective substrates. Additionally, the catalyst can be easily removed and recycled for at least 6 cycles without losing its activity. Following this work, Li *et al.* conducted a series of investigations on catalytic oxidation for the desulfurization process in ILs using different catalytic systems [68 - 71]. They investigated iron chloride-based ILs with H_2O_2 as an oxidant for the removal of benzothiophene (BT), dibenzothiophene (DBT), and 4,6-dimethyl dibenzothiophene (4,6-DMDBT) [68]. They further explored the efficiency of a catalytic system containing $Na_2MoO_4 \cdot 2H_2O$, H_2O_2, and [bmim]. BF_4 in desulfurization and in a subsequent study a combination of H_2O_2 and V_2O_5 in [bmim].BF_4 for desulfurization of fuels [69 - 71]. Zhao and the group conducted a detailed investigation on oxidative desulfurization of DBT using N-methyl-pyrrolidonium tetrafluoroborate ([hnmp].BF_4), a Brønsted acidic IL as a catalyst, and as well as the solubilizing media for the oxidative process in the presence of H_2O_2 as an oxidant [72, 73]. Later, Rafiee *et al.* synthesized a series of tungstophosphate anion-based POM-ILs, including [BuPyPS]PW, [PhPyPS]PW, and [BzPyPS]PW, and employed in the oxidation of methyl phenyl sulfide using H_2O_2. They further designed and synthesized POM-IL containing a longer alkyl group attached to the N atom of pyridinium, which was found to have the superior catalytic ability for oxidative desulfurization of a synthetic mixture of model oil [74, 75]. Several thermoregulated ILs such as imidazolium IL combined with POM as catalyst center and poly(ethylene oxide) (PEO) as thermoregulated structure and ammonium oxidative thermoregulated ILs showed good activity towards oxidative desulfurization.

3.1.2. Epoxidation

Several efforts have been undertaken to escalate the research concerning the use of ILs for epoxidation. Basheer *et al.* utilized [bmim]BF_4, IL as the solubilizing media for the epoxidation of cyclohexene using manganese(II) and copper(II) complexes of Schiff and reduced Schiff bases as catalysts and hydrogen peroxide as the oxidant [76]. Bortolini and the group reported epoxidation of electrophilic alkenes whose structures are closely related to Vitamin-K in 1 [bmim]PF_6 and [bmim]BF_4 IL utilizing aqueous solutions of hydrogen peroxide as an oxygen source in the presence of NaOH as the catalyst [77]. Basic catalysts such as Na_2CO_3, $NaHCO_3$, or NaOH, combined with hydrogen peroxide as an oxygen source, have been utilized for the epoxidation of several α,β-unsaturated carbonyls in IL/water systems [78]. Several authors have exemplified the use of different molybdenum compounds as catalysts for the epoxidation of olefin [78 - 80]. Valente *et al.* investigated the solubility of the Mo (VI) complex (Fig. **23**) in various solvents counting ILs, and their results showed that the polar IL N-buty--3-methylpyridinium tetrafluoroborate ([bmPy]BF_4) led to a higher conversion of trans-methyl styrene (31%) compared to the N-butyl-3-methylpyridinium hexafluorophosphate ([bmPy].PF_6) (1%). Additionally, it was reported that the desired epoxide has 100% selectivity in ILs [79]. Also, Gago *et al.* investigated different molybdenum (VI) dioxo complexes for the epoxidation of cyclooctene in different solvents (DCE, [bmim].BF_4, [bmim].PF_6, [bmPy].BF_4, and [bmPy].PF_6) at 55 °C [80]. Among the three catalysts, 1 was soluble in all solvents, while 2 was preferably more soluble in ILs having BF_4 anion and 3 was slightly less soluble in all the investigated solvents, leading to considerable differences in their catalytic activity. The reaction was found to be heterogeneous in nature, as depicted by the yellowish tinge in the IL phase, indicating the formation of active species while the organic phase remained colorless. Furthermore, the authors also evaluated the effect of different oxidants on the cyclooctene oxidation process. Tsang and group reported the selective oxidation of styrene to acetophenone using H_2O_2 as oxidant and Palladium-based catalysts in different multicarboxylic acid-affixed imidazolium ILs as the solubilizing media [81]. It was described that treating $PdCl_2$ with ILs, results in the formation of active catalytic species containing the $PdCl_4^{2-}$ or $PdCl_2Br_2^{2-}$ anion. The resultant system showed higher activity than the neat $PDCl_2$. Based on the investigation of various combinations, the authors concluded that the reaction rate is primarily dependent on temperature rather than the cation and anion. The selectivity and activity of the catalyst were found to be persistent even after 10 cycles.

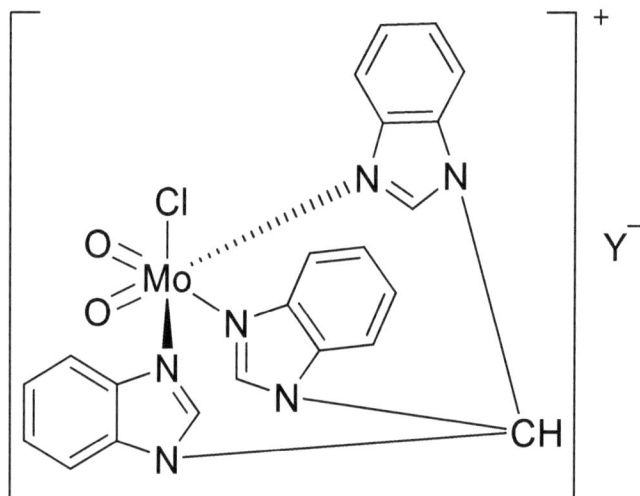

Fig. (23). Mo Catalyst (1) Y = Cl, (2) Y= BF_4 and (3) Y = PF_6.

Protic and PEG-functionalized imidazolium polyoxometalates−ILs (POM-ILs) have shown potential as thermo-regulated catalysts for epoxidation [82 - 84]. The catalytic material obtained following the immobilization of a tungsten peroxo complex on imidazole-functionalized silica was exploited for the hydroxymethylation of olefins [82]. The immobilization process was highly temperature-dependent since, at the reaction temperature, the complex would dissociate and diffuse into the liquid phase, however, on reducing the temperature to 0 °C, it would again bind with the imidazole-functionalized silica *via* H-bonding. This temperature-dependent behavior helps in recycling the catalyst as well as brands the reaction system as homogeneous. Similar to the tungsten peroxo complex, the catalysts formed by Au nanoparticles appended on PEG-based imidazolium ILs exhibited a distinct self-separating nature depending on the temperature in the epoxidation of styrene [83, 84].

3.1.3. Oxidation of Alcohols

The researchers have undertaken several efforts to study the oxidation of alcohols in various IL systems. In 2007, Shen group investigated the selective oxidation of alcohols to corresponding carbonyl compounds with molecular oxygen as oxidant and the copper-*bis*-isoquinoline as a catalyst in ionic liquids [bmim]PF$_6$, 1-octy--3-methylimidazolium ([omim]BF$_4$), and 1-n-hexyl-3-methylimidazolium. ([hmim] BF$_4$), as a solvent [85]. It was observed that ILs aid in improving the catalytic activity of the system. The authors further stated that the yield and selectivity of the system were higher than 80%. In the same year, Liu and the group explored the catalytic oxidation of secondary alcohols to respective ketones employing TBHP as an oxygen source and Cu(acac)$_2$ as a catalyst in different ILs under varying reaction conditions [86, 87]. The comparative study in different ILs demonstrated that the imidazolium-type ILs offered the best results. The authors afterward verified the recyclability of the system using the 1-phenylpropan-1-ol oxidation to the corresponding propiophenone as a test run. It was noted that the yield dropped from 91% to 84% in the 5th cycle. Bhat and coworkers reported that the catalytic oxidation of alcohols employing NaOCl as an oxidant and Ni(II)-Schiff-base as catalysts in an [emim]-based IL ensued >61% yields only after 15 minutes at room temperature [88]. Shaabani group examined the aerobic oxidation of alcohols and alkyl arenes to the subsequent carbonyl compounds in a series of solvents- [bmim]PF$_6$, [bmim]Br, and [bmim]Cl and tetrame-thylguanidinium trifluoroacetate (TMGT) in neutral, acidic, and basic conditions using Co (II) phthalocyanine as a catalyst [89]. The authors perceived that the best results were achieved with [bmim]Br in terms of reaction work up as well as the yield. The yields for the model oxidation reaction-benzoin to benzil in different solvents are presented in Table **1** and oxidation of different alcoholic substrates in [bmim]Br is presented in Table **2** [89]. It was inferred that the dual behavior of IL as a solvent and a phase transfer catalyst is responsible for the increase in yields in ILs compared to conventional organic solvents. Lei and group used N-chlorosuccinimide (NCS)/NaBr/TEMPO-IL as a catalytic system for the chemoselective oxidation of benzylic alcohols in a [bmim].PF$_6$–H$_2$O-mixture as the solubilizing media in the presence of aliphatic alcohols [90]. The system compromised the catalyst TEMPO-IL and solvent [bmim]. PF$_6$ was found to highly effective and easy to recycle up to ten successive runs while retaining the selectivity and yield of the product.

Table 1. Solvent effect on the aerobic oxidation of fluorene and benzoin using Co-Pc as a catalyst.

S. No.	Ionic Liquid	Yield (%)/time (h)	Yield (%)/time (h)
1	[bmim]Br	91 (6)	92 (1)
2	[bmim]Cl	60 (6)	80 (1)
3	[bmim]PF$_6$	30 (6)	50 (1)
4	TMG/TFA 1:1	40 (6)	60 (1)
5	TMG/TFA 1:5	50 (6)	70 (1)
6	TMG/TFA 1:5:1	30 (6)	70 (1)

(Table 1) cont.....

S. No.	Ionic Liquid	Yield (%)/time (h)	Yield (%)/time (h)
7	MeOH	10 (24)	30 (1)
8	Water	0 (24)	10 (1)
9	Acetonitrile	0 (24)	0 (1)
10	Xylene	10 (24)	40 (12)

Table 2. Oxidation of alcohols to carbonyl compounds using Co-Pc/[bmim]Br at 70 °C.

Entry	Substrate	Product	Time (min)	Yield (%)
1	HO—cyclohexane	O=cyclohexane	160	80
2	HO—cyclooctane	O=cyclooctane	120	89
3	HO—cyclodecane	O=cyclodecane	120	90
4	Me—C₆H₄—CH₂OH	Me—C₆H₄—CHO	80	83
5	MeO—C₆H₄—CH₂OH	MeO—C₆H₄—CHO	90	80
6	Cl—C₆H₄—CH₂OH	Cl—C₆H₄—CHO	80	81
7	O₂N—C₆H₄—CH₂OH	O₂N—C₆H₄—CHO	90	87
8	O₂N—C₆H₄—CH(OH)CH₃	O₂N—C₆H₄—C(O)Me	170	80
9	indanol	indanone	120	91
10	diphenylmethanol	benzophenone	100	84
11	Ph—CH(OH)—C(O)—Ph	Ph—C(O)—C(O)—Ph	60	92, 88, 91
12	(MeO-C₆H₄)CH(OH)—C(O)(C₆H₄-OMe)	(MeO-C₆H₄)C(O)—C(O)(C₆H₄-OMe)	70	92
13	fluorenyl-CH₂OH	fluorenyl-CHO	110	81

Miao *et al.* studied catalytic oxidation of aliphatic, allylic, heterocyclic, and benzylic alcohols to the corresponding carbonyl compounds using molecular oxygen and a three-component catalytic system, including a 2,2,6,6-tetramethylpiperidine-1-oxyl functionalized imidazolium salt ([imim-TEMPO]$^+$ X$^-$), a carboxylic acid substituted imidazolium salt ([imim-COOH]$^+$ X$^-$), and sodium nitrite (NaNO$_2$) in the presence of an optimum amount of water [91]. The designed catalytic system showed a higher preference for the aerobic oxidation of primary alcohols to aldehydes in comparison with the oxidation of secondary alcohols to ketones. The system was highly effective, reusable for at least 4 times, and exhibited excellent selectivity up to >99%.

3.2. Hydrogenation

ILs have shown promising application in hydrogenation reactions. The solubility of the active catalysts in the solvent is one of the controlling factors. In most cases, the hydrogen solubility in ILs is more comparable to water rather than the organic solvents [92]. Despite the low solubility of H$_2$, ILs have been very useful in hydrogenation. Several evidences indicate higher solubility of imidazolium appended chiral or achiral Rh and Ru salts catalysts in ILs [93].

3.2.1. Hydrogenation of Olefins

In a communication, Han and group have shown that phenanthroline ligand-protected palladium nanoparticles in [bmim]PF$_6$ are highly efficient and selective for hydrogenation of hex-1-ene, cyclohexene, cyclohexadiene [94]. In this system, IL has been used to synthesize the nanoparticles and also as the solvent medium for the hydrogenation of olefins. The authors further stated that the ligands inhibit the aggregation of nanoparticles in IL, thus the nanoparticle/IL system can be applied multiple times without any reduction in its selectivity and activity. Dupont and coworker prepared Pd (0) nanoparticles by reducing Pd(acac)$_2$ in [bmim]BF$_4$ and [bmim]PF$_6$ with molecular hydrogen at 75°C, which in turn were used for the hydrogenation of 1,3-butadiene to but-1-ene [95]. These Pd nanoparticles showed high selectivity towards but-1-ene, leading to the formation of but-1-ene >thrice more than but-2-ene. The selectivity was strongly reliant on the presence of IL as well as the anion, for example, using PF$_6$ instead of BF$_4$ results in the formation of butane. Later, Dupont and group used reduced Pd nanoparticles for partial-hydrogenation of a series of alkynes, including diphenyl acetylene, phenylacetylene, 4-chloro-butyne, ethynyl-methyl-phenyl amine, and 3-hexyne to alkenes in ILs under mild conditions while the same system leads to the formation of alkanes under higher hydrogen pressure [96]. The results showed good recyclability without loss in its activity up to 4 cycles. Fan *et al.* prepared silica-supported nickel catalysts synthesized from nickel nitrate and tetraethyl orthosilicate in imidazolium-type ILs as media. The authors compared the catalytic activity of Ni/SiO$_2$ for the hydrogenation of cinnamaldehyde to hydrocinnamaldehyde [97]. The best results were achieved with the catalyst prepared in 1-(2-hydroxyethyl)-3-methylimidazolium tetrafluoroborate IL. Ruta *et al.* employed a structured supported ionic liquid-phase (SSILP) catalysis method for selective gas-phase hydrogenation of 1,3-cyclohexadiene to cyclohexene [98]. The SSLIP affords the use of IL as solvents for homogeneous catalysis while providing benefits associated with heterogeneous catalysis. The SSLIP catalyst was prepared by constraining the IL with the Rh complex bicyclo[2.2.1]hepta-2,--diene–rhodium(I) chloride dimer, [Rh(nbd)Cl]$_2$) to the surface of a modified sintered metal fibers (SMF) as structured support. The Rh-based SSILP catalysts were found to be active only in the presence of acid. Moores and co-workers investigated the catalytic performance of rhodium nanoparticles stabilized by phosphine functionalized ILs using H$_2$ as reducer as Rh(allyl)$_3$ precursor for hydrogenation of arenes [99]. The biphasic catalyst successfully reduced toluene, styrene, and xylene. These functionalized ILs based catalysts showed higher catalytic activity and recyclability. The effect of bis(trifluo-romethanesulfonyl)amide anion (NTf$_2$) type-ILs ([bmim]NTf$_2$ and [hmim]NTf$_2$) on the catalytic activity of alumina-supported ruthenium catalyst (Ru/Al$_2$O$_3$) for liquid-phase hydrogenation of citral was investigated by Claus and group [100]. The results indicate that in the presence of [bmim]NTf$_2$ and [hmim]NTf$_2$ the citral conversion to geraniol and nerol was favored. The decrease in the selectivity of citronellal in the presence of ILs as additives was attributed to the low hydrogen solubility in comparison to organic solvents. The same group explored the effect of dicyanamide-based IL [bmim]DCA as a solvent, additive, or coated catalyst for selective hydrogenation of citral with Pd/C as catalysts [101]. The authors stated that the use of IL led to a generation of citronellal as a selective hydrogenation product. It was reported that even a small quantity of IL can have a significant effect on the selectivity profile of the reaction. The results for the application of IL *via* different pathways in the reaction are presented in Table **3** [101]. The study revealed that outcomes of IL-coated catalysts were similar to IL-free catalysts and also avoids the tedious process of product extraction while using IL as a solvent. The use of IL coated catalyst offers a green pathway for selective hydrogenation of citral with >99% conversion. To further understand the influence of the support on the catalytic activity of Palladium, the reaction was also carried out using Pd/SiO$_2$ as a catalyst. The results for both Pd/C and Pd/SiO$_2$ were in agreement, thus excluding the effect of support on the selective hydrogenation of citral.

Virtanen *et al.* explored the influence of different ILs on silica-supported IL catalyst (SILCA) for selective hydrogenation of citral and cinnamaldehyde [102]. The study demonstrated that the IL layer residing on heterogeneous support can have a substantial effect on the reaction pattern or the selectivity of the product.

The synchronous hydrogenation of cyclohexene and acetone using Pt/Al_2O_3 as a catalyst in a mixture of RTIL (2-hydroxy ethylammonium formate) with propan-1-ol and ethanol was investigated by the Gholami group [103]. The increase in mole fraction of IL accelerates the reaction rate which was attributed to the increased reactant tendency towards the catalyst surface. It was concluded that the reaction rate constant and the selectivity for the catalytic hydrogenation in the solution of RTIL and conventional organic solvents are concurrent with the solvatochromic parameters of the solubilizing media. The catalytic system was preferably more selective for hydrogenation of cyclohexene.

Table 3. Hydrogenation of citral under different reaction conditions with Pd/C as catalyst and [bmim]DCA as IL.

Reaction Type	Conversion (%)	Selectivity		
		Cal	DHC	Others
IL Free	100	41	49	10
IL Coated	100	>99	<1	<1
Additive	42	>99	<1	<1
Bulk Solvent	100	97	1	2

3.2.2. Hydrogenation of Carbonyls

Liu *et al.* synthesized an IL-regulated ruthenium complex [PEG-4000-omim][-u-TsDPENDS] from thermoregulated IL [PEG-4000-omim][TsDPENDS] [104]. The authors stated that the IL cation imparts the thermoregulated nature to the complex while the anion affords chirality to Ru. The thermoregulated catalyst was employed for the asymmetric transfer hydrogenation of several ketone substrates presented in Table 4 [104]. The authors further investigated the recyclability of the system using acetophenone as a model substrate. It was concluded that the catalytic system retains its activity and can be reused up to five cycles owing to the easy thermoregulated phase-separation. Chan and group studied the asymmetric hydrogenation of α-ketoesters and β-ketoesters in room temperature IL ionic as media and transition metal complex as catalysts [105] It was reported that the hydrogenation of α-ketoesters, pyruvate using $[Ru((R)-P-Phos)Cl_2]$ catalyst was very slow and provided low yield in both the investigated ILs, $[bmim]PF_6$ and $[bmim]BF_4$.

Table 4. Asymmetric Transfer Hydrogenation of ketones with catalyst [PEG-4000-C8MIM][-u-TsDPENDS] in EtOAc.

$$\underset{Ar}{\overset{O}{\|}}Me \quad \xrightarrow[\text{HCOOH-NEt}_3]{\text{[PEG-4000-omim][RuTsDPENDS]}} \quad \underset{Ar}{\overset{OH}{|}}Me$$

S. No.	Ar	Time (h)	Conv. (%)	Ee (%)
1	C_6H_5	10	97.5	95
2	4-BrC_6H_4	6.5	99.4	93
3	4-ClC_6H_4	6.5	99.5	92
4	$4NO_2C_6H_4$	6.5	99.7	86
5	4-Thienyl	10	80.5	97
6	4-Furyl	10	96.5	97
7	2-Naphthyl	10	99.5	96
8	4-MeOC_6H_4	16	66.3	95
9	4-MeOC_6H_4	16	55.0	92

However, in presence of MeOH as cosolvent, higher enantioselectivities and conversion were obtained in both the ILS, especially in [bmim]BF$_4$. It was noted that although the high conversion was obtained employing Ru-BINAP as a catalyst, however, the enantioselectivity was low. The hydrogenation of different α-ketoesters examined is presented in Table **5** [105]. The authors further investigated the hydrogenation of several β-ketoesters in a mixture of ILs and MeOH using [Ru((R)-P-Phos)Cl$_2$] and Ru-BINAP as catalysts. The results showed high enantioselectivity and good conversion were achieved for most of the substrates (presented in Table **6**) [105]. Moreover, the [Ru((R)-P-Phos)Cl$_2$] catalyst showed good recyclability in both ILs. It was observed that the catalytic activity declines in [bmim]BF$_4$ after the second run while in [bmim]PF$_6$, the catalytic system can be certainly reused up to nine cycles without showing any drop in its activity.

4. COUPLING REACTIONS

Coupling reactions have been studied extensively owing to the ease of unification of the two different hydrocarbons utilizing a metal complex as a catalyst. Suzuki coupling, Sonogashira coupling, and Heck coupling reactions are some of the most popular examples of coupling reactions. Gholap *et al.* developed a copper and ligand-free method for aryl iodides/bromides coupling reaction with alkynes in acetone and [bbim]BF$_4$ under ultrasound irradiation [106]. This method allows performing highly chemoselective Sonogashira coupling using Pd nanoparticles and triethylamine as a base. The outcomes revealed that IL offers a higher yield than acetone and the added benefit of recyclability. ILs have been used widely used in Suzuki coupling reactions for a wide-ranging substrate. Welton group studied the Suzuki coupling of bromobenzene and tolylboronic acid in an extensive series of solvents, including ILs comprising different sets of cations and anions [107]. The reaction was conducted in the presence of various palladium imidazolium complexes as catalyst precursors. The authors concluded that the functioning species, $(mim)_2PdCl_2$, is generated from the reaction between the ILs and the palladium imidazolium complexes leading to a recyclable and air-stable catalytic system for Suzuki reactions of iodo- and bromo arenes. It was reported that the stability and the activity of the catalytic system are strongly dependent on the imidazole-based ligand and the cation/anion of the ionic liquid. Similarly, the Heck reaction is also an important organic transformation and has been the focus of many research groups. Brønsted acid-base ILs based on guanidine and acetic acid were reported to act as solvent media, base, and stabilizing ligand for Pd species acting as a catalyst for coupling reaction between an aryl halide and olefins [108]. Rafiee and Kahrizi explored the application of thermoregulated IL as a co-catalyst to enhance the efficiency of the Heck reaction [109]. They successfully employed molybdovanadate-based ionic liquid salts as co-catalyst with palladium-deposited oleic acid-coated-Fe_3O_4 nanoparticles to catalyze the aryl halides and styrene reaction. This heterogeneous catalytic system is highly efficient and recyclable for up to 4 cycles. Handy and Zhang performed Stille coupling of α and β-iodoenones and aryl iodides with vinyltributyltin and phenyltributyltin using $PdCl_2(PhCN)_2$, Ph_3As, and CuI catalytic system in [bmim]BF_4 [110]. This system affords easy recovery of solvent and the catalyst system, which can be further exercised at least 5 times. Similarly, Henry coupling and Sakurai reactions have also been investigated in ILs [111, 112].

Table 5. Asymmetric hydrogenation of α-keto esters in ionic liquids.

Entry	Substrate	RTIL	ee (%)	Conv. (%)
1		[bmim]BF_4	47	34

(Table 5) cont.....

Entry	Substrate	RTIL	ee (%)	Conv. (%)
2		[bmim]PF$_6$	60	46
3		[bmim]BF$_4$	74	37
4		[bmim]PF$_6$	76	63
5		[bmim]BF$_4$	90	18
6		[bmim]PF$_6$	93	65
7		[bmim]BF$_4$	55	26
8		[bmim]PF$_6$	38	20

4.1. Polymerization

Over the years, ILs have also been actively used in the field of polymerization. Haddleton group was the first to investigate the copper(I)-mediated living radical polymerization in [bmim]PF$_6$ to synthesize poly (methyl methacrylate) PMMA [113]. The authors reported that the reaction proceeds at a faster rate in IL as compared to the traditional solvents at room temperature. Additionally, the immobility of IL with toluene allows an efficient pathway for polymer extraction. It was observed that at 50 and 30 °C, maximum conversion is obtained in 90 and 300 min., respectively. The 1-ethyl-3-methyl imidazolium bis(trifluoromethane sulfonyl)imide, [emim]TFSI has been utilized for *in-situ* free radical polymerization of a set of vinyl monomers by Susan *et al.* [114]. The study showed that polymers synthesized from MMA, and methyl acrylate had good compatibility in the IL, while polymers synthesized from styrene, acetonitrile, acrylamide, and 2-hydroxyethyl methacrylate were not compatible with the IL. The binary system of IL and compatible polymers form an ion gel giving a novel electrolyte series. Ma and coworkers investigated the effect of alkyl group chain length of imidazolium-based IL cation on AIBN/CuCl$_2$/bipyridine initiated reverse atom-transfer radical polymerization of MMA [115]. It was inferred that the polymerization was more controlled in 1-dodecyl-3-methylimidazolium tetrafluoroborate ([dodecmim]BF$_4$) as compared to [bmim]BF$_4$.

Table 6. Asymmetric hydrogenation of β-keto esters in ionic liquids.

Entry	Substrate	RTIL	ee (%)	Conv. (%)
1	H$_3$C—C(=O)—CH$_2$—C(=O)—OEt	[bmim]BF$_4$	>99	>99
2	H$_3$C—C(=O)—CH$_2$—C(=O)—OEt	[bmim]PF$_6$	97	>99
3	H$_3$C—C(=O)—CH$_2$—C(=O)—OtBu	[bmim]BF$_4$	>99	86

(Table 6) cont.....

Entry	Substrate	RTIL	ee (%)	Conv. (%)
4	H_3C — (COCH$_2$CO) — OtBu	[bmim]PF$_6$	>99	92
5	ClH$_2$C — (COCH$_2$CO) — OEt	[bmim]BF$_4$	54	91
6	ClH$_2$C — (COCH$_2$CO) — OEt	[bmim]PF$_6$	58	93
7	F$_3$C — (COCH$_2$CO) — OEt	[bmim]BF$_4$	50	21
8	F$_3$C — (COCH$_2$CO) — OEt	[bmim]PF$_6$	21	78

4.1.1. IL-Based Molecularly Imprinted Polymers

In recent years, ILs have made significant progress in every dynamic of research, including molecular imprinting technology, where they have been utilized as solvent media, monomers, templates, modifier and pyrogens, among others. Molecular Imprinted Polymers (MIP) are rapidly gaining attention in various fields, for instance, protein sensing, peptide recognition and food safety detection. Booker *et al.*, reported that [bmim]PF$_6$ and [bmim]BF$_4$ considerably accelerate the polymerization rates of MIP generated from trans-aconitic acid and cocaine [120]. In 2018, Li and group successfully incorporated 1-vinyl-3-butylimidazolium tetrafluoroborate [bvim]BF$_4$ and 1,4-butanediyl-3,3'-bis-l-vinylimidazolium dibromideas functional monomer and cross-linker within a MIP framework. This novel IL-composite MIP was used for electrochemical sensing purposes. Polymeric ionic liquids (PIL) are also widely used in catalysis processes [121]. Liu and Chen reported the preparation of a PIL-supported catalyst which is utilized to effectively catalyze the conversion of glucose or cellulose into HMF. The catalysis process was not only effective but also readily recyclable up to many cycles. They employed a poly(3-butyl-1-vinylimidazolium chloride) P[bvim]Cl PIL-based catalyst (P[bvim]Cl-CrCl$_2$) for the biomass conversion and observed that the efficiency of the reaction was much higher than the monomeric [bvim]Cl-CrCl$_2$ catalyst [122]. Similarly, Han *et al.* reported the conversion of benzene to phenol *via* oxidation using H2O2 and PIL-supported vanadium oxide catalyst [123].

CONCLUSION AND FUTURE PERSPECTIVES

The application of ionic liquids (ILs) is at an incredibly exciting stage, gaining increased attention from researcher's day-by-day from various interdisciplinary research areas. The dual nature (both as solvent and electrolyte) of these unique materials has enhanced their potential applications in many fields of science and technology, such as synthesis, catalysis, nanoscience, pharmaceutics, spectroscopy, and electrochemistry, to mention just a few. Currently, ionic liquids are being vastly used as solvents for a wide range of synthetic and catalytic reactions. ILs have shown tremendous potential both as solvents and catalysts for a wide range of reactions that take place more efficiently than molecular solvents. This book chapter highlights the importance of ILs used for a variety of synthetic and catalytic reactions.

REFERENCES

[1]   Greaves, T.L.; Drummond, C.J. Protic ionic liquids: properties and applications. *Chem. Rev.,* **2008,** *108*(1), 206-237.
        [http://dx.doi.org/10.1021/cr068040u] [PMID: 18095716]

[2]   Hallett, J.P.; Welton, T. Room-temperature ionic liquids: solvents for synthesis and catalysis. 2. *Chem. Rev.,* **2011,** *111*(5), 3508-3576.
        [http://dx.doi.org/10.1021/cr1003248] [PMID: 21469639]

[3]   Estager, J.; Holbrey, J.D.; Swadźba-Kwaśny, M. Halometallate ionic liquids – revisited. *Chem. Soc. Rev.,* **2014,** *43*(3), 847-886.
        [http://dx.doi.org/10.1039/C3CS60310E] [PMID: 24189615]

[4]   Pandey, S. Analytical applications of room-temperature ionic liquids: A review of recent efforts. *Anal. Chim. Acta,* **2006,** *556*(1), 38-45.
        [http://dx.doi.org/10.1016/j.aca.2005.06.038] [PMID: 17723329]

[5]   van Rantwijk, F.; Sheldon, R.A. Biocatalysis in ionic liquids. *Chem. Rev.,* **2007,** *107*(6), 2757-2785.
        [http://dx.doi.org/10.1021/cr050946x] [PMID: 17564484]

[6]   Andrushko, V.; Andrushko, N. Stereoselective Synthesis of Drugs and Natural Products. John Wiley & Sons, **2013.**
        [http://dx.doi.org/10.1002/9781118596784]

[7]   Neuhaus, W.C.; Bakanas, I.J.; Lizza, J.R.; Boon, C.T., Jr; Moura-Letts, G. Novel biodegradable protonic ionic liquid for the Fischer indole synthesis reaction. *Green Chem. Lett. Rev.,* **2016,** *9*(1), 39-43.
        [http://dx.doi.org/10.1080/17518253.2016.1149231]

[8]   Duan, X.; Sun, G.; Sun, Z.; Li, J.; Wang, S.; Wang, X.; Li, S.; Jiang, Z. A heteropolyacid-based ionic liquid as a thermoregulated and environmentally friendly catalyst in esterification reaction under microwave assistance. *Catal. Commun.,* **2013**, *42*, 125-128.
[http://dx.doi.org/10.1016/j.catcom.2013.08.014]

[9]   Joni, J.; Schmitt, D.; Schulz, P.; Lotz, T.; Wasserscheid, P. Detailed kinetic study of cumene isopropylation in a liquid–liquid biphasic system using acidic chloroaluminate ionic liquids. *J. Catal.,* **2008**, *258*(2), 401-409.
[http://dx.doi.org/10.1016/j.jcat.2008.06.018]

[10]   Brausch, N.; Metlen, A.; Wasserscheid, P. New, highly acidic ionic liquid systems and their application in the carbonylation of tolueneElectronic supplementary information (ESI) available: experimental details. See http://www.rsc.org/suppdata/cc/b4/b403464c/. *Chem. Commun. (Camb.),* **2004**, *10*(13), 1552-1553.
[http://dx.doi.org/10.1039/b403464c] [PMID: 15216376]

[11]   Aschauer, S.; Schilder, L.; Korth, W.; Fritschi, S.; Jess, A. Liquid-Phase Isobutane/Butene-Alkylation Using Promoted Lewis-Acidic IL-Catalysts. *Catal. Lett.,* **2011**, *141*(10), 1405-1419.
[http://dx.doi.org/10.1007/s10562-011-0675-2]

[12]   Liu, Z.C.; Meng, X.H.; Zhang, R.; Xu, C.M. Friedel-Crafts Acylation of Aromatic Compounds in Ionic Liquids. *Petrol. Sci. Technol.,* **2009**, *27*(2), 226-237.
[http://dx.doi.org/10.1080/10916460701700898]

[13]   Joni, J.; Haumann, M.; Wasserscheid, P. Continuous gas-phase isopropylation of toluene and cumene using highly acidic Supported Ionic Liquid Phase (SILP) catalysts. *Appl. Catal. A Gen.,* **2010**, *372*(1), 8-15.
[http://dx.doi.org/10.1016/j.apcata.2009.09.048]

[14]   Baldelli, S. Interfacial Structure of Room-Temperature Ionic Liquids at the Solid–Liquid Interface as Probed by Sum Frequency Generation Spectroscopy. *J. Phys. Chem. Lett.,* **2013**, *4*(2), 244-252.
[http://dx.doi.org/10.1021/jz301835j] [PMID: 26283429]

[15]   Peñalber, C.Y.; Baker, G.A.; Baldelli, S. Sum frequency generation spectroscopy of imidazolium-based ionic liquids with cyano-functionalized anions at the solid salt-liquid interface. *J. Phys. Chem. B,* **2013**, *117*(19), 5939-5949.
[http://dx.doi.org/10.1021/jp4019074] [PMID: 23650965]

[16]   Zanatta, M.; Girard, A.L.; Simon, N.M.; Ebeling, G.; Stassen, H.K.; Livotto, P.R.; dos Santos, F.P.; Dupont, J. The formation of imidazolium salt intimate (contact) ion pairs in solution. *Angew. Chem. Int. Ed.,* **2014**, *53*(47), 12817-12821.
[http://dx.doi.org/10.1002/anie.201408151] [PMID: 25257391]

[17]   Giernoth, R. Task-specific ionic liquids. *Angew. Chem. Int. Ed.,* **2010**, *49*(16), 2834-2839.
[http://dx.doi.org/10.1002/anie.200905981] [PMID: 20229544]

[18]   Chen, X.; Souvanhthong, B.; Wang, H.; Zheng, H.; Wang, X.; Huo, M. Polyoxometalate-based Ionic liquid as thermoregulated and environmentally friendly catalyst for starch oxidation. *Appl. Catal. B,* **2013**, *138-139*, 161-166, 161-166.
[http://dx.doi.org/10.1016/j.apcatb.2013.02.028]

[19]   Ma, Y.; Qing, S.; Wang, L.; Islam, N.; Guan, S.; Gao, Z.; Mamat, X.; Li, H.; Eli, W.; Wang, T. Production of 5-hydroxymethylfurfural from fructose by a thermo-regulated and recyclable Brønsted acidic ionic liquid catalyst. *RSC Advances,* **2015**, *5*(59), 47377-47383.
[http://dx.doi.org/10.1039/C5RA08107F]

[20]   Pagni, R.M.; Mamantov, G.; Mamantov, C.B.; Braunstein, J. *Advances in Molten Salt Chemistry.,* Elsevier, Oxford., **1987**, *6*, 211-349.

[21]   Smith, G.P.; Pagni, R.M. *Molten Salt Chemistry, An introduction to selected Applications*; Mamantov, G.; Marassi, R., Eds.; D. Reidel Publishing Co.: Dordrecht, **1987**, pp. 383-416.

[http://dx.doi.org/10.1007/978-94-009-3863-2_18]

[22]   Studt, P. Notizüber die Synthese von Indeno[1,2,3-*cd*]pyren. *Lebigs Ann. Chem,* **1978**, 528-529.

[23]   Buchanan, A.C., III; Chapman, D.M.; Smith, G.P. Redox *vs.* Lewis acid catalysis. The chemistry of 1,2-diarylethanes in antimony trichloride-rich molten salt media. *J. Org. Chem.,* **1985**, *50*(10), 1702-1711.
        [http://dx.doi.org/10.1021/jo00210a026]

[24]   Smith, G.P.; Dworkin, A.S.; Pagni, R.M.; Zingg, S.P. Broensted superacidity of hydrochloric acid in a liquid chloroaluminate. Aluminum chloride - 1-ethyl-3-methyl-1H-imidazolium chloride (55.O m/o AlCl3). *J. Am. Chem. Soc.,* **1989**, *111*(2), 525-530.
        [http://dx.doi.org/10.1021/ja00184a020]

[25]   Pagni, R.M. *Advances in Molten Salt Chemistry*; Mamantov, G.; Braunstein, J., Eds.; Elsevier: Oxford, **1987**, Vol. 6, pp. 211-346.

[26]   Green, L.; Hemeon, I.; Singer, R.D. 1-Ethyl-3-methylimidazolium halogenoaluminate ionic liquids as reaction media for the acylative cleavage of ethers. *Tetrahedron Lett.,* **2000**, *41*(9), 1343-1346.
        [http://dx.doi.org/10.1016/S0040-4039(99)02289-3]

[27]   Jaeger, D.A.; Tucker, C.E. Diels-Alder reactions in ethylammonium nitrate, a low-melting fused salt. *Tetrahedron Lett.,* **1989**, *30*(14), 1785-1788.
        [http://dx.doi.org/10.1016/S0040-4039(00)99579-0]

[28]   Earle, M.J.; McCormac, P.B.; Seddon, K.R. Diels–Alder reactions in ionic liquids. *Green Chem.,* **1999**, *1*(1), 23-25.
        [http://dx.doi.org/10.1039/a808052f]

[29]   Ludley, P.; Karodia, N. Phosphonium tosylates as solvents for the Diels–Alder reaction. *Tetrahedron Lett.,* **2001**, *42*(10), 2011-2014.
        [http://dx.doi.org/10.1016/S0040-4039(01)00064-8]

[30]   Abbott, A.P.; Capper, G.; Davies, D.L.; Rasheed, R.K.; Tambyrajah, V. Quaternary ammonium zinc- or tin-containing ionic liquids: water insensitive, recyclable catalysts for Diels–Alder reactions. *Green Chem.,* **2002**, *4*(1), 24-26.
        [http://dx.doi.org/10.1039/b108431c]

[31]   Chiappe, C.; Malvaldi, M.; Pomelli, C.S. The solvent effect on the Diels–Alder reaction in ionic liquids: multiparameter linear solvation energy relationships and theoretical analysis. *Green Chem.,* **2010**, *12*(8), 1330-1339.
        [http://dx.doi.org/10.1039/c0gc00074d]

[32]   Olah, G.A. *Friedel–Crafts and Related Reactions*; Interscience: New York, **1963**.

[33]   Olah, G.A. *Friedel Crafts Chemistry*; Wiley–Interscience: New York, **1973**.

[34]   Baddeley, G.; Williamson, R. 434. Preparation and isomerisation of aromatic ketones and keto-acids. *J. Chem. Soc.,* **1953**, 2120-2125.
        [http://dx.doi.org/10.1039/jr9530002120]

[35]   Baddeley, G.; Holt, G.; Makar, S.M.; Ivinson, M.G. 688. Interaction of naphthalene and maleic anhydride through the agency of aluminium chloride. *J. Chem. Soc.,* **1952**, 3605-3607.
        [http://dx.doi.org/10.1039/jr9520003605]

[36]   Boon, J.A.; Levisky, J.A.; Pflug, J.L.; Wilkes, J.S. Friedel-Crafts reactions in ambient-temperature molten salts. *J. Org. Chem.,* **1986**, *51*(4), 480-483.
        [http://dx.doi.org/10.1021/jo00354a013]

[37]   Piersma, B.J.; Merchant, M. Proceedings of the 7th International Symposium on Molten Salts. *(Hussey, C. L.; Flengas, S. N.; Wilkes, J. S.; Ito, Y. eds.), The Electrochemical Society, Pennington,* **1990**, , pp. 805-821.

[38]   Keim, W.; Korth, W.; Wasserscheid, P. Ionic liquids. *World Patent, WO ,* **2000**, *0016902*.

[39]   Song, C.E.; Roh, E.J.; Shim, W.H.; Choi, J.H. Scandium(iii) triflate immobilised in ionic liquids: a novel and recyclable catalytic system for Friedel–Crafts alkylation of aromatic compounds with alkenes. *Chem. Commun. (Camb.),* **2000**, (17), 1695-1696.
       [http://dx.doi.org/10.1039/b005335j]

[40]   DeCastro, C. P.; Sauvage, E.; Valkenberg, M. H.; Hölderich, W. F. W.F. Immobilized ionic liquids. *World Patent, WO ,* **2001**, (0132308).

[41]   DeCastro, C.; Sauvage, E.; Valkenberg, M.H.; Hölderich, W.F. Immobilised Ionic Liquids as Lewis Acid Catalysts for the Alkylation of Aromatic Compounds with Dodecene. *J. Catal.,* **2000**, *196*(1), 86-94.
       [http://dx.doi.org/10.1006/jcat.2000.3004]

[42]   Scholl, R.; Meyer, K.; Donat, J. Vom pyren in das gebiet hoeher anellierter ring systeme. *Ber. Dtsch. Chem. Ges. B,* **1937**, *70*(11), 2180-2189.
       [http://dx.doi.org/10.1002/cber.19370701104]

[43]   Bruce, D.B.; Sorrie, A.J.S.; Thomson, R.H. 489. Reactions in fused aluminium chloride–sodium chloride. *J. Chem. Soc.,* **1953**, *0*(0), 2403-2406.
       [http://dx.doi.org/10.1039/JR9530002403]

[44]   Boon, J.A.; Lander, S.W., Jr; Levisky, J.A.; Pflug, J.L. Proceedings of the Joint International Symposium on Molten Salts. *6th ed.,* **1987**, 979-990.

[45]   Earle, M.J.; Seddon, K.R.; Adams, C.J.; Roberts, G. Friedel–Crafts reactions in room temperature ionic liquids. *Chem. Commun. (Camb.),* **1998**, (19), 2097-2098.
       [http://dx.doi.org/10.1039/a805599h]

[46]   Davey, P.N.; Earle, M.J.; Newman, C.P.; Seddon, K.R. Imrovements in or relating to friedel-crafts reactions. *World Patent WO.,* **1999**. 99 19288.

[47]   Ross, J.; Xiao, J. Friedelâ€"Crafts acylation reactions using metal triflates in ionic liquid. *Green Chem.,* **2002**, *4*(2), 129-133.
       [http://dx.doi.org/10.1039/b109847k]

[48]   Earle, M.J.; McCormac, P.B.; Seddon, K.R. Regioselective alkylation in ionic liquids. *Chem. Commun. (Camb.),* **1998**, (20), 2245-2246.
       [http://dx.doi.org/10.1039/a806328a]

[49]   Adams, C.J.; Earle, M.J.; Seddon, K.R. Catalytic cracking reactions of polyethylene to light alkanes. *Green Chem.,* **2000**, *2*(1), 21-24.
       [http://dx.doi.org/10.1039/a908167d]

[50]   Li, K.; Choudhary, H.; Mishra, M.K.; Rogers, R.D. Enhanced Acidity and Activity of Aluminum/Gallium-Based Ionic Liquids Resulting from Dynamic Anionic Speciation. *ACS Catal.,* **2019**, *9*(11), 9789-9793.
       [http://dx.doi.org/10.1021/acscatal.9b03132]

[51]   Gmouh, S.; Yang, H.; Vaultier, M. Activation of bismuth(III) derivatives in ionic liquids: novel and recyclable catalytic systems for Friedel-Crafts acylation of aromatic compounds. *Org. Lett.,* **2003**, *5*(13), 2219-2222.
       [http://dx.doi.org/10.1021/ol034529n] [PMID: 12816413]

[52]   Nara, S.J.; Harjani, J.R.; Salunkhe, M.M. Friedel-Crafts sulfonylation in 1-butyl-3-methylimidazolium chloroaluminate ionic liquids. *J. Org. Chem.,* **2001**, *66*(25), 8616-8620.
       [http://dx.doi.org/10.1021/jo016126b] [PMID: 11735546]

[53]   Chica, A.; Corma, A. Hydroisomerization of Pentane, Hexane, and Heptane for Improving the Octane Number of Gasoline. *J. Catal.,* **1999**, *187*(1), 167-176.
       [http://dx.doi.org/10.1006/jcat.1999.2601]

[54]   Setiabudi, H.D.; Jalil, A.A.; Triwahyono, S.; Kamarudin, N.H.N.; Jusoh, R. Ir/Pt-HZSM5 for n-

pentane isomerization: Effect of Si/Al ratio and reaction optimization by response surface methodology. *Chem. Eng. J.,* **2013**, *217*, 300-309.
[http://dx.doi.org/10.1016/j.cej.2012.12.011]

[55] Arribas, M.A.; Marquez, F.; Martínez, A. Activity, Selectivity, and Sulfur Resistance of Pt/WOx–ZrO2 and Pt/Beta Catalysts for the simultaneous hydroisomerization of n-Heptane and Hydrogenation of Benzene. *J. Catal.,* **2000**, *190*, 309-319.
[http://dx.doi.org/10.1006/jcat.2000.2768]

[56] Liu, P.; Wang, J.; Zhang, X.; Wei, R.; Ren, X. Catalytic performances of dealuminated Hβ zeolite supported Pt catalysts doped with Cr in hydroisomerization of n-heptane. *Chem. Eng. J.,* **2009**, *148*(1), 184-190.
[http://dx.doi.org/10.1016/j.cej.2008.12.016]

[57] Xu, D.Z.; Liu, Y.; Shi, S.; Wang, Y. A simple, efficient and green procedure for Knoevenagel condensation catalyzed by [C₄dabco][BF₄] ionic liquid in water. *Green Chem.,* **2010**, *12*(3), 514-517.
[http://dx.doi.org/10.1039/b918595j]

[58] Forsyth, S.A.; Fröhlich, U.; Goodrich, P.; Gunaratne, H.Q.N.; Hardacre, C.; McKeown, A.; Seddon, K.R. Functionalised ionic liquids: synthesis of ionic liquids with tethered basic groups and their use in Heck and Knoevenagel reactions. *New J. Chem.,* **2010**, *34*(4), 723-731.
[http://dx.doi.org/10.1039/b9nj00729f]

[59] Jiang, T.; Ma, X.; Zhou, Y.; Liang, S.; Zhang, J.; Han, B. Solvent-free synthesis of substituted ureas from CO_2 and amines with a functional ionic liquid as the catalyst. *Green Chem.,* **2008**, *10*(4), 465-469.
[http://dx.doi.org/10.1039/b717868a]

[60] Feng, L.; Ye, R.; Yuan, T.; Zhang, X.; Lu, G.; Zhou, B. A concerted addition mechanism in [Hmim]Br-triggered thiol–ene reactions: a typical "ionic liquid effect" revealed by DFT and experimental studies. *New J. Chem.,* **2019**, *43*(15), 5752-5758.
[http://dx.doi.org/10.1039/C8NJ05674A]

[61] Xu, J.M.; Wu, Q.; Zhang, Q.Y.; Zhang, F.; Lin, X.F. A Basic Ionic Liquid as Catalyst and Reaction Medium: A Rapid and Simple Procedure for Aza-Michael Addition Reactions. *Eur. J. Org. Chem.,* **2007**, *2007*(11), 1798-1802.
[http://dx.doi.org/10.1002/ejoc.200600999]

[62] Rebeiro, G.L.; Khadilkar, B.M. Chloroaluminate Ionic Liquid for Fischer Indole Synthesis. *Synthesis,* **2001**, *2001*(3), 0370-0372.
[http://dx.doi.org/10.1055/s-2001-11441]

[63] Kumar, A.; Pawar, S.S. Catalyzing Henry reactions in chloroaluminate ionic liquids. *J. Mol. Catal. Chem.,* **2005**, *235*(1-2), 244-248.
[http://dx.doi.org/10.1016/j.molcata.2005.04.010]

[64] Gilbert, A.; Haines, R.S.; Harper, J.B. Understanding the effects of ionic liquids on a unimolecular substitution process: correlating solvent parameters with reaction outcome. *Org. Biomol. Chem.,* **2019**, *17*(3), 675-682.
[http://dx.doi.org/10.1039/C8OB02460J] [PMID: 30601540]

[65] Schaffarczyk McHale, K.S.; Haines, R.S.; Harper, J.B. Ionic Liquids as Solvents for S_N 2 Processes. Demonstration of the Complex Interplay of Interactions Resulting in the Observed Solvent Effects. *ChemPlusChem,* **2018**, *83*(12), 1162-1168.
[http://dx.doi.org/10.1002/cplu.201800510] [PMID: 31950706]

[66] Keaveney, S.T.; Haines, R.S.; Harper, J.B. Ionic liquid effects on a multistep process. Increased product formation due to enhancement of all steps. *Org. Biomol. Chem.,* **2015**, *13*(33), 8925-8936.
[http://dx.doi.org/10.1039/C5OB01214G] [PMID: 26214746]

[67] Venkat Reddy, C.; Verkade, J.G. An advantageous tetrameric titanium alkoxide/ionic liquid as a recyclable catalyst system for the selective oxidation of sulfides to sulfones. *J. Mol. Catal. Chem.,*

2007, *272*(1-2), 233-240.
[http://dx.doi.org/10.1016/j.molcata.2007.02.053]

[68]   Li, H.; Zhu, W.; Wang, Y.; Zhang, J.; Lu, J.; Yan, Y. Deep oxidative desulfurization of fuels in redox ionic liquids based on iron chloride. *Green Chem.,* **2009**, *11*(6), 810-815.
[http://dx.doi.org/10.1039/b901127g]

[69]   Zhu, W.; Li, H.; Jiang, X.; Yan, Y.; Lu, J.; He, L.; Xia, J. Commercially available molybdic compound-catalyzed ultra-deep desulfurization of fuels in ionic liquids. *Green Chem.,* **2008**, *10*(6), 641-646.
[http://dx.doi.org/10.1039/b801185k]

[70]   Chao, Y.; Li, H.; Zhu, W.; Zhu, G.; Yan, Y. Deep Oxidative Desulfurization of Dibenzothiophene in Simulated Diesel with Tungstate and H_2O_2 in Ionic Liquids. *Petrol. Sci. Technol.,* **2010**, *28*(12), 1242-1249.
[http://dx.doi.org/10.1080/10916460903030441]

[71]   Xu, D.; Zhu, W.; Li, H.; Zhang, J.; Zou, F.; Shi, H.; Yan, Y. Oxidative Desulfurization of Fuels Catalyzed by V_2O_5 in Ionic Liquids at Room Temperature. *Energy Fuels,* **2009**, *23*(12), 5929-5933.
[http://dx.doi.org/10.1021/ef900686q]

[72]   Wang, J.; Zhao, D.; Li, K. Oxidative Desulfurization of Dibenzothiophene Catalyzed by Brønsted Acid Ionic Liquid. *Energy Fuels,* **2009**, *23*(8), 3831-3834.
[http://dx.doi.org/10.1021/ef900251a]

[73]   Zhao, D.; Wang, J.; Zhou, E. Oxidative desulfurization of diesel fuel using a Brønsted acid room temperature ionic liquid in the presence of H_2O_2. *Green Chem.,* **2007**, *9*(11), 1219-1222.
[http://dx.doi.org/10.1039/b706574d]

[74]   Rafiee, E.; Mirnezami, F. Temperature regulated Brønsted acidic ionic liquid-catalyze esterification of oleic acid for biodiesel application. *J. Mol. Struct.,* **2017**, *1130*, 296-302.
[http://dx.doi.org/10.1016/j.molstruc.2016.10.049]

[75]   Rafiee, E.; Eavani, S. A new organic–inorganic hybrid ionic liquid polyoxometalate for biodiesel production. *J. Mol. Liq.,* **2014**, *199*, 96-101.
[http://dx.doi.org/10.1016/j.molliq.2014.08.034]

[76]   Basheer, C.; Vetrichelvan, M.; Suresh, V.; Lee, H.K. Ionic-liquid supported oxidation reactions in a silicon-based microreactor. *Tetrahedron Lett.,* **2006**, *47*(6), 957-961.
[http://dx.doi.org/10.1016/j.tetlet.2005.11.147]

[77]   Bortolini, O.; Conte, V.; Chiappe, C.; Fantin, G.; Fogagnolo, M.; Maietti, S. Epoxidation of electrophilic alkenes in ionic liquids. *Green Chem.,* **2002**, *4*(2), 94-96.
[http://dx.doi.org/10.1039/b110009m]

[78]   Wang, B.; Kang, Y.R.; Yang, L.M.; Suo, J.S. Epoxidation of α,β-unsaturated carbonyl compounds in ionic liquid/water biphasic system under mild conditions. *J. Mol. Catal. Chem.,* **2003**, *203*(1-2), 29-36.
[http://dx.doi.org/10.1016/S1381-1169(03)00282-6]

[79]   Neves, P.; Gago, S.; Pereira, C.C.L.; Figueiredo, S.; Lemos, A.; Lopes, A.D.; Gonçalves, I.S.; Pillinger, M.; Silva, C.M.; Valente, A.A. Catalytic Epoxidation and Sulfoxidation Activity of a Dioxomolybdenum(VI) Complex Bearing a Chiral Tetradentate Oxazoline Ligand. *Catal. Lett.,* **2009**, *132*(1-2), 94-103.
[http://dx.doi.org/10.1007/s10562-009-0065-1]

[80]   Gago, S.; Balula, S.S.; Figueiredo, S.; Lopes, A.D.; Valente, A.A.; Pillinger, M.; Gonçalves, I.S. Catalytic olefin epoxidation with cationic molybdenum(VI) cis-dioxo complexes and ionic liquids. *Appl. Catal. A Gen.,* **2010**, *372*(1), 67-72.
[http://dx.doi.org/10.1016/j.apcata.2009.10.016]

[81]   Li, X.; Geng, W.; Zhou, J.; Luo, W.; Wang, F.; Wang, L.; Tsang, S.C. Synthesis of multicarboxylic acid appended imidazolium ionic liquids and their application in palladium-catalyzed selective

oxidation of styrene. *New J. Chem.,* **2007**, *31*(12), 2088-2094.
[http://dx.doi.org/10.1039/b702573d]

[82]   Li, H.; Hou, Z.; Qiao, Y.; Feng, B.; Hu, Y.; Wang, X.; Zhao, X. Peroxopolyoxometalate-based room temperature ionic liquid as a self-separation catalyst for epoxidation of olefins. *Catal. Commun.,* **2010**, *11*(5), 470-475.
[http://dx.doi.org/10.1016/j.catcom.2009.11.025]

[83]   Zeng, A.; Li, Y.; Su, S.; Li, D.; Hou, B.; Yu, N. Gold nanoparticles stabilized by task-specific ionic complexes: Quasi-homogeneous catalysts with self-separating nature for aerobic epoxidation of styrene. *J. Catal.,* **2014**, *319*, 163-173.
[http://dx.doi.org/10.1016/j.jcat.2014.08.012]

[84]   Chen, J.; Hua, L.; Zhu, W.; Zhang, R.; Guo, L.; Chen, C.; Gan, H.; Song, B.; Hou, Z. Polyoxometalate anion-functionalized ionic liquid as a thermoregulated catalyst for the epoxidation of olefins. *Catal. Commun.,* **2014**, *47*, 18-21.
[http://dx.doi.org/10.1016/j.catcom.2014.01.003]

[85]   Shen, H.Y.; Ying, L.Y.; Jiang, H.L.; Judeh, Z. Efficient Copper-bisisoquinoline-based Catalysts for Selective Aerobic Oxidation of Alcohols to Aldehydes and Ketones. *Int. J. Mol. Sci.,* **2007**, *8*(6), 505-512.
[http://dx.doi.org/10.3390/i8060505]

[86]   Liu, C.; Han, J.; Wang, J.A. Simple, Efficient and Recyclable Copper(II) Acetylacetonate Catalytic System for Oxidation of sec-Alcohols in Ionic Liquid. *Synlett,* **2007**, *4*, 643-645.

[87]   Rong, M.; Liu, C.; Han, J.; Sheng, W.; Zhang, Y.; Wang, H. Catalytic Oxidation of Alcohols by a Novel Copper Schiff Base Ligand Derived from Acetylacetonate and l-Leucine in Ionic Liquids. *Catal. Lett.,* **2008**, *125*(1-2), 52-56.
[http://dx.doi.org/10.1007/s10562-008-9486-5]

[88]   Ramakrishna, D.; Bhat, B.R.; Karvembu, R. Catalytic oxidation of alcohols by nickel(II) Schiff base complexes containing triphenylphosphine in ionic liquid: An attempt towards green oxidation process. *Catal. Commun.,* **2010**, *11*(5), 498-501.
[http://dx.doi.org/10.1016/j.catcom.2009.12.011]

[89]   Shaabani, A.; Farhangi, E.; Rahmati, A. Aerobic oxidation of alkyl arenes and alcohols using cobalt(II) phthalocyanine as a catalyst in 1-butyl-3-methyl-imidazolium bromide. *Appl. Catal. A Gen.,* **2008**, *338*(1-2), 14-19.
[http://dx.doi.org/10.1016/j.apcata.2007.12.014]

[90]   Hu, R.; Lei, M.; Wei, H.; Wang, Y. Ionic liquid-H_2O Resulting in a Highly Chemoselective Oxidation of Benzylic Alcohols in the Presence of Aliphatic Analogues Catalyzed by Immobilized TEMPO. *Chin. J. Chem.,* **2009**, *27*(3), 587-592.
[http://dx.doi.org/10.1002/cjoc.200990096]

[91]   Miao, C.X.; He, L.N.; Wang, J.Q.; Wang, J.L. TEMPO and Carboxylic Acid Functionalized Imidazolium Salts/Sodium Nitrite: An Efficient, Reusable, Transition Metal-Free Catalytic System for Aerobic Oxidation of Alcohols. *Adv. Synth. Catal.,* **2009**, *351*(13), 2209-2216.
[http://dx.doi.org/10.1002/adsc.200900285]

[92]   Dyson, P.J.; Laurenczy, G.; André Ohlin, C.; Vallance, J.; Welton, T. Determination of hydrogen concentration in ionic liquids and the effect (or lack of) on rates of hydrogenation. *Chem. Commun. (Camb.),* **2003**, (19), 2418-2419.
[http://dx.doi.org/10.1039/B308309H] [PMID: 14587710]

[93]   Favre, F.; Olivier-Bourbigou, H.; Commereuc, D.; Saussine, L. Hydroformylation of 1-hexene with rhodium in non-aqueous ionic liquids : how to design the solvent and the ligand to the reaction. *Chem. Commun. (Camb.),* **2001**, (15), 1360-1361.
[http://dx.doi.org/10.1039/b104155j]

[94]   Huang, J.; Jiang, T.; Han, B.; Gao, H.; Chang, Y.; Zhao, G.; Wu, W. Hydrogenation of olefins using

ligand-stabilized palladium nanoparticles in an ionic liquid. *Chem. Commun. (Camb.),* **2003**, (14), 1654-1655.
[http://dx.doi.org/10.1039/b302750c]

[95] Umpierre, A.P.; Machado, G.; Fecher, G.H.; Morais, J.; Dupont, J. Selective Hydrogenation of 1,3-Butadiene to 1-Butene by Pd(0) Nanoparticles Embedded in Imidazolium Ionic Liquids. *Adv. Synth. Catal.,* **2005**, *347*(10), 1404-1412.
[http://dx.doi.org/10.1002/adsc.200404313]

[96] Venkatesan, R.; Prechtl, M.H.G.; Scholten, J.D.; Pezzi, R.P.; Machado, G.; Dupont, J. Palladium nanoparticle catalysts in ionic liquids: synthesis, characterisation and selective partial hydrogenation of alkynes to Z-alkenes. *J. Mater. Chem.,* **2011**, *21*(9), 3030-3036.
[http://dx.doi.org/10.1039/c0jm03557b]

[97] Fan, Q.; Liu, Y.; Zheng, Y.; Yan, W. Preparation of Ni/SiO$_2$ catalyst in ionic liquids for hydrogenation. *Front. Chem. Eng. China,* **2008**, *2*(1), 63-68.
[http://dx.doi.org/10.1007/s11705-008-0013-4]

[98] Ruta, M.; Yuranov, I.; Dyson, P.; Laurenczy, G.; Kiwiminsker, L. Structured fiber supports for ionic liquid-phase catalysis used in gas-phase continuous hydrogenation. *J. Catal.,* **2007**, *247*(2), 269-276.
[http://dx.doi.org/10.1016/j.jcat.2007.02.012]

[99] Stratton, S.A.; Luska, K.L.; Moores, A. Rhodium nanoparticles stabilized with phosphine functionalized imidazolium ionic liquids as recyclable arene hydrogenation catalysts. *Catal. Today,* **2012**, *183*(1), 96-100.
[http://dx.doi.org/10.1016/j.cattod.2011.09.016]

[100] Arras, J.; Ruppert, D.; Claus, P. Supported ruthenium catalysed selective hydrogenation of citral in presence of [NTf$_2$]$^-$ based ionic liquids. *Appl. Catal. A Gen.,* **2009**, *371*(1-2), 73-77.
[http://dx.doi.org/10.1016/j.apcata.2009.09.034]

[101] Arras, J.; Steffan, M.; Shayeghi, Y.; Claus, P. The promoting effect of a dicyanamide based ionic liquid in the selective hydrogenation of citral. *Chem. Commun. (Camb.),* **2008**, (34), 4058-4060.
[http://dx.doi.org/10.1039/b810291k] [PMID: 18758625]

[102] Virtanen, P.; Karhu, H.; Kordas, K.; Mikkol, J-P. The effect of ionic liquid in supported ionic liquid catalysts (SILCA) in the hydrogenation α,β-unsaturated aldehydes. *Chem. Eng. Sci.,* **2007**, *62*(14), 3660-3671.
[http://dx.doi.org/10.1016/j.ces.2007.03.029]

[103] Khodadadi-Moghaddam, M.; Habibi-Yangjeh, A.; Gholami, M.R. Solvent effects on the reaction rate and selectivity of synchronous heterogeneous hydrogenation of cyclohexene and acetone in ionic liquid/alcohols mixtures. *J. Mol. Catal. Chem.,* **2009**, *306*(1-2), 11-16.
[http://dx.doi.org/10.1016/j.molcata.2009.02.018]

[104] Liu, X.; Chen, C.; Xiu, Y.; Chen, A.; Guo, L.; Zhang, R.; Chen, J.; Hou, Z. Asymmetric transfer hydrogenation of ketones catalyzed by thermoregulated ionic liquid-regulating ruthenium complexes. *Catal. Commun.,* **2015**, *67*, 90-94.
[http://dx.doi.org/10.1016/j.catcom.2015.04.013]

[105] Lam, K.H.; Xu, L.; Feng, L.; Ruan, J.; Fan, Q.; Chan, A.S.C. Ruthenium catalyzed asymmetric hydrogenation of α- and β-keto esters in ionic liquids using chiral P-Phos ligand. *Can. J. Chem.,* **2005**, *83*(6-7), 903-908.
[http://dx.doi.org/10.1139/v05-102]

[106] Gholap, A.R.; Venkatesan, K.; Pasricha, R.; Daniel, T.; Lahoti, R.J.; Srinivasan, K.V. Copper- and ligand-free Sonogashira reaction catalyzed by Pd(0) nanoparticles at ambient conditions under ultrasound irradiation. *J. Org. Chem.,* **2005**, *70*(12), 4869-4872.
[http://dx.doi.org/10.1021/jo0503815] [PMID: 15932333]

[107] Mathews, C.; Smith, P.J.; Welton, T. N-donor complexes of palladium as catalysts for Suzuki cross-coupling reactions in ionic liquids. *J. Mol. Catal. Chem.,* **2004**, *214*(1), 27-32.

[http://dx.doi.org/10.1016/j.molcata.2003.11.030]

[108] Li, S.; Lin, Y.; Xie, H.; Zhang, S.; Xu, J. Brønsted guanidine acid-base ionic liquids: novel reaction media for the palladium-catalyzed Heck reaction. *Org. Lett.,* **2006**, *8*(3), 391-394.
[http://dx.doi.org/10.1021/ol052543p] [PMID: 16435842]

[109] Rafiee, E.; Kahrizi, M. Mechanistic investigation of Heck reaction catalyzed by new catalytic system composed of Fe$_3$O$_4$@OA-Pd and ionic liquids as co-catalyst. *J. Mol. Liq.,* **2016**, *218*, 625-631.
[http://dx.doi.org/10.1016/j.molliq.2016.02.055]

[110] Handy, S.T.; Zhang, X. Organic synthesis in ionic liquids: the Stille coupling. *Org. Lett.,* **2001**, *3*(2), 233-236.
[http://dx.doi.org/10.1021/ol0068849] [PMID: 11430042]

[111] Howarth, J.; James, P.; Dai, J. An exploration of the catalytic Sakurai reaction in the moisture stable ionic liquids [bmim]PF$_6$ and [bmim]BF$_4$. *J. Mol. Catal. Chem.,* **2004**, *214*(1), 143-146.
[http://dx.doi.org/10.1016/j.molcata.2003.09.036]

[112] Jiang, T.; Gao, H.; Han, B.; Zhao, G.; Chang, Y.; Wu, W.; Gao, L.; Yang, G. Ionic liquid catalyzed Henry reactions. *Tetrahedron Lett.,* **2004**, *45*(12), 2699-2701.
[http://dx.doi.org/10.1016/j.tetlet.2004.01.129]

[113] Carmichael, A.J.; Haddleton, D.M.; Bon, S.A.F.; Seddon, K.R. Copper(i) mediated living radical polymerisation in an ionic liquid. *Chem. Commun. (Camb.),* **2000**, (14), 1237-1238.
[http://dx.doi.org/10.1039/b003335i]

[114] Susan, M.A.B.H.; Kaneko, T.; Noda, A.; Watanabe, M. Ion gels prepared by *in situ* radical polymerization of vinyl monomers in an ionic liquid and their characterization as polymer electrolytes. *J. Am. Chem. Soc.,* **2005**, *127*(13), 4976-4983.
[http://dx.doi.org/10.1021/ja045155b] [PMID: 15796564]

[115] Ma, H.; Wan, X.; Chen, X.; Zhou, Q.F. Reverse atom transfer radical polymerization of methyl methacrylate in imidazolium ionic liquids. *Polymer (Guildf.),* **2003**, *44*(18), 5311-5316.
[http://dx.doi.org/10.1016/S0032-3861(03)00531-7]

[116] Ding, S.; Lyu, Z.; Niu, X.; Zhou, Y.; Liu, D.; Falahati, M.; Du, D.; Lin, Y. Integrating ionic liquids with molecular imprinting technology for biorecognition and biosensing: A review. *Biosens. Bioelectron.,* **2020**, *149*, 111830.
[http://dx.doi.org/10.1016/j.bios.2019.111830] [PMID: 31710919]

[117] Claus, J.; Sommer, F.O.; Kragl, U. Ionic liquids in biotechnology and beyond. *Solid State Ion.,* **2018**, *314*, 119-128.
[http://dx.doi.org/10.1016/j.ssi.2017.11.012]

[118] Comtet, J.; Niguès, A.; Kaiser, V.; Coasne, B.; Bocquet, L.; Siria, A. Nanoscale capillary freezing of ionic liquids confined between metallic interfaces and the role of electronic screening. *Nat. Mater.,* **2017**, *16*(6), 634-639.
[http://dx.doi.org/10.1038/nmat4880] [PMID: 28346432]

[119] Wen, W.; Yan, X.; Zhu, C.; Du, D.; Lin, Y. Recent Advances in Electrochemical Immunosensors. *Anal. Chem.,* **2017**, *89*(1), 138-156.
[http://dx.doi.org/10.1021/acs.analchem.6b04281] [PMID: 28105820]

[120] Booker, K.; Bowyer, M.C.; Lennard, C.J.; Holdsworth, C.I.; McCluskey, A. Molecularly Imprinted Polymers and Room Temperature Ionic Liquids: Impact of Template on Polymer Morphology. *Aust. J. Chem.,* **2007**, *60*(1), 51-56.
[http://dx.doi.org/10.1071/CH06284]

[121] Zhu, X.; Zeng, Y.; Zhang, Z.; Yang, Y.; Zhai, Y.; Wang, H.; Liu, L.; Hu, J.; Li, L. A new composite of graphene and molecularly imprinted polymer based on ionic liquids as functional monomer and cross-linker for electrochemical sensing 6-benzylaminopurine. *Biosens. Bioelectron.,* **2018**, *108*, 38-45.
[http://dx.doi.org/10.1016/j.bios.2018.02.032] [PMID: 29499557]

[122] Liu, D.D.J.; Chen, E.Y-X. Polymeric ionic liquid (PIL)-supported recyclable catalysts for biomass conversion into HMF. *Biomass Bioenergy,* **2013**, *48*, 181-190.
[http://dx.doi.org/10.1016/j.biombioe.2012.11.020]

[123] Han, H.; Jiang, T.; Wu, T.; Yang, D.; Han, B. VxOy supported on hydrophobic poly(ionic liquid)s as an efficient catalyst for direct hydroxylation of benzene to phenol. *ChemCatChem,* **2015**, *7*(21), 3526-3532.
[http://dx.doi.org/10.1002/cctc.201500639]

CHAPTER 5

Zinc Oxide Nanomaterials for Biomedical Applications

Mohammad Ruhul Amin Bhuiyan[1,*], Hayati Mamur[2] and Ömer Faruk Dilmaç[3]

[1] *Department of Electrical and Electronic Engineering, Islamic University, Bangladesh*

[2] *Department of Electrical and Electronics Engineering, Manisa Celal Bayar University, Turkey*

[3] *Department of Chemical Engineering, Çankiri Karatekin University, Turkey*

Abstract: Semiconducting metal oxide nanomaterials are the future potential materials for biomedical applications. Zinc oxide (ZnO) nanomaterials are developed by using the organic synthesis process for excellent biocompatibility, selectivity, sensitivity, good chemical stability, non-toxicity, and fast electron transfer properties. They have a high surface-to-volume ratio that performs proper contouring on the human body to feel comfortable. Recent advanced studies on these nanomaterials show that they are promising materials for effective antibacterial and antifungal agents against a variety of microbes. They also promise to provide advanced technology for biomedical applications that can be used to destroy several types of malignant cells in the human body. Moreover, they can be used as antibacterial agents in the human body. This chapter briefly discusses the cost-effective approach to organically synthesizing ZnO nanomaterials. Moreover, these ideas can be developed to characterize these materials as biomaterials to perform easily upscaled in biomedical applications.

Keywords: Antibacterial and Antifungal Agent, Band-gap Energy, Biomaterial, Biomedical Applications, Cost-effective, Nanoparticles, Nanostructure, Organic Synthesis, Semiconductor, ZnO Nanomaterial.

INTRODUCTION

Zinc oxide (ZnO) is the II–VI metal oxide semiconductor material with outsized exciting-binding energy of 60 meV. It has a high dielectric constant, broad band-gap energy of 3.37 eV at 300 K, and a small Bohr exciting radius of 2.34 nm. It can exhibit nanostructures in forms that possess significant optical, electrical, and semiconducting performance. The ZnO nanoparticles can be studied for extensive utilization. In the synthesis process to develop ZnO nanoparticles, these processes

* **Corresponding author Mohammad Ruhul Amin Bhuiyan:** Department of Electrical and Electronic Engineering, Islamic University, Bangladesh; E-mail: mrab_eee@iu.ac.bd

Shazia Anjum (Ed.)

are generally classified into three categories: first the physical, second the chemical, and third, the green synthesis process. The physical and chemical synthesis processes generally are involved in laser/vapor deposition, spray-pyrolysis, epitaxy, thermal evaporation, sol-gel, sonochemical, electro-deposition, solvothermal, and hydrothermal [1 - 10]. The other synthesis process is a green synthesis that involves plants and herbal extracts [11 - 15]. Among the various metal oxides, ZnO nanoparticles have been accepted by researchers because of their exceptional electrical and optical characteristics. ZnO is a low-cost specimen in electrical appliances, such as an excellent sensor and a more acceptable agent for targeted drug delivery purposes. It has revolutionized industrial sectors like agriculture, drug delivery, and the food industry [16]. It is also widely used in ethanol gas sensors, UV light-emitting devices, pharmaceuticals, photo-catalysts, and cosmetic industries. These nanoparticles are also used in sunscreen because they absorb ultraviolet light effectively. Compared to other metal oxide nanoparticles, it has inexpensive and relatively low toxic properties that can exhibit excellent application in biomedical sectors, such as bioimaging, antibacterial, antimicrobial, anticancer, and drug delivery purposes.

According to our previous work, some of the materials are silver gallium di-selenide ($AgGaSe_2$) thin films [17 - 21], bismuth telluride (Bi_2Te_3) nanostructure materials [22 - 26] that are more acceptable for device manufacturing purposes, unlike thin films. Recently, ZnO nanoparticles have attracted more attention from scientists for their remarkable biological properties and biomedical uses. It shows the prospect and promise of a biomedical application field by developing the perfect size of nanomaterials. Zinc is an important trace element that extensively exists in human body tissues, along with the muscle, brain, skin, and bone marrow. It plays a crucial role in the nucleic acid synthesis, protein, neurogenesis, and hematopoiesis as a prominent part of different enzyme systems. It also takes part in the body's metabolism [27 - 30]. Low-dimensional ZnO nanoparticles can make zinc more easily absorbed by the human body. ZnO nanoparticles are a proficient platform for biomedical applications [31 - 35]. Nanoparticles exhibit improved characteristics that are supported by particle size, distribution, and morphology. They consist of particles that have nano-scale dimensions. These nanosized elements can enhance the reactivity of catalytic processes and thermal and non-linear optical activity [36 - 38]. These nanoparticles have received more consideration in biomedical applications for these characteristics. The ZnO nanoparticles can perform excellent biomedical applications compared to other metal oxide nanoparticles. Recently, in medical science, ZnO nanoparticles have started to be thought of in order to be used as nano-antibiotics for their antimicrobial activities. One of the reasons behind acute consent is nanoparticles consisting of a particle size less than 100 nm. The typical diameter of many human cells is around 7 µm, so the nanoparticles are comparable to naturally occurring proteins and bio-molecules of the cell. The contraction of particle size to the nano-scale formation can frequently change their structural, morphological, electrical, optical, magnetic, and chemical characteristics. The nano-scale particles have been making them for physical transport into the internal structures of cells and interacting with cell bio-molecules in unique ways. The particles typically possess a large atomic percentage on the main exterior of an element, which can easily lead to enhanced surface reactivity. It can also enhance their capability to be loaded with therapeutic agents to deliver them to target bimolecular cells. By using proper mixing dilation, these particles can acquire the ability to target selectively. It has emerged as a promising candidate for antibacterial fields [39 - 43]. It has a very effective way of destroying the cancer cells in the human body [44 - 48]. Strain is the most important parameter in the biomedical sensing arena. Flexibility and incompatibility in the biomedical field are two major issues with the use of biosensors [49 - 53]. ZnO nonmaterial may be used for biosensor manufacturing purposes. Recently, many researchers have shown that ZnO nanostructures are suitable materials for manufacturing biosensors. There is ongoing research to investigate the bio-imaging sensing function. It has shown positive results only in mice so far. The requirements of this sensor are non-allergenic, non-carcinogenic, high sensitivity, good reproducibility, and non-toxic characteristics. It can help in diagnosing the early stages of cancer germs. Moreover, sensing is important for monitoring and tracking the disease of a patient.

There are various synthesis processes that have been employed to develop ZnO nanostructures due to their vast areas of applications. It has drawn researchers' interest because of its various applications in the fields of electrical, optical, and biomedical sectors. The ZnO nanoparticles can be prepared easily, inexpensively, safely, and securely. Generally, ZnO is a safe metal oxide material recognized and enlisted by the US FDA. At ambient conditions, crystalline ZnO has a wurtzite crystal structure. It has two lattice constants, a & c, with a hexagonal structure. It belongs to the C_{6v}^4 or $P6_3mc$ space group. Fig. (**1**) shows the hexagonal wurtzite ZnO crystal structure. The structure of ZnO wurtzite occurs along the c-axis within energy polar exteriors that exist as O^{2-} terminated, or Zn^{2+} is relatively high.

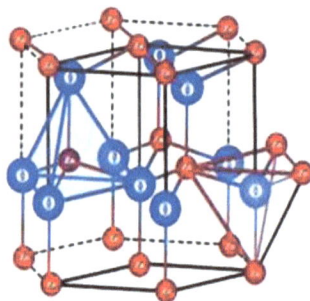

Fig. (1). Hexagonal wurtzite ZnO crystal structure; adapted from Kayode Adesina Adegoke *et al.* 2015 [54].

It is simply explained schematically as a layer to layer movement along the c-axis. It belongs to the tetrahedral coordinates of Zn^{2+} and O^{2-} inorganic compound semiconductors. The non-centrosymmetric structure is given rise by tetrahedral coordination. Some physical properties of ZnO are presented in Table **1** [55].

Table 1. Some physical properties of ZnO for an ideal hexagonal structure.

Properties	Zinc Oxide, ZnO
Stable phase at 300 K	Wurtzite
Lattice parameters at 300 K, a and c	0.32495 nm and 0.52069 nm
Density	5.606 g/cm^3
Melting point	1975 °C
Thermal conductivity	0.6 W/cm.°C, 1-1.2 W/cm.°C
Static dielectric constant	8.656
Refractive index	2.008
Band gap (Room Temperature)	3.370 eV
Band gap (low Temperature)	3.437 eV
Exciton binding energy	60 meV
Electron Hall mobility at 300 K	200 cm^2/V.s
Hole Hall mobility at 300 K	5-50 cm^2/V.s
Electron effective mass	0.24
Hole effective mass	0.59

Already, discovered ZnO nanoparticles have been able to bind biological substances in nanotechnology. Its dimensions below 20 nm can display UV absorption behavior without light scattering. They can also be extensively employed as inorganic long wave ultraviolet A (UVA) and short wave ultraviolet

B (UVB) filters. It is used to kill harmful microorganisms with its antibacterial and antifungal characteristics. It shows the semiconductor nature that has wide band-gap energy, and therefore, it is used as a UV photodetector. A considerable amount of research work has been completed through different synthesis routes for ZnO nanostructures. One of the common processes to develop ZnO nanoparticles is a chemical vapor deposition (CVD) process. It can also be obtained through an aqueous solution growth process. The effect of ZnO-nanostructured material on human cells has not been clear to researchers until now. They are thinking of the way it will dissolve the Zn ions into human cells. The human cell must possess the following characteristics of antimicrobial substances.

- It has to be nontoxic.
- It shouldn't react with food.
- It has to have a good taste or it has to be tasteless.
- It shouldn't have a disagreeable smell.

ZnO nanoparticles are one such inorganic metal oxide that would achieve all the above demands. It can safely be used as a drug delivery system that shows preservation in antibacterial and antimicrobial agents [56, 57].

The ZnO nanoparticles exhibit tremendous semiconducting characteristics because of their extremely high exciton binding energy and wide energy gap. In this circumstance, its application is in a wide range of biomedical sectors. The development of appropriate particle dimensions is a challenging research task for researchers. Intensive and systematic investigations are necessary to understand and improve the process of this material's transformation to nanostructure form. This chapter represents a search for the efficient synthesis process for developing the ZnO nanomaterial. It highlights which processes would be suitable for development performance in the biomedical sector. It also presents a new way for biomaterials that will have a high impact on biomedical applications. It can be applied to human cells that offer a way to recover from some diseases. The preliminary advancement of this material is not satisfactory for biomedical applications. Therefore, the efficient use of this material doesn't have sufficient influence on the biomedical sector. The discussion of the current situation in biomedical applications is the main objective of this chapter. Moreover, this chapter also discussed ZnO nanoparticle prospects in biomedical applications. Ultimately, we will try to summarize the cost-effective synthesis process and update exciting developments in the utilization of ZnO nanoparticles in biomedical applications.

SYNTHESIS PROCESS

The nanomaterials are developed by using physical, chemical, electrochemical, and biological synthesis processes. Physical and chemical synthesis processes are carried out using a thermodynamic, solvothermal, and hydrothermal approach, whereas particle growth rates are significant and used for industrial advancement. Biological synthesis processes are performed by plants, fungus, algae, bacteria, *etc*. Plant parts like fruits, seeds, stems, leaves, and roots have been used for nanomaterial development. Their extract is rich in photochemicals, which can act as both stabilizing and reducing agents.

Physical Synthesis Processes

The physical synthesis process can be divided into various processes, such as high-energy ball milling, laser ablation, and chemical vapour deposition methods. C. Prommalikit *et al.* [58] reduced particle size from microns to the nanoscale of ultra-fine ZnO material by using a ball milling process. The authors used the starting material of commercial-grade ZnO with an average crystalline dimension of 0.8 μm. They investigated the structure, morphology, and crystalline dimension of milled specimens by using the X-ray diffractometer (XRD), scanning electron microscopy (SEM), and particle analyzer, respectively. The XRD pattern indicated the hexagonal crystalline structure. The SEM images indicated that the particle size of ZnO nanopowders distinctly decreased, whereas the milling speed increased with time. After a milling process, the particle size is found in the range between 200 and 400 nm, forming nanoparticles. These results indicated that commercial ZnO powders minimized the particle size with a specific milling speed within a reasonable time. Finally, the authors recommended that the milling parameters (speed and time) of this process show a significant influence on the reduction of particle size. Fig. (**2**) shows the typical high-energy ball milling (HEBM) process within a schematic representation of nanomaterial synthesis.

S. Amirkhanlou *et al.* [60] also investigated whether milling time had influenced particle size in a high-energy ball milling process. The authors obtained a particle dimension of from 800 to 60 nm at an 8 hour milling time. In this process, the other researcher [61] investigated milling time durations from 02 to 50 hours to develop ZnO nanoparticles. The results showed that as milling time was increased, the particle dimension decreased from 600 to 30 nm. For other ball milling processes [62 - 64], researchers obtained a particle size of between 10 and 58 nm at different parameters. The authors investigated ball milling media, which also had an influence on particle size.

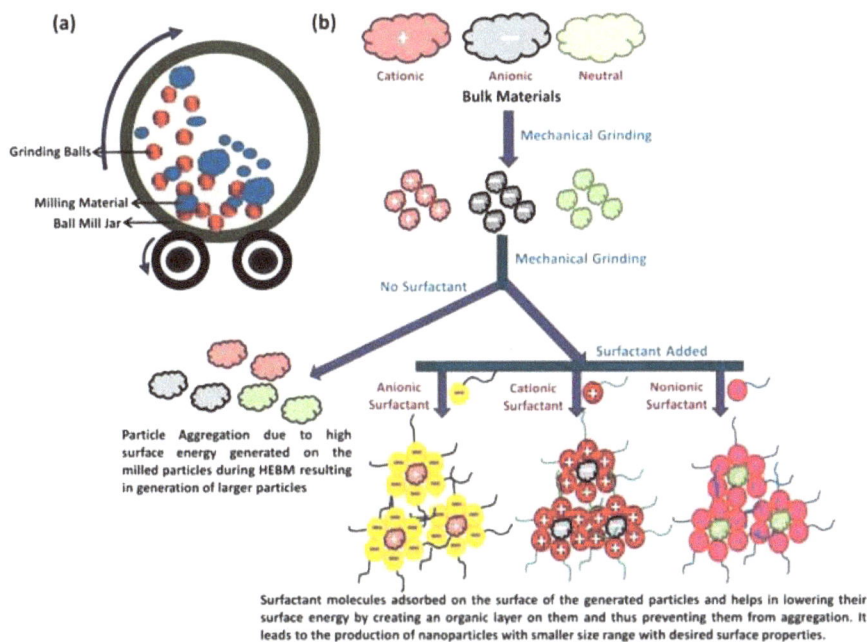

Fig. (2). (**a**) High-energy ball milling (HEBM) process; and (**b**) Schematic presentation of the nanomaterials synthesis using HEBM process; adapted from Chetna Dhand *et al.* 2015 [59].

Onur Tigli *et al.* [65] used a vapour deposition process to deposit ZnO nanowires onto silicon (Si) substrates. They used the source material of 1.38 grams of Zn powder and employed the measurements in a Lindberg horizontal multi-zone tuber furnace. The XRD and SEM characterization methods were used in the experiment. In experiments, the authors used variable reaction temperatures between 800 and 1200°C and reaction time durations between 45 and 60 min. The prepared sample has a uniform appearance. The SEM image shows aggregates of ZnO nanowires that accumulated sparsely onto the substrate. These experimental results indicate that it is possible to use a simple synthesis process to produce ZnO nanowires while eliminating all contamination. Finally, the authors reported that for a longer duration with high temperatures, a comb-like growth structure is not acceptable. Higher thermal conditions from the synthesized to the vaporization zones, combined with a rapid change in the heat of the self-catalysis mechanism, may result in premature termination. For this reason, a comb-like structure can be developed on the grown nanowires. The authors concluded that the physical vapour deposition process is an inexpensive growth technique to synthesize ZnO nanowires. Fig. (**3**) shows the typical chemical vapour deposition process for ZnO nanowires onto a Si substrate.

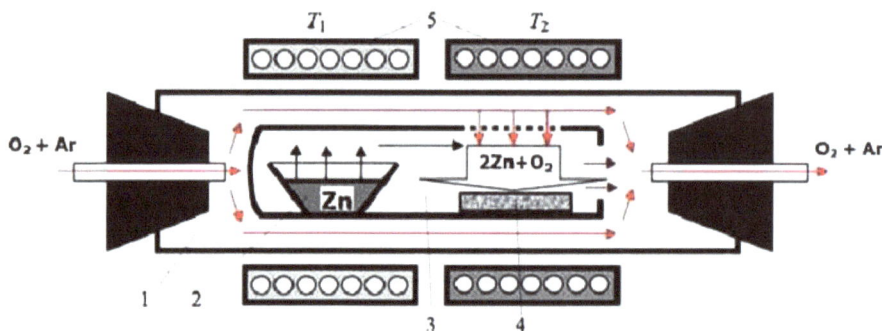

Fig. (3). Typical chemical vapour deposition process for ZnO nanowires onto Si substrate: (**1**) Flowing quartz reactor, (**2**) Zn source, (**3**) Internal quartz retort, (**4**) Substrates, (**5**) Electric heaters; adapted from Oleg Lupan *et al.* 2010 [66].

Oleg Lupan *et al.* synthesized the ZnO nanowires on silicon substrates by using a chemical vapour deposition process at 650°C with low pressure. They used the starting reactants of a high purity metallic Zn and 15 vol.% of O_2 within the Ar mixture. The XRD, X-ray photoelectron spectroscopy (XPS), photoluminescence (PL), Raman spectroscopy (RS), SEM, and transmission electron microscopy (TEM) characterization tools have been employed in the experimental investigation. The chemical and micro-structural quality of these prepared samples were justified through these experiments. According to JCPDS number 36-1451, the experimental results showed that nanowires have a hexagonal crystal structure within diameters ranging between 50 and 200 nm. The authors recommended that the obtained nanowires be transferable by pre-patterned external contacts to another substrate. They also fabricated a single ZnO nanosensor using a focused ion beam. They contacted metal electrode patterns for every end of a single nanowire by using this lithography process. Finally, the authors conclude that nanowires can be used as a sensor having a higher response to the H_2 atmosphere in the room temperature environment. Other researchers also developed ZnO nanostructures by using these processes. In physical vapour deposition processes [67 - 69], other authors developed the ZnO nanowires and nanosheets onto the indium-tin-oxide glass substrate that was used in dye-synthesized solar cells with an efficiency ranging between 0.1 and 0.5%. In conclusion, they recommended that its integration capability be proven for solar cell devices. Another researcher developed the ZnO nanowires in a hexagonal structure of around 60 nm at 450°C. It emitted ultraviolet light at room temperature conditions. In chemical vapour deposition processes [70, 71], other researchers developed ZnO nanorods, nanotubes, nanowalls, and nanowires depending on different growth parameters. It shows a morphologically smooth surface with uniform diameters of between 50 and 60 nm.

G. Al-Dahash *et al.* [72] prepared ZnO nanomaterials by using laser ablation within the aqueous solution of NaOH. The laser beam is focused on the pure zinc target surface within 1 mm of its diameter. The laser was placed at the bottom of a quartz cell and immersed at an 8 mm depth in the 0.1 M NaOH solution. In the experiment, UV-Visible spectroscopy, atomic force microscopy (AFM), SEM, and XRD characterization methods were used. In terms of optical characteristics, UV-Visible spectroscopy plays a part in the produced solution. It revealed red shifting peak positions within laser ablation power that exhibited optical extinction. The surface topography by AFM measurement showed a wide distribution of the particle size that lay between 80.76 and 102.54 nm. The SEM image showed that nanoparticles have a spherical shape. We estimated the crystalline size to be between 35 and 40 nm. In conclusion, the authors recommended that ZnO nanoparticles could be used as good insulating materials in reducing crystalline size by using the laser ablation process. In this process, other researchers [73 - 77] developed the ZnO nanoparticle characterized for structure, morphology, and optical performance. Most of the particles show the wurtzite hexagonal crystal structure, which is polycrystalline in nature. They can produce irregular shapes and different sizes of particles by controlling pulse energies. For this process, other researchers can control the particle size between 10 and 60 nm at different conditions. Fig. (4) shows the schematic diagram of a pulsed laser ablation process.

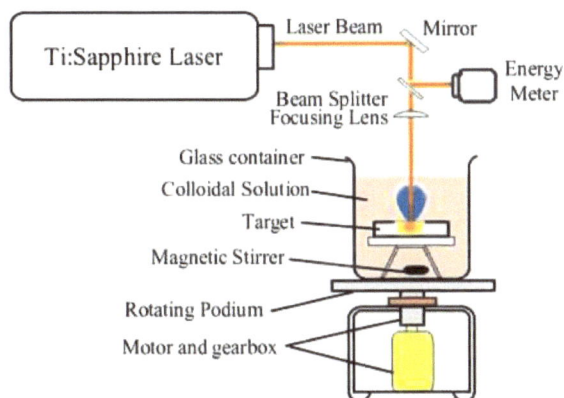

Fig. (4). Schematic diagram of a pulsed laser ablation process; adapted from Suha I. Al-Nassar *et al.* 2019 [78].

Chemical Synthesis Processes

A variety of chemical synthesis processes can be used for the development of ZnO nanoparticles. Chemical synthesis processes can generally be divided into two categories: gas and liquid phases. The gas-phase process can be divided into processes such as spray pyrolysis and gas condensation. The liquid phase process can be divided into precipitation or co-precipitation, colloidal, sol-gel, oil microemulsion, hydrothermal, and solvothermal. G.J. Lee *et al.* [79] used a spray pyrolysis process to create ZnO nanoparticles from zinc acetate dihydrate $(Zn(CH_3COOH)_2.2H_2O)$ at a 0.5 M concentration of solution within distilled H_2O. The concentrated solution is used as a precursor for spraying by sonication into a vertical quartz reactor by being allowed to flow without a carrier gas in a 900°C furnace. The authors studied the optical sensing characteristics conducted by temperature-dependent PL measurements. They observed the low-thermal PL spectra of ZnO nanoparticles that displayed a significant exciton peak with multiple sideband regions. This result attributed the bound exciton to optical phonon sidebands.

The PL peak intensity was higher, which revealed high optical sensing capabilities. Finally, the authors concluded that ZnO nanoparticles could be added to help form a strong platform for optical sensing devices as biomaterials. Other researchers [80 - 82] synthesized the ZnO particles/nanoparticles depending on various reactor temperatures, solution concentrations, and atomizing pressures. The ultraviolet-visible spectroscopy, XRD, FTIR, SEM, and TEM charac-terization tools are used in their experiments. The authors recommended that all parameters show a significant influence on the particle sizes of synthesized materials. Their experimental results showed the particle size to be between 10 and 400 nm, with uniform morphologies that indicated the wide distribution of particle size. The authors recommended ZnO nanoparticles to enhance photo-catalytic efficiency and control the crystalline dimension. Fig. (**5**) shows the schematic diagram of a spray pyrolysis process.

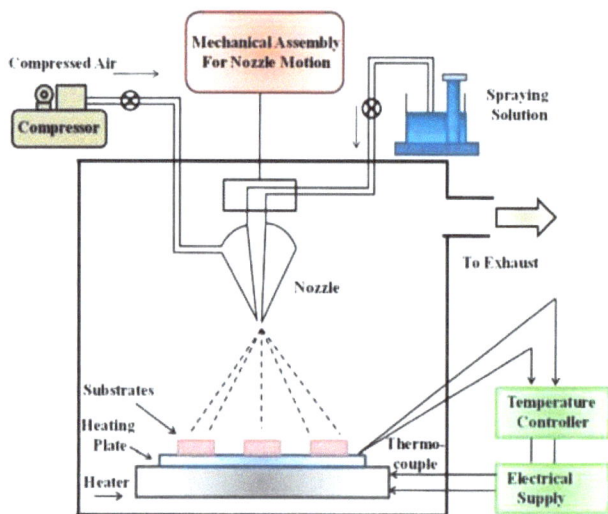

Fig. (5). Schematic diagram of spray pyrolysis process; adapted from Mangesh Waghmare *et al.* 2018 [83].

M. Vaghayenegar *et al.* [84] synthesized the ZnO nanoparticles at a high production rate by using the gas condensation process. The Zn droplet is oxidized at high temperatures by considering the atmosphere and reduced pressure within this process. They used XRD, SEM, and TEM measurements to characterize the prepared samples. They investigated the effects of temperature, carrier gas type, oxygen content, and pressure of the reactor on morphology, measuring the particle size distribution of samples. Finally, they recommended that all growth parameters show a significant influence on the samples. The authors estimated the optimum O_2/He–0. 2 Ar molar gas ratio; and synthesized the oxidized materials that found 0.18 and 0.21 ratios, having an average length/width between 80/35 and 83/27 nm, respectively, within atmospheric and reduced pressure conditions. In this process, other researchers [85, 86] synthesized the ultra-fine ZnO nanoparticles. In the experimental results, the authors estimated the particle size at between 30 and 40 nm with a web-like hexagonal crystalline structure. Fig. (**6**) shows a schematic illustration of the equipment for levitational gas condensation.

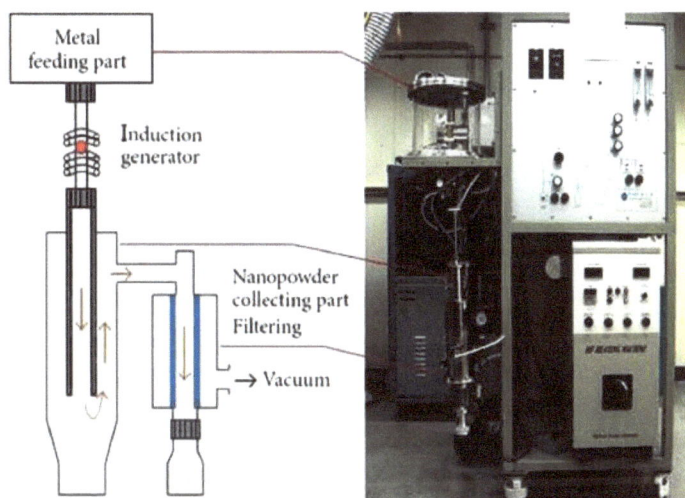

Fig. (6). Schematic illustration of the equipment for levitational gas condensation; adapted from Young Rang Uhm 2017 [87].

Monalisha Goswami *et al.* [88] synthesized ZnO nanostructure particles by the chemical process at a low-temperature condition. The authors used the $Zn(NO_3)_2.6H_2O$ and NaOH as precursors at different annealing temperatures ranging between 200 and 600°C for 2 hours. They observed different characteristics by using UV-Vis spectroscopy, XRD, FTIR, SEM, and PL measurements. XRD results revealed that prepared samples exhibited a hexagonal crystal structure, and the mean crystalline dimension is proportional to the annealing temperatures. SEM images revealed the formation of the crystal, such as a spherical shape with slight agglomeration. All samples exhibited the peak, which corresponds to a Zn-O stretching band observed in FTIR spectra. The UV–Vis absorption spectra showed the optical energy gap changing between 3.56 and 3.84 eV with annealing temperature. PL spectra show an excitation wavelength that increases in UV emission intensity at room temperature. The authors recommended annealing temperatures to influence the improvement of the crystal quality and control the crystalline dimension. Fig. (7) shows the schematic representation of a chemical precipitation process for developing ZnO nanoparticles.

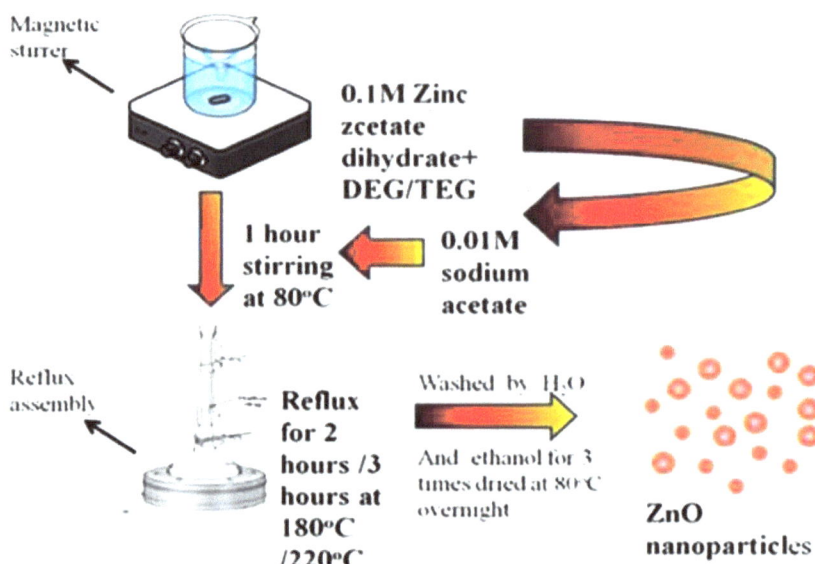

Fig. (7). Schematic representation of a chemical precipitation process; adapted from Pranjali P. Mahamuni *et al.* 2019 [89].

Rania E. Adam *et al.* [90] prepared ZnO particles by using the precipitation process in the low-temperature conditions for solar photo-degradation. The authors performed XRD, SEM, and photocatalytic measurements for the produced samples. The experimental results revealed that the nanoparticles indicated no impurity phases, and they had hexagonal wurtzite structures with uniform distribution. They investigated the sun-driven photocatalytic activity under different pH values of the aqueous solutions. The authors saw that solar-driven photocatalytic methods to abolish organic toxic from aqueous solutions were very prominent due to their environmental benefits. The ZnO nanoparticles were used as a catalyst to degrade Congo red dye from aqueous solutions at different pH values. Finally, the authors found a higher photodegradation efficiency in the acidic-medium solutions. In conclusion, the authors recommended that ZnO nanoparticles could perform with considerable degradation efficiency after different usage cycles. In these processes, other researchers [91 - 95] prepared the ZnO nanoparticles at different temperatures ranging between 100 and 600°C for 4 hours. They studied XRD, FTIR, SEM, Energy Dispersive X-ray Spectroscopy (EDX), Electron Spin Resonance Spectroscopy (ESR), TEM, UV-Vis spectroscopy, Brunauer–Emmet–Teller (BET) analysis, and thermal analysis TG-DTA for prepared ZnO nanostucture particles. Prepared samples possessed hexagonal crystal structures within crystalline dimensions ranging from 3 to 45 nm.

H. Wang *et al.* [96] prepared the colloidal ZnO nanoparticles in ethanol solutions. Subsequently, the authors annealed the prepared samples at different temperatures ranging from 150 to 500°C. They examined different characteristics of these specimens by using XRD, TEM, UV–vis absorption spectrum, and FTIR measurements. The TEM image indicated the sample possessed a narrow size distribution within 90 nm. These nanoparticles have a significantly higher surface-to-volume ratio than the bulk material. The morphologies of the samples mainly showed an irregular shape with aggregates of each other. Crystalline dimension is significantly influenced by annealing heat, and it rapidly increases with a wide dimension distribution in nature. The average dimension increases from 10 to 90 nm with the temperature. Finally, the authors recommended that the stability and photocatalytic activity of these materials could be adjusted by utilizing the size effect. Fig. (**8**) shows a typical preparation of the colloidal ZnO nanoparticles. The average dimension of these particles is -25 nm within the hexagonal crystal structure form.

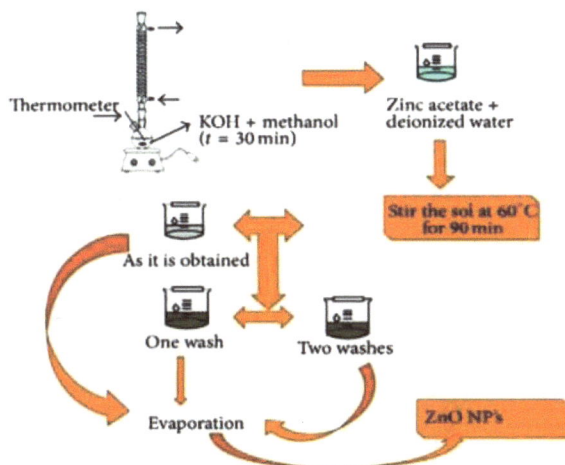

Fig. (8). A typical preparation of the colloidal ZnO nanoparticles; adapted from Jose Alberto Alvarado *et al.* 2013 [97].

A. Nagar *et al.* [98] developed ZnO nanomaterials by using the sol-gel process for a transparent element in transistors. The authors used $ZnC_4H_6O_4$ and distilled water as precursors. The experimental characterization has been performed by using the field emission scanning electron microscope (FESEM), XRD, and high-resolution transmission electron microscope (HRTEM) for synthesized samples. The FESEM micrographs showed a nanoflower structure of morphology that was uniform and highly dense in nanoparticles. This behaviour enhanced the field emission characteristics of a transistor. According to JCPDS No 01–1136, the nanoflowers indicated a wurtzite hexagonal crystalline phase structure. The HRTEM micrographs revealed the ZnO nanoparticles are closely stacked with each other. The authors estimated the diameter distribution of these nanoparticles to be in a range between 30 and 60 nm. This flower-type nanoparticle has a high current density and a significant field enhancement factor to conform to the field emission study. In conclusion, the authors recommended that ZnO nanoparticles might be effectively used in different liquid-crystal and field emission display devices. Fig. (**9**) shows the methodology of sol–gel auto-combustion synthesis of ZnO nanomaterials.

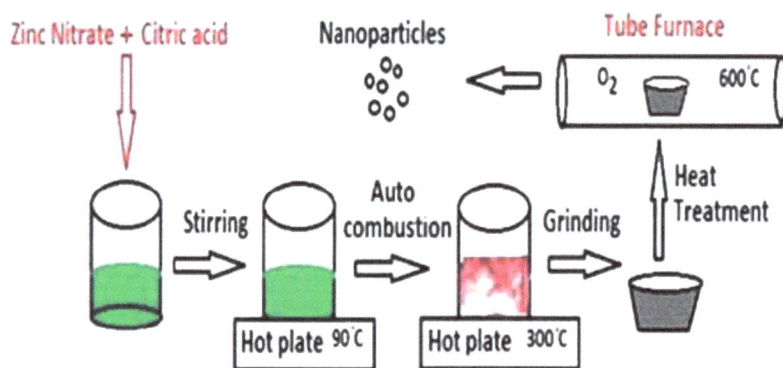

Fig. (9). Methodology of sol-gel auto-combustion synthesis process; adapted from Hassan Soleimani *et al.* 2018 [99].

In this process, other researchers [100 - 103] synthesized this nanoparticle for antibacterial effects, dye-sensitized solar cells, and humidity sensing purposes. The synthesized samples were calcined at different temperatures ranging between 70 and 800°C. The characterization of the developed samples was studied by using XRD, HRSEM, HRTEM, AFM, FT-IR, and UV-Vis spectroscopy with different calcination temperatures. The authors recommended that temperature played a significant influence in controlling the crystalline dimension and using the nanoparticlesfor different purposes. They estimated the particle size ranged between 10 and 40 nm having a crystalline wurtzite hexagonal structure.

Ana M. Pineda-Reyes *et al.* [104] synthesized the ZnO nanopowder by using an oil microemulsion, which is a human and environmentally friendly process. This process consists of water-in-oil (w/o) microemulsions, where continuous or discontinuous phases are emu oil, and 0.5 M $ZnC_4H_6O_4$, respectively, in the solution that is stabilized in molecules. After that, the precipitation was prepared by adding 1.0 M sodium hydroxide to this aqueous solution. Finally, the resultant precipitates were calcined at 800°C for 2h and then dried in the air at 100°C. The prepared nanopowders are characterized by using XRD, SEM, and TEM, and the average 31.2 nm dimension is estimated within the wurtzite hexagonal structure. Fig. (**10**) shows a typical water-in-oil microemulsion process.

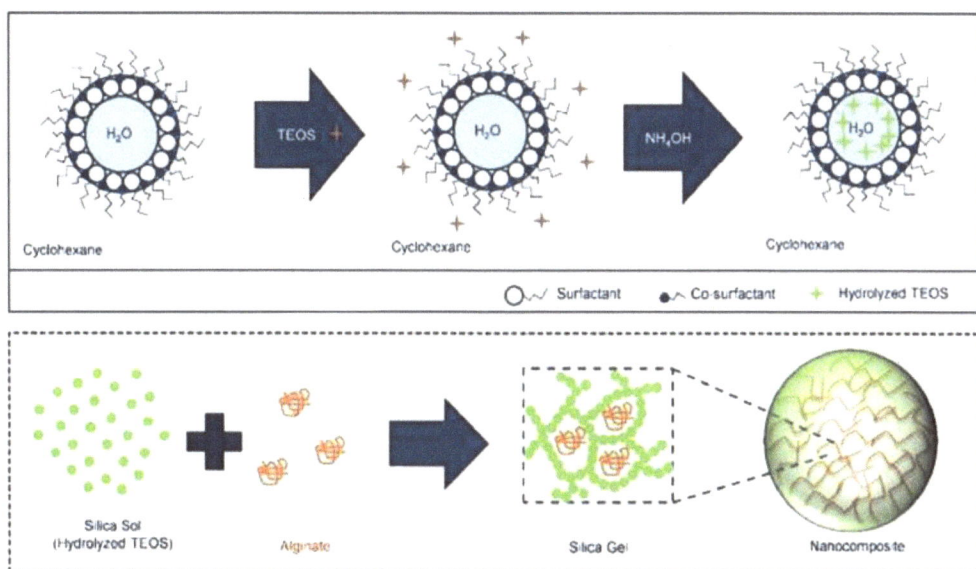

Fig. (10). Illustration of a water-in-oil microemulsion process; adapted from Xin Fan *et al.* 2019 [105].

In this process, other researchers [106 - 108] synthesized ZnO nanoparticles of different shapes and particle dimensions at the reaction (40 to 70°C) and calcination (300 to 500°C) temperatures. The reaction temperature did not considerably affect the average particle size, whereas the calcination temperature significantly affected the particle size. The authors used $Zn(NO_3)_2$, $ZnSO_4$, and zinc chloride ($ZnCl_2$) to create the Zn^{2+} source. Finally, the authors suggested that $Zn(NO_3)_2$ perform the best activity during the synthesis process. They estimated the average crystalline dimension ranged from 2.1 to 24 nm having a wurtzite hexagonal structure. The SEM and TEM images showed the particle-like nanorod shape with a diameter and length ranging between 22 and 28 nm and 66 and 72 nm, respectively.

S. Agarwal *et al.* [109] synthesized the ZnO nanostructures like flowers or rods by using a hydrothermal process. The authors investigated their morphology-dependent gas sensing characteristics. They used SEM, XRD, TEM, UV-Vis spectroscopy, and PL measurements for the prepared samples. The SEM images formed two kinds of floral structures, flower-like structures having a composition of nanoparticles were formed at short-time reactions, whereas at long-time reactions, floral assemblies of nanorods were formed. XRD spectra reveal a hexagonal crystal structure of the samples according to JCPDS card No. 01-07--0205. The crystalline dimensions were found to be 21 and 43 nm for nanoflowers and nanorods, respectively. The energy gap of the prepared specimen is found to be 3.0 and 3.19 eV for nanoflowers and nanorods, respectively, by using UV–Vis absorption spectra. The presence of oxygen vacancies is confirmed in both samples for the PL spectra. Finally, the authors investigated the gas sensing characteristics with morphology for different gases at operating temperatures. In conclusion, the authors recommended good sensitivity to nitrogen dioxide (NO_2) gas with the response of prepared samples. Fig. (**11**) shows a typical hydrothermal process for developing ZnO nanostructure.

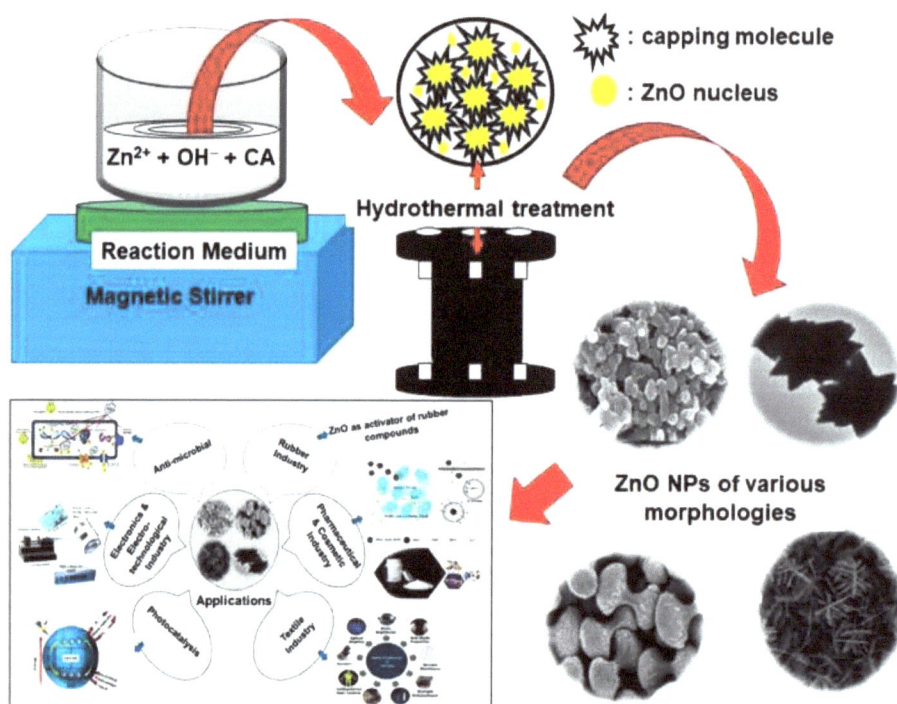

Fig. (11). A typical hydrothermal process for developing ZnO nanostructure; adapted from Parita Basnet *et al.* 2020 [110].

In this process, other researchers [111 - 114] synthesized ZnO nanoparticles, nanopleates, nanorods, and nanoflowers for gas sensing and photoconductive applications. The authors used structural and optical measurements for the characterization of the developed samples. They observed gas sensing characteristics towards ethanol at different operating temperatures. The authors estimated the average crystalline dimension to range from 6 to 64 nm having a wurtzite hexagonal crystallographic structure.

Ankica Šarić *et al.* [115] developed ZnO nanosize particles from zinc acetylacetonate within the presence of various alcoholic solvents and triethanolamine (TEA) at 170°C. The authors monitored the structural and optical properties with XRD, ultraviolet-visible spectroscopy, FT-IR, and FE-SEM investigations for the developed samples. The authors proposed the nucleation and the formation of this nanoparticle in various alcoholic solvents and TEA systems. The experimental findings indicated that the alcoholic solvents, ethanol, and TEA would dominate the crucial influences on particle size. The authors concluded that the impact of surface interactions between developed nanostructures and molecules of a solvent with TEA on the way of growth and aggregation could control structural characteristics. Finally, the authors estimated the particle size at 10 nm. In this process, ZnO nanoparticles are synthesized in different solvents, temperatures, and time durations. The authors used $Zn(CH_3COO)_2.2H_2O$ and KOH as a precursor to the developing samples. The authors recommended that every growth parameter has a significant influence on controlling the particle size. The authors estimated the average crystalline dimension to range from 10 to 76 nm having a wurtzite hexagonal structure in the different shapes.

Biological Synthesis Processes

A variety of biological synthesis processes have been used to develop ZnO nanoparticles. These processes have been performed through waste materials. The mediation of microbes such as fungi, algae, viruses, actinomyces, and bacteria and the mediation of plants such as roots, shoots, leaves, and stems are performed. A. Deep *et al.* [116] improved a process for developing pure ZnO nanostructure particles from the waste electrode of used batteries of alkaline $Zn–MnO_2$. Authors compiled used batteries to find out the waste electrode. After that, they were dismantled to collect Zn and Mn materials. In an aqueous solution, the authors added 5M HCl and 0.1M Cyanex 923 at 250°C for 30 minutes to form the Zn-Cyanex 923 complex. The authors used ethanol for centrifugation to separate the organic phase. Finally, pure ZnO nanoparticles were developed, having about a 5 nm particle diameter.

There is viable importance in developing environmentally-friendly ZnO nanoparticles that don't produce toxic elements. The mediation of many microbes, including fungi, algae, viruses, actinomyces, and bacteria, can be used to synthesize the ZnO nanoparticle by using the intracellular or extracellular process. This kind of synthesis process is a green, environmentally friendly process, which is a blessing to biological nature. Nowadays, microbe mediation is a simple and one of the most attractive sources of green synthesis processes for developing ZnO nanomaterials. Heba K. Abdelhakim *et al.* [117] successfully synthesized the ZnO nanoparticles by using an eco-friendly and cost-effective process of culture filtrate of the endophytic fungus Alternaria tenuissima. For this synthesis purpose, the authors used zinc sulfate $ZnSO_4.7H_2O$ and 100mL of the Alternaria tenuissima cell-free culture filtrate (ATCF). To prepare 2mM concentrations of an aqueous solution, a volume of 100mL of ATCF was taken in a flask and mixed with 100mL of $ZnSO_4.7H_2O$. The reaction was maintained for 20 minutes of stirring at room temperature, and they observed that it looked like a white precipitate. The precipitate is separated, washed, and dried at 50°C in the open air to prepare the ZnO nanoparticle. The prepared powder was dissolved in ethanol for the ultrasonically treated dispersion, and finally, they developed the fine powders to be used for characterizations. Synthesized samples are investigated by using UV–Vis spectroscopy, XRD, dynamic light scattering (DLS), TEM, and FT-IR measurements. The authors rapidly completed this synthesis process and confirmed surface plasmon resonance by UV–Vis spectroscopy. The XRD spectrum revealed the hexagonal crystal structure of ZnO nanoparticles having a lattice parameter of 0.323420Å according to JCPDS card No. 361451. The DLS investigation demonstrated the crystalline dimension distribution was in the range between 10 and 30 nm. This distribution is confirmed by TEM micrographs, and it also showed that particles are monodisperse and spherical in shape. The FTIR spectra showed the prominent bands of symmetric/asymmetric vibrations of ATCF and ZnO nanoparticles. In conclusion, from experimental results, the authors found that low dimensional ZnO nanoparticles showed promising activities, which could be better utilized for biomedical applications. Finally, the authors recommended that excellent biomedical potential could be achieved with an innovative and alternate approach for the revelation of ZnO nanostructure particles by using microbial platforms. It would open up a new way for biomedical applications. In accordance with the other reviews and research articles [118 - 122], the production of ZnO nanoparticles from microbes' mediation confirmed a promising utilization for biomedical applications. The overview of the microbe's mediated extract for ZnO nanoparticles is summarized in Table **2**.

Table 2. Overview of microbe's mediated extract for ZnO nanoparticles.

Microbe Part	Morphology	Size (nm)	Reference
Fungus *Alternaria tenuissima*	Hexagonal, Spherical	10-30	[117]
Endophytic fungi	Hexagonal, Quasi-spherical	16-78	[118]
Fusarium keratoplasticum, Aspergillus niger	Hexagonal, Nanorods	8-42	[119]
Clonorchis sinensis, Candida albicans, Serratia ureilytica	Hexagonal	16-37	[120]
Saccharomyces cerevisiae	Quasi-spherical	-10	[121]
Aeromonas hydrophila	Spherical	- 57.72	[122]

For this process, the authors used marine microbes and microorganisms to produce ultra-fine, uniform, and well-dispersed ZnO nanoparticles for various biological and photocatalytic activities. Finally, the authors concluded that this process was time-saving, and it was eco-friendly to produce ZnO nanoparticles using these sources. Several production processes, like the physical and chemical processes of these nanoparticles, have a high rate of toxicity and extreme environmental issues. In these circumstances, the green synthesis processes of plants, fungus, bacteria, and algae are used. Fig. (**12**) shows the green synthesis process to synthesize ZnO nanoparticles to increase popularity.

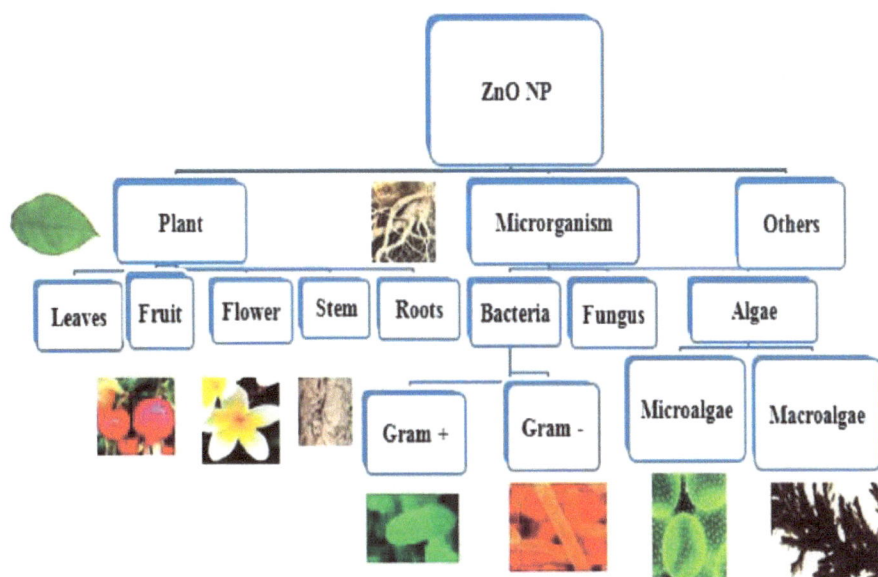

Fig. (12). Green synthesis process to synthesize ZnO nanoparticles; adapted from Happy Agarwal *et al.* 2017 [123].

Multifunction inorganic ZnO nanomaterials are very significant for their uses in the biomedical sector. Due to their constructional view, potential applications of the ZnO nanoparticle can play a dominating role in this sector day by day. Using plant mediation is the green synthesis approach that is a reliable, cost-effective, eco-friendly, and biocompatible process.

The plant mediation synthesis process has an easy step in the sequence for the development of eco-friendly nanoparticles. This process can minimize the hazards associated with the use of these processes. To achieve the extract, the first step of washing plants mediate is followed by boiling in distilled water. Then other subsequent steps followed, such as the compounding of Zn with this extract and the separation of liquid, leading to the aggregation of nanoparticles. Fig. (**13**) shows the symmetric diagram of the development of ZnO nanoparticles from plant mediation.

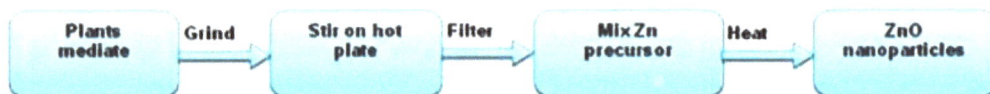

Fig. (13). Symmetric diagram of the development of ZnO nanoparticles from plant mediation.

M. Rafique *et al.* [124] have synthesized the ZnO nanoparticles from Syzygium cumini (S. cumini) plants. The authors declared that their report was a facile and cost-effective green synthesis process to develop ZnO nanoparticles from this plant's leaves by using an extract. The authors prepared the aqueous solution with 5mL of S. cumini leaf extract. It was added drop-wise to 0.05M, 25mL $ZnC_4H_6O_4$ under constant stirring at 60°C for 2 hours. The reduction of $ZnC_4H_6O_4$ into ZnO nanoparticles showed a reaction to this mixture with a colour change after constant stirring. Then the prepared solution was centrifuged at 4500rpm for 15 minutes, and the resulting sample was dried at 60°C overnight. Finally, the synthesized sample was calcinated at 400°C for 4h, and again a colour change occurred, meaning ZnO nanoparticles were ready for characterization. The authors observed their characteristics by adjusting the consolidation of leaves' extract between 10 and 25mL at constant molarity of 25 mL. The synthesized samples were characterized by UV–Vis spectroscopy, XRD, SEM, and FT-IR measurements. The absorption spectra exhibited the exciton band at 320 to 350 nm; it represents the robust absorption of a wurtzite hexagonal phase structure. The other peak is not identified in these spectra, so it is associated with impurities and structural defects within the specimen. The formation is confirmed to develop ultra-pure crystalline ZnO nanoparticles. The extract concentration had a significant influence on the bandgap energy that varies between 2.22 and 3.00 eV. The decrease in band-gap energy occurred with increases in extract concentrations in the prepared ZnO nanoparticles. According to JCPDS 76-0704 and 75-1533, the XRD pattern revealed the hexagonal crystal structure of the prepared sample having a crystalline dimension of 16.40 nm. SEM micrographs indicated uniformly dispersed spherical-like ZnO nanoparticles, and that particle size decreased from 78 to 64 nm with increases in extract concentration. The FTIR spectra indicated the oxidation occurrence and proved that the synthesised samples had pure form. Prepared samples are used to increase seed germination; they exhibit 60% advancement by controlling particle size. ZnO nanoparticles are also used in polluted water purification by using the RhB dye. The influence of different growth parameters on dye removal is analyzed under visible light irradiation. They estimated a maximum of 98% performance. In conclusion, the authors concluded that ZnO nanoparticles had potential applications in technology enhancement. Other researchers have synthesized ZnO nanoparticles from other plants mediated [125 - 139], and they declared they can be used as antimicrobial agents in biomedical applications. The overview of plant-mediated extracts for ZnO nanoparticles is summarized in Table **3**.

Table 3. Overview of plants-mediated extracts for ZnO nanoparticles.

Plant Part	Morphology	Size (nm)	Reference
Syzygium cumini plant	Hexagonal, Spherical	16-78	[124]
Star fruit	Hexagonal, Nanoflakes	19.6	[125]
Beta vulgaris, Cinnamomum tamala, Cinnamomum verum, Brassica oleracea var. Italica	Hexagonal, Spherical, Nanotubes	14-30	[126]
Radish root	Hexagonal, Spherical, Nanorods	15-25	[127]
Punica granatum fruit	Spherical, Hexagonal	33-82	[128]
Rhamnus virgata	Hexagonal	-20	[129]
Achyranthes aspera and *Couroupita guianensis* leaf	Hexagonal, Spherical, Nanoflakes	5-40	[130]
Andrographis paniculata leaf	Hexagonal, Spherical	57-115	[131]
Lagerstroemia speciosa leaf	Hexagonal	-40	[132]
Caralluma fimbriata	Hexagonal, Spherical, Nanoflakes	-30	[133]
Pongamia pinnata leaf	Hexagonal, Nanorods	10-120	[134]
Plectranthus amboinicus leaf	Hexagonal, Nanorods	-88	[135]
Aspalathus linearis	Hexagonal, Quasi-spherical	-12.5	[136]
Cassia fistula plant	Hexagonal	5-15	[137]
Ocimum basilicum leaf	Hexagonal	-50	[138]
Aloe barbadensis miller leaf	Spherical, Hexagonal	-35	[139]

The authors used different plants' mediation to produce ultra-fine, uniform, and well-dispersed ZnO nanoparticles for various biomedical applications. The authors estimated an average crystalline dimension ranging from 5 to 120 nm, having a wurtzite hexagonal structure in different shapes.

Fig. (14) shows different shapes of one-dimensional ZnO nanoparticles developed from plant mediation. Finally, the authors concluded that this process consisted of a time-saving, eco-friendly, and cost-effective way to produce hexagonal ZnO nanoparticles by using these sources.

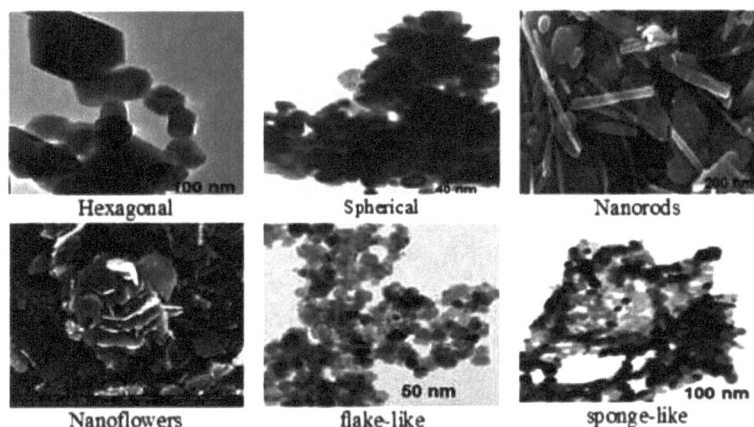

Fig. (14). Symmetric diagram of the development of ZnO nanoparticles from plant mediation; adapted from Sohini Chakraborty *et al.* 2020 [125], Siti Nur Amalina Mohamad Sukri *et al.* 2019 [128], J. Duraimurugan *et al.* 2019 [130], V. Sai Saraswathi *et al.* 2017 [132], Doddavenkatanna Suresh *et al.* 2015 [137], Hasna Abdul Salam *et al.* 2014 [138].

In the above discussion, it is clear that physical and chemical synthesis strategies are expensive to develop the ZnO nanoparticle. Chemical agents added for precipitation and reduction purposes require a lot of source material that occurs in-depth labour and more time. Most of the chemicals are naturally poisonous, and their synthesis by these products does not behave in an eco-friendly way. However, some physical and chemical processes have the potential to exceed a short time period to synthesize more quantities of ZnO nanoparticles. Moreover, some of the chemical solution-based synthesis processes are also effective, such as chemical precipitation or co-precipitation, sol-gel, solvothermal, and hydro-thermal. The biological synthesis process is used to make a non-toxic, eco-friendly operation, and it secures reagents.

Nowadays, the fast universal advance of microorganism resistance to antibiotics is causing a serious hazard to worldwide public fitness. Nano-medicine is a promising research field that combines research knowledge about nanotechnology with medicine. It presents nano-formed materials of antimicrobial agents utilized in biomedical applications.

Fast technology is nanotechnology that low dimensional nano-scale particle size can create with advanced characteristics for multifunction. Biological synthesis is significantly utilized in biomedical applications and emerging nanotechnology. An overview of these synthesis processes to prepare ZnO nanoparticles is shown in Table **4**.

Table 4. Summarize of synthesis processes for ZnO nanoparticles.

Synthesis Process	Main Morphology	Size (nm)
Physical synthesis process		
High energy ball milling	Hexagonal	10-400
Physical/chemical vapour deposition	Hexagonal	50-200
Laser ablation	Hexagonal	10-103
Chemical synthesis process		
Spray pyrolysis	Hexagonal	10-400
Gas condensation	Hexagonal	30-40
Precipitation or co precipitation	Hexagonal	03-45
Colloidal	Hexagonal	10-90
Sol-gel	Hexagonal	10-60
Oil microemulsion	Hexagonal	02-30
Hydrothermal	Hexagonal	06-64
Solvothermal	Hexagonal	10-76
Biological synthesis process		
Microbes mediated extract	Hexagonal	08-78
Plants mediated extract	Hexagonal	05-120

The biological synthesis process, mainly based on plant mediation, has more advantages over the physical and chemical synthesis proceses for synthesized ZnO nanosize particles for biomedical applications. The advantages of the plant-mediated ZnO nanoparticles synthesis process are as given below:

• In plant mediation, the synthesis process can easily collect plant material for its availability and presence of bio-component activity, which performs as the capping and reducing agent.
• It is possible to develop the large-scale nanostructure particles by using plant mediation for biomedical applications.

• Nanoparticle synthesis by plants' mediation would not require a well-designed laboratory room.
• It can synthesize at normal pressure and room temperature conditions without necessarily undergoing difficult energy processes.
• The produced sample is mostly stable and safe due to natural plant application.

Synthesis of ZnO nanoparticles from plant mediation is stable, safe, eco-friendly, and cost-effective for biomedical applications.

BIOMEDICAL APPLICATIONS

Nanotechnology has revealed novel insights into ZnO nanoparticles in the advancement of material sciences. The nanoparticles are known as a safe material that is utilized in sunscreen products and food additives.

Moreover, because of the luminescent characteristics of ZnO nanostructure materials, they can be used in various biomedical sectors. It is also well known as a semiconductor material, which has the potential for biomedical applications. The materials are extensively used in environmental remediation and healthcare utilization to perform their different characteristics. For these circumstances, conventional Cd-related biology purposes can easily be replaced by ZnO nanoparticles. Biodegradability is a prominent feature of ZnO nanostructure materials. The zinc ion is involved in different perspectives of metabolism. In the chemical aspect, the ZnO surface is rich in -OH. It has been functionalized through surface decorating molecules. Finally, in this chapter, we summarized the current state of growth and application of ZnO nanomaterials in biomedical sectors. The ZnO nanoparticles have biomedical acceptance due to potential applications in bioimaging, biosensors, antibacterial, antimicrobial, anticancer, and drug delivery. These nanoparticles are ubiquitous in numerous biological fields within consumer products and can play a crucial role in advances in assessing their toxicological effects. Fig. (15) shows the pictorial view of the green synthesis process, toxicity assessment, and utilization in biomedical areas of ZnO nanoparticles.

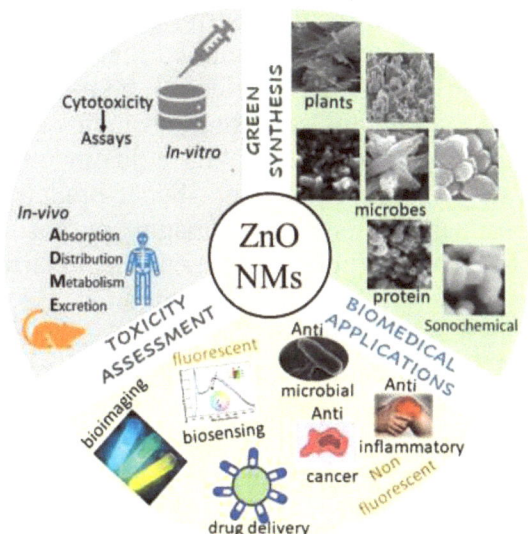

Fig. (15). Pictorial view of the green synthesis process, toxicity assessment, and utilization in biomedical areas of ZnO nanoparticles; adapted from Rajni Verma *et al.* 2021 [140].

Bioimaging

Recently, different shapes of ZnO nanoparticles have been recognized by researchers to be utilized as bioimaging materials. Fluorescence imaging is widely used in the preclinical research arena because it is inexpensive and convenient. ZnO nanoparticles are performing a powerful exit in blue and near-ultraviolet emission. It can also have a green luminescence related to O_2 vacancies. Other literature reports on the use of ZnO nanoparticles for cellular imaging. The advantage of intrinsic fluorescence can be achieved by using the human skin image. For safety concerns, most of the ZnO nanostructure materials stayed within the stratum corneum. Biocompatible ZnO nanocrystals encapsulated in a non-linear core of phospholipid micelles are conjugated to folic acid. The nanocrystals are reported for cancer cell imaging with low toxicity. Doping in other materials can be used to tune the behavior of ZnO nanoparticles. In our previous report, ZnO nanoparticles and thin films were doped with cations such as Ni, Al, and Sb that stabilized in aqueous solutions [141 - 143]. It was suggested that ZnO nanoparticles could penetrate into the cell nucleus. The most widely significant nanostructure materials are QDs due to their desirable characteristics for optical imaging purposes. CdSe/CdTe QDs are commonly used to have potential toxicity in biological systems. ZnO-based QDs improved the less toxic fluorescent nanomaterials, which have relatively weak emission and stability in aqueous solutions. Every imaging process has advantages or disadvantages; nanoparticles function to be detectable by multiple imaging modalities that can provide synergistic facilities. There are several literature reports on ZnO nanomaterial utilization for bioimaging purposes. G. Lei *et al.* [144] prepared ZnO QDs with blue fluorescence and discussed their utilization for bioimaging purposes. The surfaces of ZnO nanoparticles can be modified conveniently, and they can stay stable in an aqueous solution, and their quantum dots (QDs) are enhanced by around thirty percent after the adjustment. Finally, the authors recommend that water-soluble ZnO with hyper-branched polyethylenimine compounds has the best performance for bioimaging purposes. J.E. Eixenberger *et al.* [145] have investigated the pure n-type ZnO nanoparticles for bioimaging purposes by using fluorescent microscopy methods. Generally, to produce emissions, most of the nanoparticles require UV excitation sources. Herein, the authors demonstrated the reducing energy gap that excited the nanostructure materials to detect their emissions. The research lays the foundation for the utilization of these nanoparticles for different bioimaging purposes. It enables researchers to experiment with n-type ZnO nanoparticles on human cells through fluorescence-based imaging methods. Fig. (**16**) shows the developing process of n-type ZnO nanoparticles for bioimaging purposes.

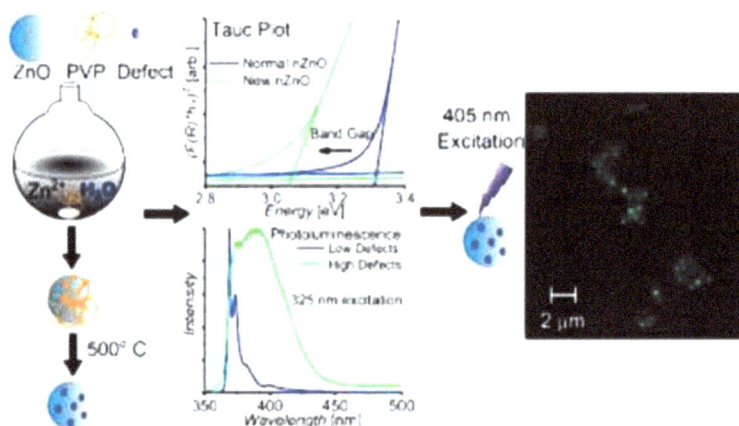

Fig. (16). Chemical solution synthesis process to produce ZnO nanoparticles for bioimaging applications; adapted from J.E. Eixenberger *et al.* 2019 [145].

In other research reports [146 - 150], the production of ZnO nanoparticles from different synthesis processes is a promising utilization for bioimaging applications.

Biosensors

Biosensors are used in chemical or biological analysis, for healthcare purposes, environmental monitoring, and in the food industry. Nanoparticles are biologically active substances that attract increasing attention.

It can provide a better platform for the advantages of high-performance biosensors due to its suitable performance. The high surface-to-volume ratio of nanoparticles can immobilize various bio-molecules that are antibodies, enzymes, and other proteins. Moreover, it can allow electron transfer to the electrode of the bio-molecules. It exhibits desirable traits for bio-sensing, which are high catalytic efficiency and strong adsorption capability. Bio-sensing is better for the adsorption of enzymes and antibodies. Moreover, nanoparticles are promising for biosensors for their low toxicity, high surface area, good biocompatibility, and high electron transfer capability performance. Classification of biosensors: typical biosensors consisting of different systems are shown as follows in Figs. (**17a & b**) [151].

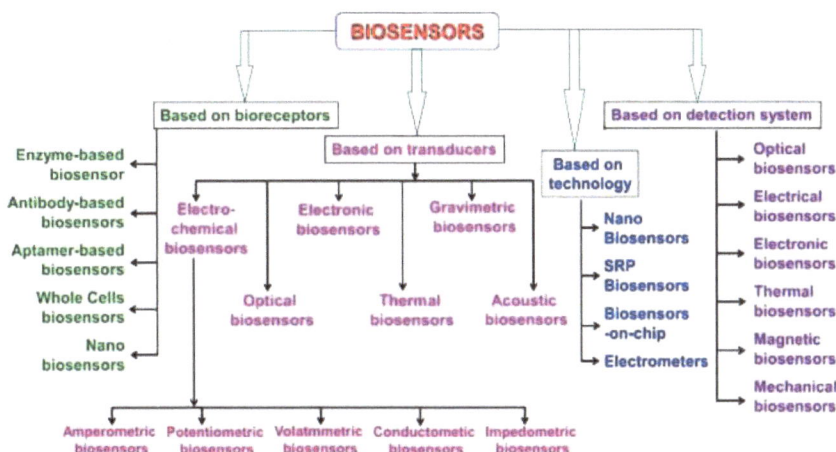

Fig. 17(a). Classification of biosensors.

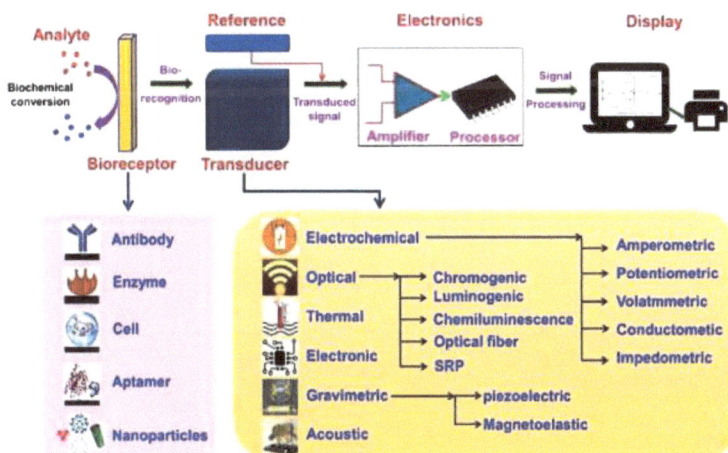

Fig. 17(b). A schematic diagram of typical biosensors.

Generally, two types of biosensors are available: catalytic and non-catalytic biosensors. An innovative biochemical reaction product is the analyte-bioreceptor interaction that improves the catalytic biosensor. Enzymes, tissues, micro-organisms, and whole cells are included in biosensors. There is no innovative biochemical reaction product developed in a non-catalytic biosensor that analyzes the bound to the receptor irreversibly and during the interaction. The sensor comprises cell receptors, antibodies, and nucleic acids as the targets for detection.

Antibacterial

Recent achievements will fulfill future public health demands, which could help biomedical areas by using organic medicinal drug agents. We believe that the effectiveness of ZnO nanoparticles will increase biomedical applications for antibacterial activity worldwide. In addition, we consequently observed that the crystalline dimension variation with exterior space to volume ratio of ZnO nanoparticles was the cause of a promising antimicrobial function. I. Kim *et al.* [152] discussed strong antibacterial activity against low concern bacteria (gram-negative and -positive). The influence on their antibacterial activity is generated by the oxidative stress interacting with them by forming Zn^{2+} ions from ZnO nanoparticles. It constrained the functions of respiratory enzymes. The authors demonstrate the capability of these nanoparticles to destroy the cell membrane and develop reactive oxygen species (ROS). Thus, the bacterial cell membrane can be damaged by leading to ROS production by the free radicals that absorb the Zn^{2+} ions. In this circumstance, the actions of oxidative stress and cell death inhibit the enzyme. The authors recommend that ZnO nanoparticles can be prominently utilized in biomedical applications that will solve the next generation's public health issues. Fig. (**18**) shows that oxidative stress influence their antibacterial activity.

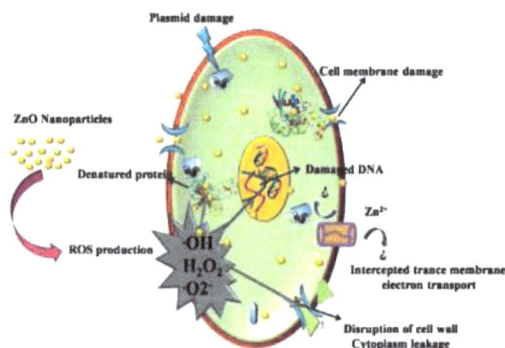

Fig. (18). Schematic diagram of the oxidative stress influence on their antibacterial activity; adapted from Insoo Kim *et al.*2020 [152].

Photodynamic therapy is a promising alternative for the non-invasive treatment of cancer. Photo sanitizers for cancer cells are within the light of a suitable wavelength, and dosage can generate ROS to induce cell death. Fig. (**19**) shows the possible mechanism of ROS production by ZnO nanorods.

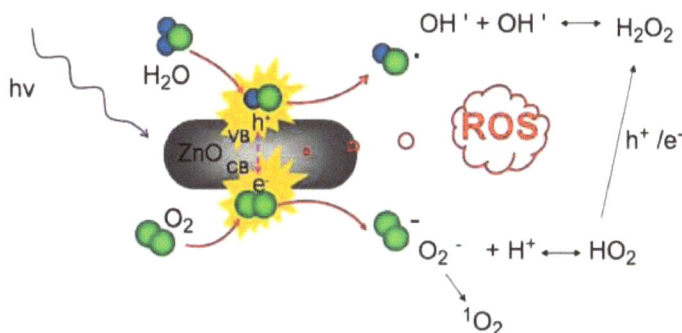

Fig. (19). Possible mechanism of ROS production by ZnO nanorods; adapted from Y. Zhang *et al.* 2013 [153].

Other researchers [154 - 157] developed the ZnO nanoparticles by using different synthesis processes to confirm a promising application for antibacterial activity.

Antimicrobial

The substitute strategy to manage the advance of detrimental microorganisms in the human body can be introduced by the ZnO nanoparticle as an antimicrobial agent. It has been investigated in microscale to nanoscale form as an associated medicament agent. The experimental results reveal that it has been indicated as an antimicrobial activity when nano-dimensional particles are involved within metal oxide materials. Although the actual structure of antimicrobial activity has not been consequently explained, it's been promptly comprehended that the major cause of cell swelling depends on the surface and dimension of specimens, Zn ions, and nanoparticle acquisition position. S. Akbar *et al.* [158] have discussed the antimicrobial potential of ZnO nanoparticles by synthesizing them from the plants' mediation. Furthermore, the authors recommend that the antimicrobial mechanism behavior of ZnO nanoparticles be studied using chemical luminescence, oxygen electrode analysis, and electrical energy.M. Batool *et al.* [159] have synthesized and stabilized ZnO nanoparticles by using an eco-friendly method based on plant extracts. This process is utilized as an antimicrobial agent for mouse skin wound healing treatment. Fig. (**20**) shows the schematic diagram of ZnO nanoparticles as antimicrobial activity.

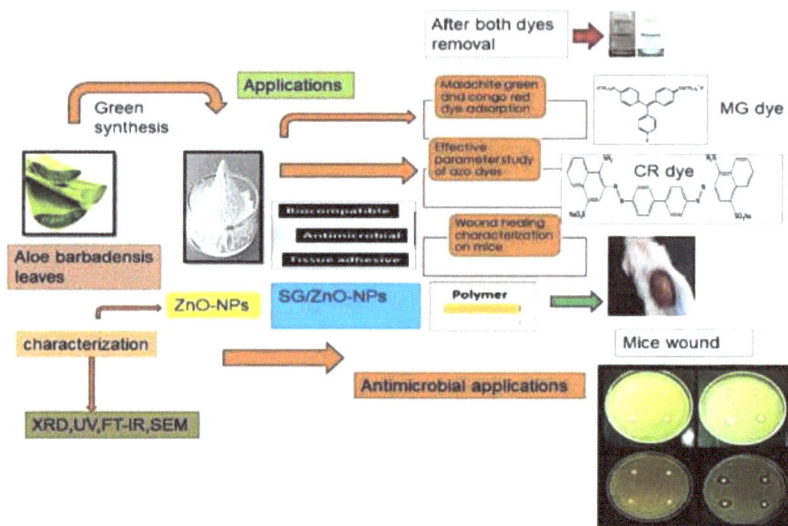

Fig. (20). Schematic diagram of different steps of ZnO nanoparticles as antimicrobial activity; adapted from M. Batool *et al.* 2021 [159].

The authors demonstrated that synthesized nanoparticles have strong antimicrobial capabilities against different bacterial strains of *Bacillus subtilis, Klebsiella pneumonia, Bacillus licheniformis, and Escherichia coli.* It also has antimicrobial activity against the fungi strains of *Aspergillus niger* and *Candida albicans.* Other researchers [160 - 162] developed the ZnO nanoparticles by using different synthesis processes to confirm a promising application for antimicrobial activity.

Anticancer

QDs are encapsulated in bio-degradable polymers like Chitosan that are appropriate for the targeted delivery of anticancer drugs. Encapsulating within Chitosan prevents drug release before attaching to the targeted cell membrane. The effect of this size on emission spectra made it beneficial for simultaneous multi-colour imaging of different parts of the body. Inorganic and organic materials can generally be used as medicinal drug agents. At high temperatures, inorganic drug agents are more suitable than organic medicinal drug agents. As a result, organic medicinal drug agents are more widely used in low-temperature conditions. However, the phenomenon of medicinal drug activity has not been accurately found. Recently, ZnO nanoparticles have been demonstrated to be a significant medicinal drug agent in a variety of forms for therapeutic use. The nano-dimension particle sizes between 5 and 100 nm have enormous exterior spaces that can be used to adjoin for diagnoses such as optical, radioisotopic, and therapies such as anticancer agents. Recent developments can be made in cancer treatment to integrate nano-devices for early cancer detection. We recommend that this advancement can increase the possibilities for generalized cancer treatment in biomedical applications. E.A. Elsayed *et al.* [163] have advanced the anticancer potential of ZnO nanoparticles against liver and breast cancer. The authors observed the experimental results that cancer cells might be affected by ZnO nanoparticles depending on the applied concentration. Authors declare that in cancer cells, serious structural change can be the obstacle to nanoparticle exposure. Finally, ZnO nanoparticles can lead to cancer cell death. The authors conclude that ZnO nanoparticles can be effectively utilized in different cancer cells to inhibit their development and generation.

Over the last several decades, gene therapy has attracted considerable interest from researchers. The advancement of safe gene vectors can protect deoxyribonucleic acid (DNA). The safe gene can degrade and enable the DNA with high efficiency, which is one of the major challenges of gene therapy. Recently, researchers have been investigating a variety of nanomaterials for gene delivery and therapy uses. ZnO nanomaterials have shown promise for these purposes. T.A. Singh *et al.* [164] discussed details about the anticancer activity of ZnO nanoparticles as well as health risks. Other literature reports [165 - 167] developed the ZnO nanoparticles by using different synthesis processes to confirm a promising application for anticancer activity. Fig. (**21**) shows the possible anticancer mechanisms of ZnO nanoparticles.

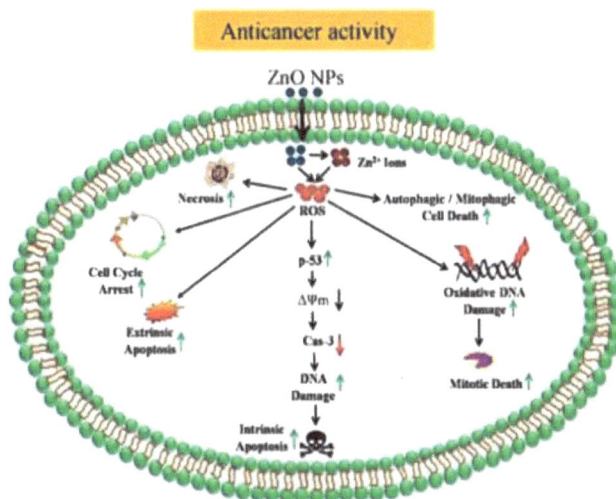

Fig. (21). Anticancer mechanisms of ZnO nanoparticles; adapted from Th. Abhishek Singh *et al.* 2020 [164].

Drug Delivery

ZnO nanoparticles have nanostructure platforms for drug delivery purposes. The purpose is supported by their large surface area, versatile surface chemistry, and phototoxic effect. The main concern in cancer treatment is minimizing the side effects of chemotherapy through localized drug delivery. For target drug delivery, there are many ways, such as pH and temperature control and optical and ultrasonic wave utilization through a magnetic field. Within intrinsic blue fluorescence, QDs of ZnO nanoparticles are decorated with special Chitosan through electrostatic interaction. It would be loaded with doxorubicin (DOX) for the chemotherapy drug at high efficiency. DOX will be entrapped through interaction with the exterior of ZnO QDs *via* H_2 bonding. The layer of external Chitosan enhanced the aqueous solution stability of ZnO QDs. However, DOX can be released quickly at the normal physiological pH value, and that needs to be improved for further studies. Different nanoparticles as the QDs have been used for target drug delivery are shown in Fig. (**22**), which represents a few targets for localized drug delivery.

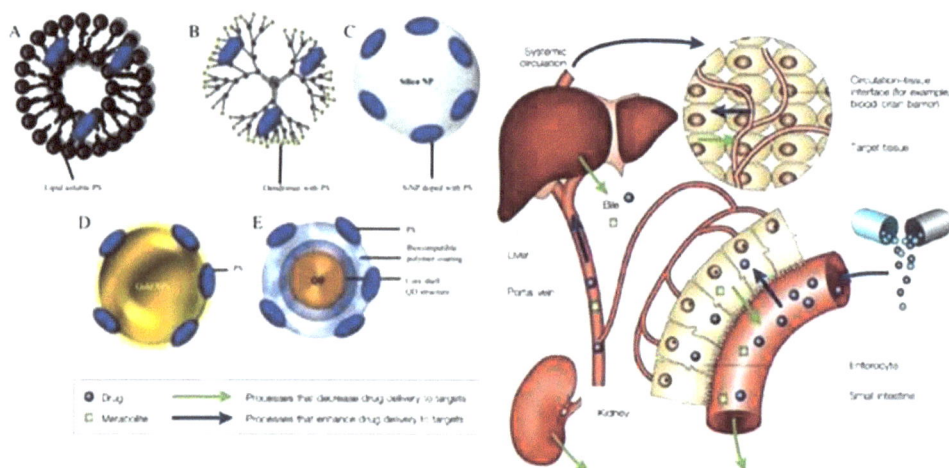

Fig. (22). Different nanoparticles as quantum dots for localized targeted drug delivery process; adapted from A. Nemati *et al.* 2020 [168].

The application of nanotechnology in cancer treatment is associated with medical science, technology, and drug delivery within deep utilizations for molecular imaging, recognition, and target for the specimen. The nanoscale particle size, such as metal oxide nanocrystals and QDs, has some prominent structural, optical, and electrical characteristics that can play a dominating role in biomedical applications. Drug delivery has emerged as an important appliance for nanomaterial advancement in the treatment of various diseases. B.A.

Fahimmunisha *et al.* [169] have synthesized eco-friendly ZnO nanoparticles for drug delivery. The authors provided an interesting explanation for the protection of bacterial infection. They also performed extensive experiments to understand the antibacterial activity of ZnO nanoparticles. The advancement of nano-medicine from plant sources is a potential agent for biomedical utilization. The authors conclude that the potent antibacterial properties of ZnO nanoparticles can be utilized as an effective bactericidal specimen for biomedical applications. From lots of studies, we can conclude that ZnO nanoparticles can be used as a suitable vehicle for drug delivery and gene silencing. Other literature reports [170 - 172] discussed the ZnO nanoparticles as a smart drug delivery system.

CONCLUSION

To witness the tremendous advancements in materials science over the last several decades, ZnO nanoparticles have had a revolutionary influence on biomedical applications. Due to their nanoscale dimension, nanoparticles of ZnO have shown various characteristics compared to the bulk form. The human cell's dimension is less than a few hundred nanometers and can exhibit the characteristics of being distinct from molecules and bulk solids. The surface and inside of cells offer unprecedented interactions with bio-molecules. It acts as a resourceful weapon against several drug-resistant micro-organisms with a talent for the substitution of antibiotics. Moreover, ZnO biosensors and bioimaging are essential for next-generation patient management. It offers synergistic advantages in providing complementary information. Using these nanoparticles has a significant role in the advancement of biomedical applications; the use of plants for synthesizing these particles is easily scaled-up, eco-friendly, and cost-effective. In particular, this is especially suitable for developing ZnO nanoparticles that can be free from contaminants in biomedical utilization. Synthesis based on biological processes can easily control particle dimensions and structural morphology. In biomedical applications, these particles can be used as agents of antibacterial and antimicrobial for targeted drug delivery and clinical diagnostics in cancer cell treatment. The chapter shows that the green synthesis of ZnO nanoparticles is safer and more environmentally friendly than other processes. It acts as a reducing agent to develop nanoparticles for controlling dimension and structure. This chapter recommends that ZnO nanoparticles from plants can be a cost-effective synthesis process for biomedical applications. We expect research on ZnO nanomaterials in biomedical applications will continue to flourish over the next generation. The productivity increase of plants in nanoparticle synthesis processes by their genetic modification with enhanced metal accumulation capabilities and metal tolerance could be the future approach. To conclude, future aspects of this study have been foreseen that will be helpful for enhancing further research in the innovative methodological characterization and clinical correlations.

REFERENCES

[1]   Camarda, P.; Vaccaro, L.; Sciortino, A.; Messina, F.; Buscarino, G.; Agnello, S.; Gelardi, F.M.; Popescu, R.; Schneider, R.; Gerthsen, D.; Cannas, M. Synthesis multi-color lLuminescent ZnO nanoparticles by ultra-short pulsed laser ablation. *Appl. Surf. Sci.,* **2020**, *506*, 144954.
[http://dx.doi.org/10.1016/j.apsusc.2019.144954]

[2]   Hu, P.; Han, N.; Zhang, D.; Ho, J.C.; Chen, Y. Highly formaldehyde-sensitive, transition-metal doped ZnO nanorods prepared by plasma-enhanced chemical vapor deposition. *Sens. Actuators B Chem.,* **2012**, *169*, 74-80.
[http://dx.doi.org/10.1016/j.snb.2012.03.035]

[3]   Zhang, C.; Han, T.; Wang, W.; Zhang, J. Dried plum-like ZnO assemblies consisted of ZnO nanoparticles synthesized by ultrasonic spray pyrolysis. *Int. J. Modern Phys. B,* **2020**. 34(01n03), 2040005.
[http://dx.doi.org/10.1142/S0217979220400056]

[4]   Sharma, S.; Chawla, S. Enhanced UV emission in ZnO/ZnS core shell nanoparticles prepared by epitaxial growth in solution. *Electron. Mater. Lett.,* **2013**, *9*(3), 267-271.
[http://dx.doi.org/10.1007/s13391-012-2222-8]

[5]   Hassan, N.K.; Hashim, M.R.; Bououdina, M. One-dimensional ZnO nanostructure growth prepared by thermal evaporation on different substrates: ultraviolet emission as a function of size and dimensionality. *Ceram. Int.,* **2013**, *39*(7), 7439-7444.
[http://dx.doi.org/10.1016/j.ceramint.2013.02.088]

[6]   Chandrasekaran, P.; Viruthagiri, G.; Srinivasan, N. The effect of various capping agents on the surface modifications of sol–gel synthesised ZnO nanoparticles. *J. Alloys Compd.,* **2012**, *540*, 89-93.
[http://dx.doi.org/10.1016/j.jallcom.2012.06.032]

[7]   Nguyen, D.T.; Kim, K.S. Structural evolution of highly porous/hollow ZnO nanoparticles in sonochemical process. *Chem. Eng. J.,* **2015**, *276*, 11-19.
[http://dx.doi.org/10.1016/j.cej.2015.04.053]

[8]   Skompska, M.; Zarębska, K. Electrodeposition of ZnO nanorod arrays on transparent conducting substrates–A review. *Electrochim. Acta,* **2014**, *127*, 467-488.
[http://dx.doi.org/10.1016/j.electacta.2014.02.049]

[9]   Mamur, H.; Dilmac, O.F.; Korucu, H.; Bhuiyan, M.R.A. Cost-effective chemical solution synthesis of bismuth telluride nanostructure for thermoelectric applications. *Micro & Nano Lett.,* **2018**, *13*(8), 1117-1120.
[http://dx.doi.org/10.1049/mnl.2018.0116]

[10]   Rai, P.; Yu, Y.T. Citrate-assisted hydrothermal synthesis of single crystalline ZnO nanoparticles for gas sensor application. *Sens. Actuators B Chem.,* **2012**, *173*, 58-65.
[http://dx.doi.org/10.1016/j.snb.2012.05.068]

[11]   Modi, S.; Fulekar, M.H. Green synthesis of Zinc Oxide nanoparticles using garlic skin extract and its characterization. *J. Nanostruc.,* **2020**, *10*(1), 20-27.
[http://dx.doi.org/10.22052/JNS.2020.01.003]

[12]   Davar, F.; Majedi, A.; Mirzaei, A. Green synthesis of ZnO nanoparticles and its application in the degradation of some dyes. *J. Am. Ceram. Soc.,* **2015**, *98*(6), 1739-1746.
[http://dx.doi.org/10.1111/jace.13467]

[13]   Bopape, D.A.; Motaung, D.E.; Hintsho-Mbita, N.C. Green synthesis of ZnO: Effect of plant concentration on the morphology, optical properties and photodegradation of dyes and antibiotics in wastewater. *Optik (Stuttg.),* **2022**, *251*, 168459.
[http://dx.doi.org/10.1016/j.ijleo.2021.168459]

[14]   Hassan, S.S.; El Azab, W.I.; Ali, H.R.; Mansour, M.S. Green synthesis and characterization of ZnO nanoparticles for photocatalytic degradation of anthracene. *Adv. Nat. Sci.: Nanosci. Nanotech.,* **2015**,

6(4), 045012.
[http://dx.doi.org/10.1088/2043-6262/6/4/045012]

[15] Buazar, F.; Bavi, M.; Kroushawi, F.; Halvani, M.; Khaledi-Nasab, A.; Hossieni, S.A. Potato extract as reducing agent and stabiliser in a facile green one-step synthesis of ZnO nanoparticles. *J. Exp. Nanosci.,* **2016**, *11*(3), 175-184.
[http://dx.doi.org/10.1080/17458080.2015.1039610]

[16] Mirzaei, H.; Darroudi, M. Zinc Oxide nanoparticles: biological synthesis and biomedical applications. *Ceram. Int.,* **2017**, *43*(1), 907-914.
[http://dx.doi.org/10.1016/j.ceramint.2016.10.051]

[17] Bhuiyan, M.R.A.; Quadir, L.; Hasan, S.M. Growth of AgGaSe$_2$ thin films by a stacked elemental layer deposition technique. *Nuclear Sci. Appl.,* **2005**, *14*(1), 73-77.

[18] Bhuiyan, M.R.A.; Hasan, S.F. Optical properties of polycrystalline Ag$_x$Ga$_{2-x}$Se$_2$ (0.4≤ x≤ 1.6) thin films. *Sol. Energy Mater. Sol. Cells,* **2007**, *91*(2-3), 148-152.
[http://dx.doi.org/10.1016/j.solmat.2006.07.010]

[19] Bhuiyan, M.R.A.; Rahman, M.K.; Hasan, S.F. Valence-band characterization of AgGaSe$_2$ thin films. *J. Phys. D Appl. Phys.,* **2008**, *41*(23), 235108.
[http://dx.doi.org/10.1088/0022-3727/41/23/235108]

[20] Bhuiyan, M.R.A.; Saha, D.K.; Hasan, S.F. Effects of temperature on the structural and optical preperties of AgGaSe$_2$ thin films. *J. Bangladesh Acad. Sci.,* **2009**, *33*(2), 179-188.
[http://dx.doi.org/10.3329/jbas.v33i2.4101]

[21] Bhuiyan, M.R.A.; Saha, D.K.; Hasan, S.M. Structural and electrical properties of polycrystalline Ag$_x$Ga$_{2-x}$Se$_2$ (0.4≤ x≤ 1.6) thin films. *Indian J. Pure Appl. Phy.,* **2009**, *47*(11), 787-792.

[22] Bhuiyan, M.R.A.; Mamur, H. Review of the bismuth telluride (Bi$_2$Te$_3$) nanoparticle: Growth and characterization. *Int. J. Ener. Appl. Tech.,* **2016**, *3*(2), 27-31.

[23] Mamur, H.; Bhuiyan, M.R.A. Development of bismuth telluride nanostructure pellet for thermoelectric applications. *Hittite J. Sci. Engg.,* **2018**, *5*(4), 293-299.
[http://dx.doi.org/10.17350/HJSE19030000106]

[24] Mamur, H.; Bhuiyan, M.R.A. Bismuth telluride (Bi$_2$Te$_3$) nanostructure for thermoelectric applications. *Int. Scient. Voc. Stud. J.,* **2019**, *3*(1), 1-7.

[25] Mamur, H.; Bhuiyan, M.R.A. Characterization of Bi$_2$Te$_3$ nanostructure by using a cost-effective chemical solution route. *Iranian J. Chem. Chem. Engg.,* **2020**, *39*(3), 23-33.

[26] Bhuiyan, M.R.A.; Mamur, H.; Dilmaç, Ö.F. A Review on performance evaluation of Bi$_2$Te$_3$-based and some other thermoelectric nanostructured materials. *Curr. Nanosci.,* **2021**, *17*(3), 423-446.
[http://dx.doi.org/10.2174/1573413716999200820144753]

[27] Severo, J.S.; Morais, J.B.S.; Beserra, J.B.; Dos Santos, L.R.; de Sousa Melo, S.R.; de Sousa, G.S.; de Matos Neto, E.M.; Henriques, G.S.; do Nascimento Marreiro, D. Role of Zinc in zinc-α$_2$-glycoprotein metabolism in obesity: A review of literature. *Biol. Trace Elem. Res.,* **2020**, *193*(1), 81-88.
[http://dx.doi.org/10.1007/s12011-019-01702-w] [PMID: 30929134]

[28] Qiao, Y.; Fan, G.; Guo, J.; Gao, S.; Zhao, R.; Yang, X. Effects of adipokine zinc-α$_2$-glycoprotein on adipose tissue metabolism after dexamethasone treatment. *Appl. Physiol. Nutr. Metab.,* **2019**, *44*(1), 83-89.
[http://dx.doi.org/10.1139/apnm-2018-0165] [PMID: 29972738]

[29] Baltaci, A.K.; Yuce, K.; Mogulkoc, R. Zinc metabolism and metallothioneins. *Biolog. Trace Elem. Resear.,* **2018**, *183*(1), 22-31.
[http://dx.doi.org/10.1007/s12011-017-1119-7] [PMID: 28812260]

[30] Grüngreiff, K.; Reinhold, D.; Wedemeyer, H. The role of zinc in liver cirrhosis. *Ann. Hepatol.,* **2016**, *15*(1), 7-16.

[http://dx.doi.org/10.5604/16652681.1184191] [PMID: 26626635]

[31]   Bhuiyan, M.R.A.; Mamur, H. A brief review of the synthesis of ZnO nanoparticles for biomedical applications. *Iranian J. Mater. Sci. Engg.,* **2021**, *18*(3), 1-27.

[32]   Jiang, J.; Pi, J.; Cai, J. The advancing of Zinc Oxide nanoparticles for biomedical applications. *Bioinorg. Chem. Appl.,* **2018**, *2018*, 1062562.
[http://dx.doi.org/10.1155/2018/1062562] [PMID: 30073019]

[33]   Mishra, P.K.; Mishra, H.; Ekielski, A.; Talegaonkar, S.; Vaidya, B. Zinc oxide nanoparticles: a promising nanomaterial for biomedical applications. *Drug Discov. Today,* **2017**, *22*(12), 1825-1834.
[http://dx.doi.org/10.1016/j.drudis.2017.08.006] [PMID: 28847758]

[34]   Zhang, Z.Y.; Xiong, H.M. Photoluminescent ZnO nanoparticles and their biological applications. *Materials (Basel),* **2015**, *8*(6), 3101-3127.
[http://dx.doi.org/10.3390/ma8063101]

[35]   Zhu, P.; Weng, Z.; Li, X.; Liu, X.; Wu, S.; Yeung, K.W.K.; Wang, X.; Cui, Z.; Yang, X.; Chu, P.K. Biomedical applications of functionalized ZnO nanomaterials: from biosensors to bioimaging. *Adv. Mater. Interfaces,* **2016**, *3*(1), 1500494.
[http://dx.doi.org/10.1002/admi.201500494]

[36]   Phukan, P.; Agarwal, S.; Deori, K.; Sarma, D. Zinc Oxide nanoparticles catalysed one-pot three-component reaction: A facile synthesis of 4-Aryl-NH-1, 2, 3-Triazoles. *Catal. Lett.,* **2020**, *150*(8), 2208-2219.
[http://dx.doi.org/10.1007/s10562-020-03143-w]

[37]   Akter, M.; Khan, M.N.I.; Mamur, H.; Bhuiyan, M.R.A. Synthesis and characterisation of CdSe QDs by using a chemical solution route. *Micro & Nano Lett.,* **2020**, *15*(5), 287-290.
[http://dx.doi.org/10.1049/mnl.2019.0200]

[38]   Mojdehi, M.S.; Yunus, W.M.M.; Fhan, K.S.; Talib, Z.A.; Tamchek, N. Nonlinear optical characterization of phosphate glasses based on ZnO using the Z-scan technique. *Chin. Phys. B,* **2013**, *22*(11), 117802.
[http://dx.doi.org/10.1088/1674-1056/22/11/117802]

[39]   Awwad, A.M.; Amer, M.W.; Salem, N.M.; Abdeen, A.O. Green synthesis of Zinc Oxide nanoparticles (ZnO-NPs) using *ailanthus altissima* fruit extracts and antibacterial activity. *Chem. Int.,* **2020**, *6*(3), 151-159.

[40]   Souza, R.C.D.; Haberbeck, L.U.; Riella, H.G.; Ribeiro, D.H.; Carciofi, B.A. Antibacterial activity of Zinc Oxide nanoparticles synthesized by solochemical process. *Braz. J. Chem. Eng.,* **2019**, *36*(2), 885-893.
[http://dx.doi.org/10.1590/0104-6632.20190362s20180027]

[41]   Kadiyala, U.; Turali-Emre, E.S.; Bahng, J.H.; Kotov, N.A.; VanEpps, J.S. Unexpected insights into antibacterial activity of zinc oxide nanoparticles against methicillin resistant Staphylococcus aureus (MRSA). *Nanoscale,* **2018**, *10*(10), 4927-4939.
[http://dx.doi.org/10.1039/C7NR08499D] [PMID: 29480295]

[42]   Balraj, B.; Senthilkumar, N.; Siva, C.; Krithikadevi, R.; Julie, A.; Potheher, I.V.; Arulmozhi, M. Synthesis and characterization of Zinc Oxide nanoparticles using marine *streptomyces sp.* with its investigations on anticancer and antibacterial activity. *Res. Chem. Intermed.,* **2017**, *43*(4), 2367-2376.
[http://dx.doi.org/10.1007/s11164-016-2766-6]

[43]   Lingaraju, K.; Naika, H.R.; Manjunath, K.; Basavaraj, R.B.; Nagabhushana, H.; Nagaraju, G.; Suresh, D. Biogenic synthesis of Zinc Oxide nanoparticles using ruta graveolens (L.) and their antibacterial and antioxidant activities. *Appl. Nanosci.,* **2016**, *6*(5), 703-710.
[http://dx.doi.org/10.1007/s13204-015-0487-6]

[44]   Zhang, T.; Du, E.; Liu, Y.; Cheng, J.; Zhang, Z.; Xu, Y.; Qi, S.; Chen, Y. Anticancer effects of Zinc Oxide nanoparticles through altering the methylation status of histone on bladder cancer cells. *Int. J.*

Nanomedicine, **2020**, *15*, 1457-1468.
[http://dx.doi.org/10.2147/IJN.S228839] [PMID: 32184598]

[45]   Ruenraroengsak, P.; Kiryushko, D.; Theodorou, I.G.; Klosowski, M.M.; Taylor, E.R.; Niriella, T.; Palmieri, C.; Yagüe, E.; Ryan, M.P.; Coombes, R.C.; Xie, F.; Porter, A.E. Frizzled-7-targeted delivery of zinc oxide nanoparticles to drug-resistant breast cancer cells. *Nanoscale,* **2019**, *11*(27), 12858-12870.
[http://dx.doi.org/10.1039/C9NR01277J] [PMID: 31157349]

[46]   Ancona, A.; Dumontel, B.; Garino, N.; Demarco, B.; Chatzitheodoridou, D.; Fazzini, W.; Engelke, H.; Cauda, V. Lipid-coated Zinc Oxide nanoparticles as innovative ros-generators for photodynamic therapy in cancer cells. *Nanomaterials (Basel),* **2018**, *8*(3), 143.
[http://dx.doi.org/10.3390/nano8030143] [PMID: 29498676]

[47]   Bai, D.P.; Zhang, X.F.; Zhang, G.L.; Huang, Y.F.; Gurunathan, S. Zinc oxide nanoparticles induce apoptosis and autophagy in human ovarian cancer cells. *Int. J. Nanomedicine,* **2017**, *12*, 6521-6535.
[http://dx.doi.org/10.2147/IJN.S140071] [PMID: 28919752]

[48]   Paino, I.M.; J Gonçalves, F.; Souza, F.L.; Zucolotto, V. Zinc Oxide flower-like nanostructures that exhibit enhanced toxicology effects in cancer cells. *ACS Appl. Mater. Interfaces,* **2016**, *8*(48), 32699-32705.
[http://dx.doi.org/10.1021/acsami.6b11950] [PMID: 27934178]

[49]   Supraja, P.; Singh, V.; Vanjari, S.R.K.; Govind Singh, S. Electrospun CNT embedded ZnO nanofiber based biosensor for electrochemical detection of Atrazine: a step closure to single molecule detection. *Microsyst. Nanoeng.,* **2020**, *6*(1), 3.
[http://dx.doi.org/10.1038/s41378-019-0115-9] [PMID: 34567618]

[50]   Cao, L.; Kiely, J.; Piano, M.; Luxton, R. A copper oxide/zinc oxide composite nano-surface for use in a biosensor. *Materials (Basel),* **2019**, *12*(7), 1126.
[http://dx.doi.org/10.3390/ma12071126] [PMID: 30959878]

[51]   George, J.M.; Antony, A.; Mathew, B. Metal oxide nanoparticles in electrochemical sensing and biosensing: a review. *Mikrochim. Acta,* **2018**, *185*(7), 358.
[http://dx.doi.org/10.1007/s00604-018-2894-3] [PMID: 29974265]

[52]   Cook, A.L.; Carson, C.S.; Marvinney, C.E.; Giorgio, T.D.; Mu, R.R. Sensing trace levels of molecular species in solution *via* Zinc Oxide nanoprobe raman spectroscopy. *J. Raman Spectrosc.,* **2017**, *48*(8), 1116-1121.
[http://dx.doi.org/10.1002/jrs.5180]

[53]   Salih, E.; Mekawy, M.; Hassan, R.Y.; El-Sherbiny, I.M. Synthesis, characterization and electrochemical-sensor applications of Zinc oxide/Graphene oxide nanocomposite. *J. Nanostructure Chem.,* **2016**, *6*(2), 137-144.
[http://dx.doi.org/10.1007/s40097-016-0188-z]

[54]   Adegoke, K.A.; Iqbal, M.; Louis, H.; Jan, S.U.; Mateen, A.; Bello, O.S. Photocatalytic conversion of CO_2 using ZnO semiconductor by hydrothermal method. *Pak. J. Anal. Environ. Chem.,* **2018**, *19*(1), 1-27.
[http://dx.doi.org/10.21743/pjaec/2018.06.01]

[55]   Vaseem, M.; Umar, A.; Hahn, Y.B. ZnO nanoparticles: growth, properties, and applications. *Metal Oxide Nanostruc. Appl,* **2010**, *5*(1), 1-36.

[56]   Basnet, P.; Inakhunbi Chanu, T.; Samanta, D.; Chatterjee, S. A review on bio-synthesized zinc oxide nanoparticles using plant extracts as reductants and stabilizing agents. *J. Photochem. Photobiol. B,* **2018**, *183*, 201-221.
[http://dx.doi.org/10.1016/j.jphotobiol.2018.04.036] [PMID: 29727834]

[57]   Siddiqi, K.S.; Ur Rahman, A.; Tajuddin, ; Husen, A. Tajuddin and Husen, A. Properties of Zinc oxide nanoparticles and their activity against microbes. *Nanoscale Res. Lett.,* **2018**, *13*(1), 141.
[http://dx.doi.org/10.1186/s11671-018-2532-3] [PMID: 29740719]

[58] Prommalikit, C.; Mekprasart, W.; Pecharapa, W. Effect of milling speed and time on ultra-fine ZnO powder by high energy ball milling technique. *J. Phys. Conf. Ser.,* **2019**, *1259*(1), 012023. [http://dx.doi.org/10.1088/1742-6596/1259/1/012023]

[59] Dhand, C.; Dwivedi, N.; Loh, X.J.; Ying, A.N.J.; Verma, N.K.; Beuerman, R.W.; Lakshminarayanan, R.; Ramakrishna, S. Methods and strategies for the synthesis of diverse nanoparticles and their applications: a comprehensive overview. *RSC Advances,* **2015**, *5*(127), 10500. [http://dx.doi.org/10.1039/C5RA19388E]

[60] Amirkhanlou, S.; Ketabchi, M.; Parvin, N. Nanocrystalline/nanoparticle ZnO synthesized by high energy ball milling process. *Mater. Lett.,* **2012**, *86*, 122-124. [http://dx.doi.org/10.1016/j.matlet.2012.07.041]

[61] Salah, N.; Habib, S.S.; Khan, Z.H.; Memic, A.; Azam, A.; Alarfaj, E.; Zahed, N.; Al-Hamedi, S. High-energy ball milling technique for ZnO nanoparticles as antibacterial material. *Int. J. Nanomedicine,* **2011**, *6*, 863-869. [http://dx.doi.org/10.2147/IJN.S18267] [PMID: 21720499]

[62] Singh, J.; Sharma, S.; Soni, S.; Sharma, S.; Singh, R.C. Influence of different milling media on structural, morphological and optical properties of the ZnO nanoparticles synthesized by ball milling process. *Mater. Sci. Semicond. Process.,* **2019**, *98*, 29-38. [http://dx.doi.org/10.1016/j.mssp.2019.03.026]

[63] Chen, D.; Ai, S.; Liang, Z.; Wei, F. Preparation and photocatalytic properties of Zinc Oxide nanoparticles by microwave-assisted ball milling. *Ceram. Int.,* **2016**, *42*(2), 3692-3696. [http://dx.doi.org/10.1016/j.ceramint.2015.10.123]

[64] Balamurugan, S.; Joy, J.; Godwin, M.A.; Selvamani, S.; Raja, T.S.G. Godwin, M.A.; Selvamani, S.; Raja, T.G. ZnO nanoparticles obtained by ball milling technique: structural, micro-structure, optical and photo-catalytic properties. *AIP Conf. Proc.,* **2016**, *1731*(1), 050121. [http://dx.doi.org/10.1063/1.4947775]

[65] Tigli, O.; Juhala, J. ZnO nanowire growth by physical vapour deposition. *IEEE Int. Conf. Nanotech,* **2011**, pp. 608-611.

[66] Lupan, O.; Emelchenko, G.A.; Ursaki, V.V.; Chai, G.; Redkin, A.N.; Gruzintsev, A.N.; Tiginyanu, I.M.; Chow, L.; Ono, L.K.; Cuenya, B.R.; Heinrich, H. Synthesis and characterization of ZnO nanowires for nanosensor applications. *Mater. Res. Bull.,* **2010**, *45*(8), 1026-1032. [http://dx.doi.org/10.1016/j.materresbull.2010.03.027]

[67] Jimenez-Cadena, G.; Comini, E.; Ferroni, M.; Vomiero, A.; Sberveglieri, G. Synthesis of different ZnO nanostructures by modified PVD process and potential use for dye-sensitized solar cells. *Mater. Chem. Phys.,* **2010**, *124*(1), 694-698. [http://dx.doi.org/10.1016/j.matchemphys.2010.07.035]

[68] Lyu, S.C.; Zhang, Y.; Lee, C.J.; Ruh, H.; Lee, H.J. Low-temperature growth of ZnO nanowire array by a simple physical vapor-deposition method. *Chem. Mater.,* **2003**, *15*(17), 3294-3299. [http://dx.doi.org/10.1021/cm020465j]

[69] Kong, Y.C.; Yu, D.P.; Zhang, B.; Fang, W.; Feng, S.Q. Ultraviolet-emitting ZnO nanowires synthesized by a physical vapor deposition approach. *Appl. Phys. Lett.,* **2001**, *78*(4), 407-409. [http://dx.doi.org/10.1063/1.1342050]

[70] Xiang, B.; Wang, P.; Zhang, X.; Dayeh, S.A.; Aplin, D.P.; Soci, C.; Yu, D.; Wang, D. Rational synthesis of p-type zinc oxide nanowire arrays using simple chemical vapor deposition. *Nano Lett.,* **2007**, *7*(2), 323-328. [http://dx.doi.org/10.1021/nl062410c] [PMID: 17297995]

[71] Zhang, B.P.; Binh, N.T.; Wakatsuki, K.; Segawa, Y.; Yamada, Y.; Usami, N.; Kawasaki, M.; Koinuma, H. Pressure-dependent ZnO nanocrsytal growth in a chemical vapor deposition process. *J. Phys. Chem. B,* **2004**, *108*(30), 10899-10902.

[http://dx.doi.org/10.1021/jp048602i]

[72]   Al-Dahash, G.; Mubder Khilkala, W.; Abd Alwahid, S.N. Preparation and characterization of ZnO nanoparticles by laser ablation in NaOH aqueous solution. *Iranian J. Chem. Chem. Engg.,* **2018**, *37*(1), 11-16.

[73]   Chen, W.; Yao, C.; Gan, J.; Jiang, K.; Hu, Z.; Lin, J.; Xu, N.; Sun, J.; Wu, J. ZnO colloids and ZnO nanoparticles synthesized by pulsed laser ablation of zinc powders in water. *Mater. Sci. Semicond. Process.,* **2020**, *109*, 104918.
[http://dx.doi.org/10.1016/j.mssp.2020.104918]

[74]   Khudiar, S.S.; Nayef, U.M.; Mutlak, F.A.H. Preparation and characterization of ZnO nanoparticles *via* laser ablation for sensing NO$_2$ gas. *Optik (Stuttg.),* **2021**, *246*, 167762.
[http://dx.doi.org/10.1016/j.ijleo.2021.167762]

[75]   Zhang, W.; Guo, L.P.; Deng, Q.W.; Li, M.X. ZnO nano-sized particles preparation by laser ablation in liquids. *Optik (Stuttg.),* **2019**, *196*, 163195.
[http://dx.doi.org/10.1016/j.ijleo.2019.163195]

[76]   Yudasari, N.; Wiguna, P.A.; Suliyanti, M.M.; Imawan, C. Antibacterial activity of ZnO nanoparticles fabricated using laser ablation in solution technique. *J. Phys. Conf. Ser.,* **2019**, *1245*(1), 012035.
[http://dx.doi.org/10.1088/1742-6596/1245/1/012035]

[77]   Islam, S.; Bakhtiar, H.; Abbas, K.N.; Riaz, S.; Naseem, S.; Johari, A.R.B. Grown of highly porous ZnO-nanoparticles by pulsed laser ablation in liquid technique for sensing applications. *J. Australian Cera. Soci.,* **2019**, *55*(3), 765-771.
[http://dx.doi.org/10.1007/s41779-018-0288-y]

[78]   Al-Nassar, S.I.; Hussein, F.I.; M, A.K. The effect of laser pulse energy on ZnO nanoparticles formation by liquid phase pulsed laser ablation. *J. Mater. Res. Technol.,* **2019**, *8*(5), 4026-4031.
[http://dx.doi.org/10.1016/j.jmrt.2019.07.012]

[79]   Lee, G.J.; Choi, E.H.; Nam, S.H.; Lee, J.S.; Boo, J.H.; Oh, S.D.; Choi, S.H.; Cho, J.H.; Yoon, M.Y. Optical sensing properties of ZnO nanoparticles prepared by spray pyrolysis. *J. Nanosci. Nanotechnol.,* **2019**, *19*(2), 1048-1051.
[http://dx.doi.org/10.1166/jnn.2019.15918] [PMID: 30360198]

[80]   Ozcelik, B.K.; Ergun, C. Synthesis of ZnO nanoparticles by an aerosol process. *Ceram. Int.,* **2014**, *40*(5), 7107-7116.
[http://dx.doi.org/10.1016/j.ceramint.2013.12.044]

[81]   Lee, S.D.; Nam, S.H.; Kim, M.H.; Boo, J.H. Synthesis and photocatalytic property of ZnO nanoparticles prepared by spray-pyrolysis method. *Phys. Procedia,* **2012**, *32*, 320-326.
[http://dx.doi.org/10.1016/j.phpro.2012.03.563]

[82]   Ghaffarian, H.R.; Saiedi, M.; Sayyadnejad, M.A.; Rashidi, A.M. Synthesis of ZnO nanoparticles by spray pyrolysis method. Iran. *J. Chem. Chem. Eng.,* **2011**, *30*(1), 1-6.

[83]   Waghmare, M.; Sonone, P.; Patil, P.; Kadam, V.; Pathan, H.; Ubale, A. Spray pyrolytic deposition of zirconium oxide thin films: influence of concentration on structural and optical properties. *Engineered Sci.,* **2018**, *5*(2), 79-87.
[http://dx.doi.org/10.30919/es8d622]

[84]   Vaghayenegar, M.; Kermanpur, A.; Abbasi, M.H. Bulk synthesis of ZnO nanoparticles by the one-step electromagnetic levitational gas condensation method. *Ceram. Int.,* **2012**, *38*(7), 5871-5878.
[http://dx.doi.org/10.1016/j.ceramint.2012.04.038]

[85]   Vaghayenegar, M.; Kermanpur, A.; Abbasi, M.H.; Yazdabadi, H.G. Effects of process parameters on synthesis of Zn ultra-fine/nanoparticles by electromagnetic levitational gas condensation. *Adv. Powder Technol.,* **2010**, *21*(5), 556-563.
[http://dx.doi.org/10.1016/j.apt.2010.02.009]

[86]   Uhm, Y.R.; Han, B.S.; Lee, M.K.; Hong, S.J.; Rhee, C.K. Synthesis and characterization of

nanoparticles of ZnO by levitational gas condensation. *Mater. Sci. Eng. A,* **2007**, *449*, 813-816.
[http://dx.doi.org/10.1016/j.msea.2006.02.427]

[87]   Uhm, Y.R. Properties and Catalytic Effects of Nanoparticles Synthesized by Levitational Gas Condensation. In: *Novel Nanomaterials*; Kyzas, G., Ed.; IntechOpen, **2018**.

[88]   Goswami, M.; Adhikary, N.C.; Bhattacharjee, S. Effect of annealing temperatures on the structural and optical properties of zinc oxide nanoparticles prepared by chemical precipitation method. *Optik (Stuttg.),* **2018**, *158*, 1006-1015.
[http://dx.doi.org/10.1016/j.ijleo.2017.12.174]

[89]   Mahamuni, P.P.; Patil, P.M.; Dhanavade, M.J.; Badiger, M.V.; Shadija, P.G.; Lokhande, A.C.; Bohara, R.A. Synthesis and characterization of zinc oxide nanoparticles by using polyol chemistry for their antimicrobial and antibiofilm activity. *Biochem. Biophys. Rep.,* **2018**, *17*, 71-80.
[http://dx.doi.org/10.1016/j.bbrep.2018.11.007] [PMID: 30582010]

[90]   Adam, R.E.; Pozina, G.; Willander, M.; Nur, O. Synthesis of ZnO nanoparticles by co-precipitation method for solar driven photodegradation of Congo red dye at different pH. *Photon. Nanostructures,* **2018**, *32*, 11-18.
[http://dx.doi.org/10.1016/j.photonics.2018.08.005]

[91]   Baharudin, K.B.; Abdullah, N.; Derawi, D. Effect of calcination temperature on the physicochemical properties of Zinc Oxide nanoparticles synthesized by co-precipitation. *Mater. Res. Express,* **2018**, *5*(12), 125018.
[http://dx.doi.org/10.1088/2053-1591/aae243]

[92]   Ghorbani, H.R.; Mehr, F.P.; Pazoki, H.; Rahmani, B.M. Synthesis of ZnO nanoparticles by precipitation method. *Orient. J. Chem.,* **2015**, *31*(2), 1219-1221.
[http://dx.doi.org/10.13005/ojc/310281]

[93]   Raoufi, D. Synthesis and microstructural properties of ZnO nanoparticles prepared by precipitation method. *Renew. Energy,* **2013**, *50*, 932-937.
[http://dx.doi.org/10.1016/j.renene.2012.08.076]

[94]   Prakoso, S.P.; Saleh, R. Synthesis and spectroscopic characterization of undoped nanocrytalline ZnO particles prepared by co-precipitation. *Mater. Sci. Appl.,* **2012**, *3*(8), 21794.
[http://dx.doi.org/10.4236/msa.2012.38075]

[95]   Wang, Y.; Zhang, C.; Bi, S.; Luo, G. Preparation of ZnO nanoparticles using the direct precipitation method in a membrane dispersion micro-structured reactor. *Powder Technol.,* **2010**, *202*(1-3), 130-136.
[http://dx.doi.org/10.1016/j.powtec.2010.04.027]

[96]   Wang, H.; Xie, C. Effect of annealing temperature on the microstructures and photocatalytic property of colloidal ZnO nanoparticles. *J. Phys. Chem. Solids,* **2008**, *69*(10), 2440-2444.
[http://dx.doi.org/10.1016/j.jpcs.2008.04.036]

[97]   Alvarado, J.A.; Maldonado, A.; Juarez, H.; Pacio, M. Synthesis of colloidal ZnO nanoparticles and deposit of thin films by spin coating technique. *J. Nanomater.,* **2013**, *2013*, 903191.
[http://dx.doi.org/10.1155/2013/903191]

[98]   Nagar, A.; Kumar, A.; Parveen, S.; Kumar, A.; Dhasmana, H.; Husain, S.; Verma, A.; Jain, V.K. Zinc oxide nanoflowers synthesized by sol-gel technique for field emission displays (FEDs). *Mater. Today Proc.,* **2020**, *32*, 402-406.
[http://dx.doi.org/10.1016/j.matpr.2020.02.087]

[99]   Soleimani, H.; Baig, M.K.; Yahya, N.; Khodapanah, L.; Sabet, M.; Demiral, B.M.; Burda, M. Synthesis of ZnO nanoparticles for oil–water interfacial tension reduction in enhanced oil recovery. *Appl. Phys., A Mater. Sci. Process.,* **2018**, *124*(2), 128.
[http://dx.doi.org/10.1007/s00339-017-1510-4]

[100]   Ismail, A.M.; Menazea, A.A.; Kabary, H.A.; El-Sherbiny, A.E.; Samy, A. The influence of calcination

temperature on structural and antimicrobial characteristics of Zinc oxide nanoparticles synthesized by Sol–Gel method. *J. Mol. Struct.,* **2019**, *1196*, 332-337.
[http://dx.doi.org/10.1016/j.molstruc.2019.06.084]

[101] Musleh, H.; Zayed, H.; Shaat, S.; Tamous, H.M.; Asad, J.; Al-Kahlout, A.; Issa, A.; Shurrab, N.; AlDahoudi, N. Synthesis and characterization of ZnO nanoparticles using sol-gel technique for dye-sensitized solar cells applications. *J. Phys. Conf. Ser.,* **2019**, *1294*(2), 022022.
[http://dx.doi.org/10.1088/1742-6596/1294/2/022022]

[102] Ba-Abbad, M.M.; Kadhum, A.A.H.; Mohamad, A.B.; Takriff, M.S.; Sopian, K. The effect of process parameters on the size of ZnO nanoparticles synthesized *via* the sol–gel technique. *J. Alloys Compd.,* **2013**, *550*, 63-70.
[http://dx.doi.org/10.1016/j.jallcom.2012.09.076]

[103] Erol, A.; Okur, S.; Comba, B.; Mermer, Ö.; Arıkan, M.C. Humidity sensing properties of ZnO nanoparticles synthesized by sol–gel process. *Sens. Actuators B Chem.,* **2010**, *145*(1), 174-180.
[http://dx.doi.org/10.1016/j.snb.2009.11.051]

[104] Pineda-Reyes, A.M.; Olvera, M.D.L.L. Synthesis of ZnO nanoparticles from water-in-oil (w/o) microemulsions. *Mater. Chem. Phys.,* **2018**, *203*, 141-147.
[http://dx.doi.org/10.1016/j.matchemphys.2017.09.054]

[105] Fan, X.; Domszy, R.C.; Hu, N.; Yang, A.J.; Yang, J.; David, A.E. Synthesis of silica-alginate nanoparticles and their potential application as pH-responsive drug carriers. *J. Sol-Gel Sci. Technol.,* **2019**, *91*(1), 11-20.
[http://dx.doi.org/10.1007/s10971-019-04995-4] [PMID: 32863592]

[106] Wang, Y.; Zhang, X.; Wang, A.; Li, X.; Wang, G.; Zhao, L. Synthesis of ZnO nanoparticles from microemulsions in a flow type microreactor. *Chem. Eng. J.,* **2014**, *235*, 191-197.
[http://dx.doi.org/10.1016/j.cej.2013.09.020]

[107] Sarkar, D.; Tikku, S.; Thapar, V.; Srinivasa, R.S.; Khilar, K.C. Formation of zinc oxide nanoparticles of different shapes in water-in-oil microemulsion. *Colloids Surf. A Physicochem. Eng. Asp.,* **2011**, *381*(1-3), 123-129.
[http://dx.doi.org/10.1016/j.colsurfa.2011.03.041]

[108] Yıldırım, Ö.A.; Durucan, C. Synthesis of zinc oxide nanoparticles elaborated by microemulsion method. *J. Alloys Compd.,* **2010**, *506*(2), 944-949.
[http://dx.doi.org/10.1016/j.jallcom.2010.07.125]

[109] Agarwal, S.; Rai, P.; Gatell, E.N.; Llobet, E.; Güell, F.; Kumar, M.; Awasthi, K. Gas sensing properties of ZnO nanostructures (flowers/rods) synthesized by hydrothermal method. *Sens. Actuators B Chem.,* **2019**, *292*, 24-31.
[http://dx.doi.org/10.1016/j.snb.2019.04.083]

[110] Basnet, P.; Chatterjee, S. Structure-directing property and growth mechanism induced by capping agents in nanostructured ZnO during hydrothermal synthesis-A systematic review. *Nano-Struc. Nano-Obj.,* **2020**, *22*, 100426.

[111] Zhu, L.; Li, Y.; Zeng, W. Hydrothermal synthesis of hierarchical flower-like ZnO nanostructure and its enhanced ethanol gas-sensing properties. *Appl. Surf. Sci.,* **2018**, *427*, 281-287.
[http://dx.doi.org/10.1016/j.apsusc.2017.08.229]

[112] Mishra, S.K.; Srivastava, R.K.; Prakash, S.G.; Yadav, R.S.; Panday, A.C. Photoluminescence and photoconductive characteristics of hydrothermally synthesized ZnO nanoparticles. *Opto-Electron. Rev.,* **2010**, *18*(4), 467-473.
[http://dx.doi.org/10.2478/s11772-010-0037-4]

[113] Hu, Y.; Chen, H.J. Preparation and characterization of nanocrystalline ZnO particles from a hydrothermal process. *J. Nanopart. Res.,* **2008**, *10*(3), 401-407.
[http://dx.doi.org/10.1007/s11051-007-9264-0]

[114] Madathil, A.N.P.; Vanaja, K.A.; Jayaraj, M.K. Synthesis of ZnO nanoparticles by hydrothermal method. *Nanophot. Mater. IV,* **2007,** *6639,* 66390J.

[115] Šarić, A.; Despotović, I.; Štefanić, G. Solvothermal synthesis of zinc oxide nanoparticles: A combined experimental and theoretical study. *J. Mol. Struct.,* **2019,** *1178,* 251-260.
[http://dx.doi.org/10.1016/j.molstruc.2018.10.025]

[116] Deep, A.; Sharma, A.L.; Mohanta, G.C.; Kumar, P.; Kim, K.H. A facile chemical route for recovery of high quality zinc oxide nanoparticles from spent alkaline batteries. *Waste Manag.,* **2016,** *51,* 190-195.
[http://dx.doi.org/10.1016/j.wasman.2016.01.033] [PMID: 26851168]

[117] Abdelhakim, H.K.; El-Sayed, E.R.; Rashidi, F.B. Biosynthesis of zinc oxide nanoparticles with antimicrobial, anticancer, antioxidant and photocatalytic activities by the endophytic Alternaria tenuissima. *J. Appl. Microbiol.,* **2020,** *128*(6), 1634-1646.
[http://dx.doi.org/10.1111/jam.14581] [PMID: 31954094]

[118] Ganesan, V.; Hariram, M.; Vivekanandhan, S.; Muthuramkumar, S. Periconium sp.(endophytic fungi) extract mediated sol-gel synthesis of ZnO nanoparticles for antimicrobial and antioxidant applications. *Mater. Sci. Semicond. Process.,* **2020,** *105,* 104739.
[http://dx.doi.org/10.1016/j.mssp.2019.104739]

[119] Mohamed, A.A.; Fouda, A.; Abdel-Rahman, M.A.; Hassan, S.E.D.; El-Gamal, M.S.; Salem, S.S.; Shaheen, T.I. Fungal strain impacts the shape, bioactivity and multifunctional properties of green synthesized zinc oxide nanoparticles. *Biocatal. Agric. Biotechnol.,* **2019,** *19,* 101103.
[http://dx.doi.org/10.1016/j.bcab.2019.101103]

[120] Madhumitha, G.; Elango, G.; Roopan, S.M. Biotechnological aspects of ZnO nanoparticles: overview on synthesis and its applications. *Appl. Microbiol. Biotechnol.,* **2016,** *100*(2), 571-581.
[http://dx.doi.org/10.1007/s00253-015-7108-x] [PMID: 26541334]

[121] Sharan, C.; Khandelwal, P.; Poddar, P. Biomilling of rod-shaped ZnO nanoparticles: a potential role of Saccharomyces cerevisiae extracellular proteins. *RSC Advances,* **2015,** *5*(3), 1883-1889.
[http://dx.doi.org/10.1039/C4RA10077H]

[122] Jayaseelan, C.; Rahuman, A.A.; Kirthi, A.V.; Marimuthu, S.; Santhoshkumar, T.; Bagavan, A.; Gaurav, K.; Karthik, L.; Rao, K.V. Novel microbial route to synthesize ZnO nanoparticles using Aeromonas hydrophila and their activity against pathogenic bacteria and fungi. *Spectrochim. Acta A Mol. Biomol. Spectrosc.,* **2012,** *90,* 78-84.
[http://dx.doi.org/10.1016/j.saa.2012.01.006] [PMID: 22321514]

[123] Agarwal, H.; Kumar, S.V.; Rajeshkumar, S. A review on green synthesis of zinc oxide nanoparticles–An eco-friendly approach. *Resource-Efficient Tech.,* **2017,** *3*(4), 406-413.
[http://dx.doi.org/10.1016/j.reffit.2017.03.002]

[124] Rafique, M.; Tahir, R.; Gillani, S.S.A.; Tahir, M.B.; Shakil, M.; Iqbal, T.; Abdellahi, M.O. Plant-mediated green synthesis of zinc oxide nanoparticles from Syzygium Cumini for seed germination and wastewater purification. *Int. J. Environ. Anal. Chem.,* **2022,** *102*(1), 23-38.
[http://dx.doi.org/10.1080/03067319.2020.1715379]

[125] Chakraborty, S.; Farida, J.J.; Simon, R.; Kasthuri, S.; Mary, N.L. Averrhoe carrambola fruit extract assisted green synthesis of zno nanoparticles for the photodegradation of congo red dye. *Surf. Interfaces,* **2020,** *19,* 100488.
[http://dx.doi.org/10.1016/j.surfin.2020.100488]

[126] Pillai, A.M.; Sivasankarapillai, V.S.; Rahdar, A.; Joseph, J.; Sadeghfar, F.; Rajesh, K.; Kyzas, G.Z. Green synthesis and characterization of zinc oxide nanoparticles with antibacterial and antifungal activity. *J. Mol. Struct.,* **2020,** *1211,* 128107.
[http://dx.doi.org/10.1016/j.molstruc.2020.128107]

[127] Liu, D.; Liu, L.; Yao, L.; Peng, X.; Li, Y.; Jiang, T.; Kuang, H. Synthesis of ZnO nanoparticles using radish root extract for effective wound dressing agents for diabetic foot ulcers in nursing care. *J. Drug*

Deliv. Sci. Technol., **2020**, *55*, 101364.
[http://dx.doi.org/10.1016/j.jddst.2019.101364]

[128] Sukri, S.N.A.M.; Shameli, K.; Wong, M.M.T.; Teow, S.Y.; Chew, J.; Ismail, N.A. Cytotoxicity and antibacterial activities of plant-mediated synthesized zinc oxide (ZnO) nanoparticles using Punica granatum (pomegranate) fruit peels extract. *J. Mol. Struct.,* **2019**, *1189*, 57-65.
[http://dx.doi.org/10.1016/j.molstruc.2019.04.026]

[129] Iqbal, J.; Abbasi, B.A.; Mahmood, T.; Kanwal, S.; Ahmad, R.; Ashraf, M. Plant-extract mediated green approach for the synthesis of ZnONPs: Characterization and evaluation of cytotoxic, antimicrobial and antioxidant potentials. *J. Mol. Struct.,* **2019**, *1189*, 315-327.
[http://dx.doi.org/10.1016/j.molstruc.2019.04.060]

[130] Duraimurugan, J.; Kumar, G.S.; Maadeswaran, P.; Shanavas, S.; Anbarasan, P.; Vasudevan, V. Structural, optical and photocatlytic properties of zinc oxide nanoparticles obtained by simple plant extract mediated synthesis. *J. Mater. Sci. Mater. Electron.,* **2019**, *30*(2), 1927-1935.
[http://dx.doi.org/10.1007/s10854-018-0466-2]

[131] Rajakumar, G.; Thiruvengadam, M.; Mydhili, G.; Gomathi, T.; Chung, I.M. Green approach for synthesis of zinc oxide nanoparticles from Andrographis paniculata leaf extract and evaluation of their antioxidant, anti-diabetic, and anti-inflammatory activities. *Bioprocess Biosyst. Eng.,* **2018**, *41*(1), 21-30.
[http://dx.doi.org/10.1007/s00449-017-1840-9] [PMID: 28916855]

[132] Sai Saraswathi, V.; Tatsugi, J.; Shin, P.K.; Santhakumar, K. Facile biosynthesis, characterization, and solar assisted photocatalytic effect of ZnO nanoparticles mediated by leaves of L. speciosa. *J. Photochem. Photobiol. B,* **2017**, *167*, 89-98.
[http://dx.doi.org/10.1016/j.jphotobiol.2016.12.032] [PMID: 28056394]

[133] Mishra, P.; Singh, Y.P.; Nagaswarupa, H.P.; Sharma, S.C.; Vidya, Y.S.; Prashantha, S.C.; Nagabhushana, H.; Anantharaju, K.S.; Sharma, S.; Renuka, L. d Renuka, L. Caralluma fimbriata extract induced green synthesis, structural, optical and photocatalytic properties of ZnO nanostructure modified with Gd. *J. Alloys Compd.,* **2016**, *685*, 656-669.
[http://dx.doi.org/10.1016/j.jallcom.2016.05.044]

[134] Sundrarajan, M.; Ambika, S.; Bharathi, K. Plant-extract mediated synthesis of ZnO nanoparticles using Pongamia pinnata and their activity against pathogenic bacteria. *Adv. Powder Technol.,* **2015**, *26*(5), 1294-1299.
[http://dx.doi.org/10.1016/j.apt.2015.07.001]

[135] Fu, L.; Fu, Z. Plectranthus amboinicus leaf extract–assisted biosynthesis of ZnO nanoparticles and their photocatalytic activity. *Ceram. Int.,* **2015**, *41*(2), 2492-2496.
[http://dx.doi.org/10.1016/j.ceramint.2014.10.069]

[136] Diallo, A.; Ngom, B.D.; Park, E.; Maaza, M. Green synthesis of ZnO nanoparticles by Aspalathus linearis: structural & optical properties. *J. Alloys Compd.,* **2015**, *646*, 425-430.
[http://dx.doi.org/10.1016/j.jallcom.2015.05.242]

[137] Suresh, D.; Nethravathi, P.C.; Rajanaika, H.; Nagabhushana, H.; Sharma, S.C. Green synthesis of multifunctional zinc oxide (ZnO) nanoparticles using Cassia fistula plant extract and their photodegradative, antioxidant and antibacterial activities. *Mater. Sci. Semicond. Process.,* **2015**, *31*, 446-454.
[http://dx.doi.org/10.1016/j.mssp.2014.12.023]

[138] Salam, H.A.; Sivaraj, R.; Venckatesh, R. Green synthesis and characterization of zinc oxide nanoparticles from Ocimum basilicum L. var. purpurascens Benth.-Lamiaceae leaf extract. *Mater. Lett.,* **2014**, *131*, 16-18.
[http://dx.doi.org/10.1016/j.matlet.2014.05.033]

[139] Sangeetha, G.; Rajeshwari, S.; Venckatesh, R. Green synthesis of zinc oxide nanoparticles by aloe barbadensis miller leaf extract: Structure and optical properties. *Mater. Res. Bull.,* **2011**, *46*(12), 2560-

2566.
[http://dx.doi.org/10.1016/j.materresbull.2011.07.046]

[140] Verma, R.; Pathak, S.; Srivastava, A.K.; Prawer, S.; Tomljenovic-Hanic, S. ZnO nanomaterials: Green synthesis, toxicity evaluation and new insights in biomedical applications. *J. Alloys Compd.,* **2021**, *876*, 160175.
[http://dx.doi.org/10.1016/j.jallcom.2021.160175]

[141] Bhuiyan, M.R.A.; Rahman, M.K. Synthesis and characterization of Ni doped ZnO nanoparticles. *Int. J. Engg. Manuf.,* **2014**, *4*(1), 10-17.
[http://dx.doi.org/10.5815/ijem.2014.01.02]

[142] Bhuiyan, M.R.A.; Alam, M.M.; Momin, M.A.; Mamur, H. Characterization of Al doped ZnO nanostructures *via* an electrochemical route. *Int. J. Ener. Appl. Tech.,* **2017**, *4*(1), 28-33.

[143] Hoq, E.; Bhuiyan, M.R.A.; Begum, J. Influence of thickness on the optical properties of Sb doped ZnO thin films. *J. Bangladesh Acad. Sci.,* **2014**, *38*(1), 93-96.
[http://dx.doi.org/10.3329/jbas.v38i1.20217]

[144] Lei, G.; Yang, S.; Cao, R.; Zhou, P.; Peng, H.; Peng, R.; Zhang, X.; Yang, Y.; Li, Y.; Wang, M.; He, Y.; Zhou, L.; Du, J.; Du, W.; Shi, Y.; Wu, H. *In situ* preparation of amphibious ZnO quantum dots with blue fluorescence based on hyperbranched polymers and their application in bio-imaging. *Polymers (Basel),* **2020**, *12*(1), 144.
[http://dx.doi.org/10.3390/polym12010144] [PMID: 31935952]

[145] Eixenberger, J.E.; Anders, C.B.; Wada, K.; Reddy, K.M.; Brown, R.J.; Moreno-Ramirez, J.; Weltner, A.E.; Karthik, C.; Tenne, D.A.; Fologea, D.; Wingett, D.G. Defect engineering of ZnO nanoparticles for bioimaging applications. *ACS Appl. Mater. Interfaces,* **2019**, *11*(28), 24933-24944.
[http://dx.doi.org/10.1021/acsami.9b01582] [PMID: 31173687]

[146] Zhang, S.; Liu, L.; Ren, S.; Li, Z.; Zhao, Y.; Yang, Z.; Hu, R.; Qu, J. Recent advances in nonlinear optics for bio-imaging applications. *Opto-Elec. Advan.,* **2020**, *3*(10), 10200003.

[147] Xiong, H.M. ZnO nanoparticles applied to bioimaging and drug delivery. *Adv. Mater.,* **2013**, *25*(37), 5329-5335.
[http://dx.doi.org/10.1002/adma.201301732] [PMID: 24089351]

[148] Urban, B.E.; Neogi, P.; Senthilkumar, K.; Rajpurohit, S.K.; Jagadeeshwaran, P.; Kim, S.; Fujita, Y.; Neogi, A. Bioimaging using the optimized nonlinear optical properties of ZnO nanoparticles. *IEEE J. Sel. Top. Quantum Electron.,* **2012**, *18*(4), 1451-1456.
[http://dx.doi.org/10.1109/JSTQE.2012.2184793]

[149] Senthilkumar, K.; Senthilkumar, O.; Yamauchi, K.; Sato, M.; Morito, S.; Ohba, T.; Nakamura, M.; Fujita, Y. Preparation of ZnO nanoparticles for bio-imaging applications. *Phys. Status Solidi, B Basic Res.,* **2009**, *246*(4), 885-888.
[http://dx.doi.org/10.1002/pssb.200880606]

[150] Wu, Y.L.; Lim, C.S.; Fu, S.; Tok, A.I.Y.; Lau, H.M.; Boey, F.Y.C.; Zeng, X.T. Surface modifications of ZnO quantum dots for bio-imaging. *Nanotechnology,* **2007**, *18*(21), 215604.
[http://dx.doi.org/10.1088/0957-4484/18/21/215604] [PMID: 23619742]

[151] Naresh, V.; Lee, N. A Review on biosensors and recent development of nanostructured materials-enabled biosensors. *Sensors (Basel),* **2021**, *21*(4), 1109.
[http://dx.doi.org/10.3390/s21041109] [PMID: 33562639]

[152] Kim, I.; Viswanathan, K.; Kasi, G.; Thanakkasaranee, S.; Sadeghi, K.; Seo, J. ZnO nanostructures in active antibacterial food packaging: Preparation methods, antimicrobial mechanisms, safety issues, future prospects, and challenges. *Food Rev. Int.,* **2022**, *38*(4), 537-565.
[http://dx.doi.org/10.1080/87559129.2020.1737709]

[153] Zhang, Y.; Nayak, T.R.; Hong, H.; Cai, W. Biomedical applications of zinc oxide nanomaterials. *Curr. Mol. Med.,* **2013**, *13*(10), 1633-1645.

[http://dx.doi.org/10.2174/15665240136666131111130058] [PMID: 24206130]

[154] Pauzi, N.; Zain, N.M.; Kutty, R.V.; Ramli, H. Antibacterial and antibiofilm properties of ZnO nanoparticles synthesis using gum arabic as a potential new generation antibacterial agent. *Mater. Today Proc.,* **2021**, *41*, 1-8.
[http://dx.doi.org/10.1016/j.matpr.2020.06.359]

[155] Rambabu, K.; Bharath, G.; Banat, F.; Show, P.L. Green synthesis of zinc oxide nanoparticles using Phoenix dactylifera waste as bioreductant for effective dye degradation and antibacterial performance in wastewater treatment. *J. Hazard. Mater.,* **2021**, *402*, 123560.
[http://dx.doi.org/10.1016/j.jhazmat.2020.123560] [PMID: 32759001]

[156] Gudkov, S.V.; Burmistrov, D.E.; Serov, D.A.; Rebezov, M.B.; Semenova, A.A.; Lisitsyn, A.B. A mini review of antibacterial properties of ZnO nanoparticles. *Front. Phys. (Lausanne),* **2021**, *9*, 641481.
[http://dx.doi.org/10.3389/fphy.2021.641481]

[157] Tavakoli, S.; Kharaziha, M.; Nemati, S. Polydopamine coated ZnO rod-shaped nanoparticles with noticeable biocompatibility, hemostatic and antibacterial activity. *Nano-Struc. Nano-Obj.,* **2021**, *25*, 100639.

[158] Akbar, S.; Tauseef, I.; Subhan, F.; Sultana, N.; Khan, I.; Ahmed, U.; Haleem, K.S. An overview of the plant-mediated synthesis of zinc oxide nanoparticles and their antimicrobial potential. *Inorg. Nano-Metal Chem.,* **2020**, *50*(4), 257-271.

[159] Batool, M.; Khurshid, S.; Qureshi, Z.; Daoush, W.M. Adsorption, antimicrobial and wound healing activities of biosynthesised zinc oxide nanoparticles. *Chem. Pap.,* **2021**, *75*(3), 893-907.
[http://dx.doi.org/10.1007/s11696-020-01343-7]

[160] Umavathi, S.; Mahboob, S.; Govindarajan, M.; Al-Ghanim, K.A.; Ahmed, Z.; Virik, P.; Al-Mulhm, N.; Subash, M.; Gopinath, K.; Kavitha, C. Green synthesis of ZnO nanoparticles for antimicrobial and vegetative growth applications: A novel approach for advancing efficient high quality health care to human wellbeing. *Saudi J. Biol. Sci.,* **2021**, *28*(3), 1808-1815.
[http://dx.doi.org/10.1016/j.sjbs.2020.12.025] [PMID: 33732066]

[161] Jayakumar, A.; Radoor, S.; Nair, I.C.; Siengchin, S.; Parameswaranpillai, J.; Radhakrishnan, E.K. Lipopeptide and zinc oxide nanoparticles blended polyvinyl alcohol-based nanocomposite films as antimicrobial coating for biomedical applications. *Process Biochem.,* **2021**, *102*, 220-228.
[http://dx.doi.org/10.1016/j.procbio.2020.12.010]

[162] Nadeem, M.S.; Munawar, T.; Mukhtar, F.; Ur Rahman, M.N.; Riaz, M.; Iqbal, F. Enhancement in the photocatalytic and antimicrobial properties of ZnO nanoparticles by structural variations and energy bandgap tuning through Fe and Co co-doping. *Ceram. Int.,* **2021**, *47*(8), 11109-11121.
[http://dx.doi.org/10.1016/j.ceramint.2020.12.234]

[163] Elsayed, E.A.; Moussa, S.A.; El-Enshasy, H.A.; Wadaan, M.A. Anticancer potentials of zinc oxide nanoparticles against liver and breast cancer cell lines. *J. Sci. Ind. Res. (India),* **2020**, *79*(01), 56-59.

[164] Singh, T.A.; Das, J.; Sil, P.C. Zinc oxide nanoparticles: A comprehensive review on its synthesis, anticancer and drug delivery applications as well as health risks. *Adv. Colloid Interface Sci.,* **2020**, *286*, 102317.
[http://dx.doi.org/10.1016/j.cis.2020.102317] [PMID: 33212389]

[165] Dulta, K.; Ağçeli, G.K. Chauhan, P.; Jasrotia, R.; Chauhan, P.K. A novel approach of synthesis zinc oxide nanoparticles by bergenia ciliata rhizome extract: antibacterial and anticancer potential. *J. Inorg. Organomet. Polym. Mater.,* **2021**, *31*(1), 180-190.
[http://dx.doi.org/10.1007/s10904-020-01684-6]

[166] González, S.E.; Bolaina-Lorenzo, E.; Pérez-Trujillo, J.J.; Puente-Urbina, B.A.; Rodríguez-Fernández, O.; Fonseca-García, A.; Betancourt-Galindo, R. Antibacterial and anticancer activity of ZnO with different morphologies: A comparative study. *3 Biotech,* **2021**, *11*(2), 1-12.

[167] Nilavukkarasi, M.; Vijayakumar, S.; Prathipkumar, S. Capparis zeylanica mediated bio-synthesized

ZnO nanoparticles as antimicrobial, photocatalytic and anti-cancer applications. *Mater. Sci. Energy Technol.,* **2020**, *3*, 335-343.
[http://dx.doi.org/10.1016/j.mset.2019.12.004]

[168] Nemati, A. Quantum dots in therapeutic, diagnostic and drug delivery applications "A brief review". *Iranian J. Mater. Sci. Engg.,* **2020**, *17*(2), 1-12.

[169] Fahimmunisha, B.A.; Ishwarya, R.; AlSalhi, M.S.; Devanesan, S.; Govindarajan, M.; Vaseeharan, B. Green fabrication, characterization and antibacterial potential of zinc oxide nanoparticles using Aloe socotrina leaf extract: A novel drug delivery approach. *J. Drug Deliv. Sci. Technol.,* **2020**, *55*, 101465.
[http://dx.doi.org/10.1016/j.jddst.2019.101465]

[170] Sathishkumar, P.; Li, Z.; Govindan, R.; Jayakumar, R.; Wang, C.; Gu, F.L. Zinc oxide-quercetin nanocomposite as a smart nano-drug delivery system: Molecular-level interaction studies. *Appl. Surf. Sci.,* **2021**, *536*, 147741.
[http://dx.doi.org/10.1016/j.apsusc.2020.147741]

[171] Wang, H.; Zhou, Y.; Sun, Q.; Zhou, C.; Hu, S.; Lenahan, C.; Xu, W.; Deng, Y.; Li, G.; Tao, S. Update on nanoparticle-based drug delivery system for anti-inflammatory treatment. *Front. Bioeng. Biotechnol.,* **2021**, *9*, 630352.
[http://dx.doi.org/10.3389/fbioe.2021.630352] [PMID: 33681167]

[172] Akbarian, M.; Mahjoub, S.; Elahi, S.M.; Zabihi, E.; Tashakkorian, H. Green synthesis, formulation and biological evaluation of a novel ZnO nanocarrier loaded with paclitaxel as drug delivery system on MCF-7 cell line. *Colloids Surf. B Biointerfaces,* **2020**, *186*, 110686.
[http://dx.doi.org/10.1016/j.colsurfb.2019.110686] [PMID: 31816463]

CHAPTER 6

Superhydrophobic Polymeric Nanocomposites Coatings for Effective Corrosion Protection

Shimaa A. Higazy[1], Olfat E. El-Azabawy[1] and Mohamed S. Selim[1,2,*]

[1] Petroleum Application Department, Egyptian Petroleum Research Institute, Nasr City 11727, Cairo, Egypt

[2] Key Laboratory of Clean Chemistry Technology of Guangdong Regular Higher Education Institutions, School of Chemical Engineering and Light Industry, Guangdong University of Technology, Guangzhou, 510006, PR China

Abstract: The contemporary era of studying superhydrophobic surfaces began in 1997, when Neinhuis and Barthlott discovered the self-cleaning qualities of the lotus effect. Corrosion of steel represents an important industrial issue with well-known negative economic and environmental consequences. The protection of steel objects during service operations is inexhaustible research subject because of the steel's high demand in the industry. Anticorrosive coatings have aided in extending the life of the material without impairing its bulk qualities. The microporous structure of polymers allows corrosive ions to pass at the coating–metal interface, resulting in poor serviceability. Advanced structural modifications, such as polymeric nanocomposites, have been used to solve these disadvantages. Organic-inorganic nanocomposites are employed as outstanding anti-corrosive coatings to provide steel constructions' service longevity. Superhydrophobic nanocomposite coatings tend to be one of the most promising methods for avoiding corrosion in steel. Various nanostructured fillers have the ability to significantly improve the corrosion-barrier efficiency of polymeric coatings. Superhydrophobicity in nature will be briefly addressed to provide a comprehensive study. This chapter focuses on introducing the anticorrosive properties of superhydrophobic coatings. It gives an overview of present and advanced research developments, such as graphene nanocomposite surfaces.

Keywords: Anticorrosive Coatings, Corrosion of Steel, Graphene, Lotus Effect, Nanomaterials, Nanofillers, Organic-inorganic, Polymeric Nanocomposites, Superhydrophobic Surfaces, Self-cleaning.

** **Corresponding author Mohamed S. Selim:**Petroleum Application Department, Egyptian Petroleum Research Institute, Nasr City 11727, Cairo, Egypt & Key Laboratory of Clean Chemistry Technology of Guangdong Regular Higher Education Institutions, School of Chemical Engineering and Light Industry, Guangdong University of Technology, Guangzhou, 510006, PR China;
E-mails: moh.selim_chem2006@yahoo.com, mohamedselim@gudt.edu.cn*

1. INTRODUCTION

Nanomaterials sciences play an important role in every aspect of technology, helping to solve issues and improve our lives [1]. Developing advanced nanomaterials is necessary to satisfy the demand for betterment [2]. The negative environmental and economic consequences of metallic corrosion have prompted the development of anti-corrosion materials. Corrosion is predicted to cost the world economy $2.5 trillion annually [3]. The significant corrosion impact on different aspects of contemporary life is depicted in Fig. (**1**). Superhydrophobic surfaces have emerged as viable alternatives for a variety of engineering applications, including self-cleaning, biofouling resistance, and corrosion resistance [4]. Many natural surfaces, such as butterflies' wings as well as plants' leaves (including the Indian cress), represent superhydrophobic and self-cleaning surfaces [5]. Lotus plant leaves (Nelumbo nucifera) are the most famous hydrophobic self-cleaning example from which *"Lotus Effect"* was derived [6]. Electron microscopy revealed projecting nubs 20–40 μm apart on the lotus leaf surface coated with a nanorough waxy crystalloids [7]. The contact angle quantifies the wettability of a solid surface by a liquid *via* the Young equation. It is conventionally measured through the liquid, where a liquid–vaporinterface meets a solid surface. Water contact angles (WCA) measurement is a qualitative way to evaluate whether the surface has a hydrophobic (>90°) or hydrophilic (<90°) characteristic.

Fig. (1). Corrosion's impact on various facets of recent life.

On the other hand, the sliding angle is a measure of the mobility of a drop on the surface [8]. It is the tilt angle at which the movement of the drop across the surface begins [9]. It provides information about the adhesion conditions of a droplet on a surface [10].

Superhydrophobic surfaces exhibited (WCAs) $\geq 150°$, reduced sliding angles, and self-cleaning properties. Such processes can be facilitated by micro/nano binary roughness and low surface free energy (SFE) [11]. Superhydrophobicity is a strategy to protect sensitive surface properties [12, 13]. Controlling both oxidation and reduction reactions is necessary to prevent metal corrosion. As a physical barrier, organic coatings are often used to separate steel from its surroundings [13]. Nanoparticles (NPs) cover the spaces between large particles utilizing nanocomposite organic coatings, preventing corrosive solutions from diffusing through these rapid routes. Fabrication of superhydrophobicity surfaces entails the creation of a hierarchical rough morphology with low-energy molecules [14]. To make superhydrophobic nano-surfaces, researchers used etching, lithography, biomimetic, and stamping processes. Also, simple methods of dispersing NPs and graphene-based materials in a hydrophobic polymer matrix can develop water-repellent nanocomposite surfaces [15]. Controlled NP structures aid in the creation of a rough morphology and the generation of extra capabilities for a superhydrophobic surface [16]. Surface roughness and hydrophobicity can be increased by trapping air between the roughness and the liquid drops [17].

As air is a superhydrophobic substance (WCA of 180°) [18], the trapped air can increase the surface self-cleaning property. Superhydrophobicity is thus caused by a hierarchical micro-nano structured surface and a low SFE [19 - 24]. Superhydrophobic surfaces have a near-perfect non-wetting state with minimized contact angle hysteresis (CAH), which are widely used for corrosion-protection applications. CAH is a reflection of the activation energy required for the movement of a droplet from one metastable state to another on a surface [25]. It expresses the difference between advancing contact angle and receding contact angle and is essential to investigate the surface non-wettability and chemical heterogeneity. This chapter looks at recent developments in anticorrosion superhydrophobic surfaces for the steel coating industry, as well as their development methods. Furthermore, it covers the manufacturing approach of stable superhydrophobic coatings for corrosion-resistant coatings. It can present long-term superhydrophobic materials for steel surfaces. It is divided into five sections; the first two of them are devoted to illustrate the coatings' superhydrophobic self-cleaning notion as well as superhydrophobicity in nature. The third section discusses superhydrophobic surface manufacturing and anticorrosion coatings. The importance of polymeric nanocomposite structures for developing anticorrosion superhydrophobic surfaces is discussed in these sections.

The advancement in designing anticorrosive superhydrophobic surfaces over different substrates is discussed in the fourth section. The final remarks are outlined in the fifth section.

2. SUPERHYDROPHOBICITY IN NATURE

The incredible superhydrophobicity of some plants and animals has gained much attention [26]. Superhydrophobic surfaces can be seen on the leaves of lotuses, butterfly wings, rice leaves, water strider legs, and shark skin (Fig. **2**) [27]. Lotus leaves demonstrated no significant loss of hydrophobicity in these settings, making them a perfect example for creating self-cleaning nanosurfaces. Lotus leaf has a hierarchical micro/nano rough surface, which contributes to its superhydrophobicity and self-cleaning ability [28].

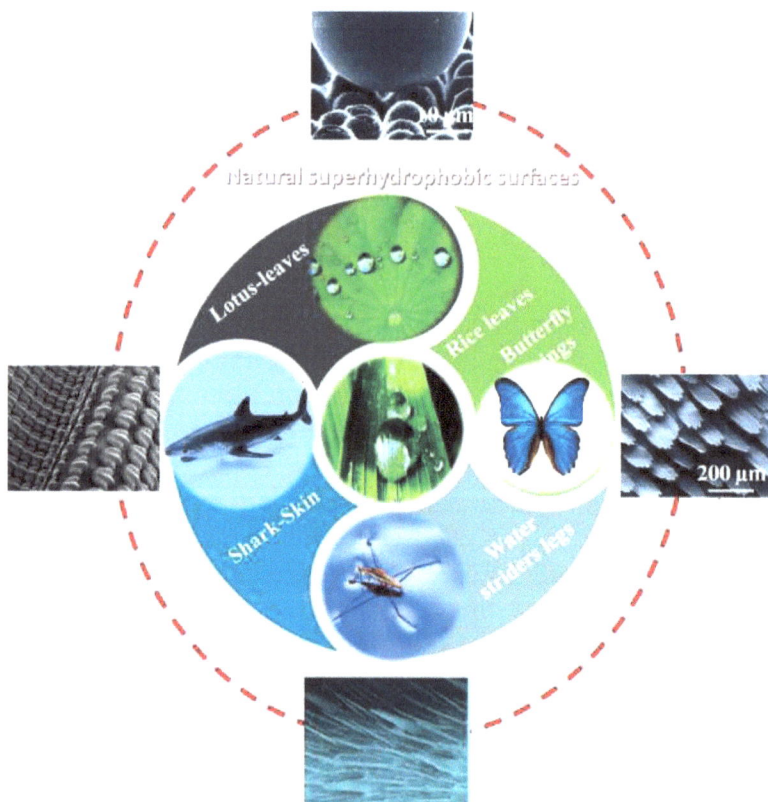

Fig. (2). Superhydrophobic natural surfaces, including the leaves of lotus plant surfaces and their SEM micrographs, rice leaf with longitudinal grooves, water strider can stay on the surface of water, the leg's SEM image of needle-shaped and grooved microsetae, and shark skin superhydrophobic structure. Copyright 2019, reproduced with permission from the RSC [23].

The surface of lotus leaves was randomly covered by roughly 10 μm sized protrusions with about 20 μm gaps between them. Many nanostructured tubes with 100 nm mean diameter and 500 nm length were also found on the surface of the protrusions. Hydrophobic structures were also discovered covering the lotus leaf's micro/nano-surface, resulting in a reduced SFE. Furthermore, superhydrophobicity is achieved when a hierarchical rough surface is coupled with the waxy structure. The WCA > 160° and a CAH of 3° are used to produce self-cleaning capability in the lotus leaf. Butterfly wings exhibit superhydrophobicity owing to their micro and nano-structured surfaces. Microgrooves on their wings produced a surface roughness with superhydrophobicity and minimized stickiness. Rice leaves have self-cleaning surfaces because of their stiffness and confinement under anisotropic flow orders along with the leaves' directions [29]. The water striders' legs have an unusual hierarchical rough structure. Large numbers of aligned tiny hairs with nanostructured grooves were reported by Gao *et al.* [30], which can produce their superhydrophobicity. Legs' superhydrophobicity (WCA of 167°) is caused by air-trapping between the nanogrooves and microsetae. Water striders are able to live on the water's surface as a result of this adaptation. According to Bechert *et al.* [31], the skin of a shark is shielded with teeth-like structures (called dermal denticles), which are ribbed with linear microgrooves. Such grooves prevent vortices from establishing an ultrasmooth surface and provide superhydrophobicity.

WCAs of around 200 plants were reported by Neinhuis and Barthlott [32], along with surface morphologies. In comparison with other plants that have a 160° WCA [33, 34], the lotus leaf has exhibited greater stability and perfection as a superhydrophobic surface. This is caused by the leaf's epidermal structure's morphology, nanoscopic waxy crystals, and superhydrophobic stability against moisture environments [35]. Rice leaves' micropapillae and butterfly wings' shingle-like microgrooves create hierarchical rough and superhydrophobic surfaces. The self-cleaning surfaces provide superhydrophobic character, reduced stickiness, and anisotropic properties. According to studies on the superhydrophobicity of mosquito eyes, the essential distance between two microstructured surfaces for achieving superhydrophobicity is ~ 100 nm. However, the superhydrophobicity (WCA of 160) and reduced adhesion (CAH of 10°) are required for attaining the self-cleaning property. In nature, superhydrophobic surfaces such as rose petals, garlic, and scallion's leaves may attach water droplets even when they are topsy-turvy termed as *"petal effect"* [36]. Water drops can enter enormous surface gaps as a result of superhydrophobicity with water adhesion.

3. SELF-CLEANING SUPERHYDROPHOBIC COATINGS FUNDAMEN-TALS

The following are the principles of surface non-wettability and the WCA essentials of a liquid:

3.1. Young Equation

Because the surface atoms and molecules have fewer links with their neighbors, they have more free energy than those in the interior. This SFE, which is measured in N/m, is the labor necessary to create a surface area at stable temperature and pressure. The angle created when a drop of liquid contacts with a solid surface, as indicated by the Young equation (Eq. 1) [37], is known as static WCA (θ) (Fig. **3a**).

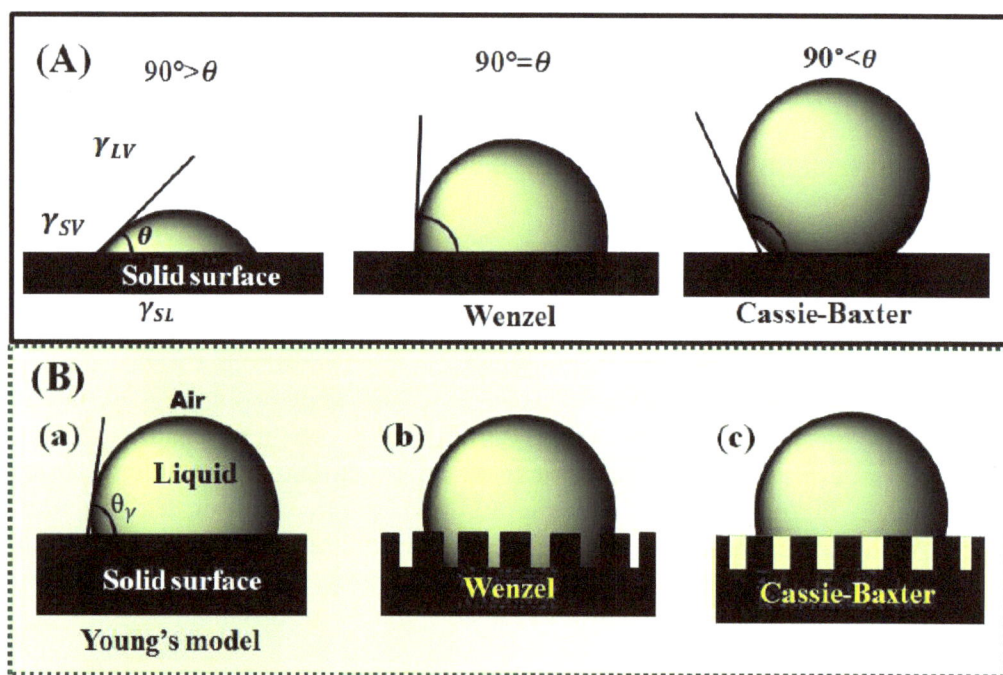

Fig. (3). (A) Represents the liquid drops forming contact angles over a solid smooth homogeneous surface, and (B) Typical wetting actions of a drop on the surface; including (a) Young, (b) Wenzel, and (c) Cassie-Baxter models. Copyright 2017, produced with permission from Elsevier [14].

$$Cos\ \theta_0 = \left(\frac{\gamma_{SA} - \gamma_{SL}}{\gamma_{LA}}\right) \tag{1}$$

Where γ_{SA}, γ_{SL}, and γ_{LA} represent the SFE of solid and air, solid and liquid, as well as liquid and air, respectively. The Young equation only works on flat surfaces, not rugged ones. WCA is improved when the differential energy between the solid–vapor and solid–liquid surfaces is minimized. The surface hydrophobicity was investigated using WCA and CAH [38]. Surface flaws accompany CAH and could be produced by chemical heterogeneity, lack of wettability, and surface roughness. CAH variations can be performed on slippery or sticky surfaces. Despite the fact that the Young model assumes a perfectly smooth surface, all surfaces exhibit imperfections. These imperfections participate in the roughness of the surface and influence the hydrophobic structure [39]. The Wenzel [40] and Cassie-Baxter [41] models have been developed to discuss such influences.

3.2. Wenzel and Cassie-Baxter Models for a Rough Surface

The WCA is critical in determining wettability because it intersects the liquid, solid, and gas-solid interfaces. Fig. (**3A**) depicts the different geometries of the contact angle (including hydrophilic, hydrophobic, and superhydrophobic characters) after dropping a liquid on a certain surface. On a rough surface (Fig. **3B**), the Young equation is no longer valid [40, 42]. Fig. (**3B**) shows Wenzel's model for describing the link between a rough surface's θ0 and a flat surface's θ. The Young equation was changed by the Wenzel model, as shown below (Eq. 2):

$$Cos\ \theta = \left(\frac{\gamma_{SA} - \gamma_{SL}}{\gamma_{LA}}\right) = r\ Cos\theta_0 \tag{2}$$

The factor of surface roughness (r) can be represented as the ratio of the actual rough surface area (A_{SL}) to the smooth fractional area (A_F) as follows (Eq. 3):

$$r = \frac{A_{SA}}{A_F} \tag{3}$$

As r is greater than 1, the hydrophobic surface's (> 90°) roughness raises the WCA and gives it a superhydrophobic character, but roughness on a hydrophilic surface (< 90°) can be reduced to 0° and give it an extra hydrophilic character. Wenzel model can merely be applied to a homogeneous solid–liquid interface and not to heterogeneous surfaces. As a result, Cassie and Baxter [41] devised a new equation for heterostructured surfaces formed from two fractions; (1) with f_1 and θ_1 fractional area and WCA, respectively; and (2) with f_2 and θ_2, where $f_1 + f_2 = 1$. WCA could be expressed as in Eq. 4:

$$Cos\theta = f_1\ Cos\theta_1 + f_1\ Cos\theta_2 \tag{4}$$

In the interface of a nanocomposite, f1 = fSL; $\theta_1 = \theta_0$ and f2 = fSL = 1 and fSL; $\theta2$ = 180°.

Eq. 5 indicates the equation of the Cassie and Baxter model:

$$Cos\theta = rCos\theta_0 - f_{LA}(rCos\theta_0 + 1) \tag{5}$$

When $\theta2 = 0°$, the Cassie equation can be applied. It reflects water-water contact over a coated rough surface with gaps filled with water rather than air (Fig. **3B**) (Eq. 6).

$$Cos\,\theta = f_{SL}(Cos\,\theta_0 - 1) + 1 \tag{6}$$

With increasing f_{LA}, the WCA of a hydrophobic surface increases, creating a superhydrophobic surface. However, on a hydrophilic surface, the WCA is enhanced as the f_{LA} increases, indicating that the surface is transmitted from hydrophilic to hydrophobic structure [43, 44]. On the definite microtextured surface, Lafuma and Quere [45] investigated the water drops' CAH in the Wenzel and Cassie models. In the Cassie model, the drops of water were slightly spread on the surface, whereas in the Wenzel model, they were condensed. Cassie's approach has significantly lower CAH and higher advancing contact angle. Furthermore, on the identical structure, the Cassie state's CAHs are weaker than those in the Wenzel state's homogeneous wetting.

4. SUPERHYDROPHOBIC SURFACE FABRICATION METHODS

Superhydrophobic surfaces could be developed by various chemical and physical methods, as mentioned in Table **1**. In these methods, the surfaces are mainly treated with chemical substances which could react with the surfaces or deposit on them. These methods could facilitate the formation of coatings with surface roughening and non-wettability [46].

Table 1. Different techniques used for developing of superhydrophobic anticorrosion nanocomposites.

Technique	Advantages	Diadvatages
Wet chemical interaction	* Simple and common method * This method was utilized to incorporate many nanofillers in different polymeric matrixes	* The used solvent should be removed by evaporation. * Several steps
Etching	* Fast and low cost * Controlling the parameters or roughness	* Choosing the proper etching agent and immersion time

(Table 1) cont.....

Technique	Advantages	Diadvatages
Hydrothermal approach	* Dense morphology	* Several steps and longtime of production * Sever conditions and need of expensive autoclaves
Electro-deposition	* High quality coatings. * Low cost and rapid deposition * Simple manufacturing process	* Environmental issues * Time consuming * Non-uniform coating
Anodization	* Easy to maintain * UV stable and will not peel or flake * Offers a wide range of high quality architectural finishes	* Use specific grades of aluminum for this process * Cannot be used on Stainless Steel * Expensive solution for small quantities * Limited color selection
Chemical and electrochemical deposition	* High quality coatings. * Thickness controllable. * Complex surface	* High temperature * High cost
Sol-gel	* Complex surfaces. * High quality films. * Good resistance to temperature	* Crackability * Thickness limits
Layer-by-layer self-assembly	* High quality coating. * Controllable thickness * High adhesive layer	* Require definite equipments * Laboratory scale. * High cost
Nanocomposite coatings	* Low cost * Intrinsic superhydrophobicity and no need to modification processes	* Environmental concern * Require definite equipments
Laser surface texturing	* Good surface finish * Dense and porous coating	* Depends on quality of powders * High cost * Low rate of production

4.1. Wet Chemical Interaction

This method entails a variety of chemical reactions that build the microstructures needed for non-wettable surfaces in a direct manner [47, 48]. Hydrophobic materials can be controlled throughout the preparation process or by later post-treatments. On a copper surface, a tetradecanoic acid solution was employed to form flower-like microstructures [47]. The wet-chemical approach offers considerable versatility because it may be applied to surfaces of various sizes and shapes [49]. The most significant constraint connected with the wet-chemical techniques for superhydrophobic coatings is the use of expensive materials with low SFE [50].

4.2. Etching

Another quick and simple way to generate a superhydrophobic coated surface with rough topology is to etch them. Plasma, chemical, and laser etching have all been employed recently [51]. To generate a superhydrophobic surface on aluminum alloy AA2024, hydrochloric acid was employed as a chemical etching solution. The roughness was demonstrated to be dependent on the chemical etching time. Microstructures can also be created by chemically etching dislocations or imperfections on steel substrate surfaces [52, 53]. This is usually followed by grafting the low SFE compounds to achieve superhydrophobicity. To etch the Mg-Li alloy, Liu and his colleagues [52] utilized 0.1 mol/L HCl. Following immersion in ethyl alcohol, including (1 wt.% fluoroalkylsilanes for 12 hours) and heating at 100 °C, the resultant surface exhibited long-term superhydrophobicity, which could be retained even after 6 months in the atmosphere of air corrosion. Lastly, as opposed to wet chemical reactions, superhydrophobic materials are etched out of bulk compounds, making them more resistant to mechanical stresses [54].

4.3. Hydrothermal Approach

Another major top-down strategy for developing anticorrosive superhydrophobic coatings is the hydrothermal method [55, 56]. High pressure and/or heat were used to develop microstructured surfaces in aqueous conditions. The hydrothermal approach is considered as eco-friendly way because the chemical employed to build a microstructured surface is merely H_2O or dilute H_2O [56]. The air cushion caused by coating non-wettability and the hydroxide layer' barrier grown on the metal substrates both contributed to the hydrothermally prepared superhydrophobic surface. This approach can provide better anticorrosive features, including lower corrosion current density and higher corrosion potential, than the etching approach.

4.4. Electrodeposition

This method can deposit a thin layer onto a conductive substrate from a solution containing ions or charged micro and NPs [57, 58]. It's a brand-new method for creating anticorrosive superhydrophobic coatings which have a lot of benefits, including great scalability and efficiency. In corrosive conditions, these superhydrophobic coatings can exhibit excellent corrosion-protection. According to Liu *et al.* [58], electrochemical deposition of a superhydrophobic coating on copper was created in ethyl alcohol including myristic acid and cerium chloride [58].

4.5. Anodization

It can be used to generate anticorrosive superhydrophobic coatings with surfaces' micro- or nano-roughness [59, 60]. The tremendous efficiency of anodization methods in creating surface micro/nanostructures is one of their most significant benefits.

4.6. Chemical and Electrochemical Deposition

These techniques were used to develop superhydrophobic coatings. The electrochemical oxidation and reduction procedures, including metals' anodization and oxidation in solution, or conducting polymers' electrodeposition, were recently reported by Darmanin *et al.* [61].

4.7. Sol-gel

Superhydrophobicity can also be engineered into inorganic-organic sol-gel coatings for corrosion-protection applications [62 - 64]. Generally, sol-gel procedures do not require high pressures or temperatures and can be applied on a variety of metallic surfaces [65]. A superhydrophobic silica covering was created by dipping a copper substrate in a sol-gel solution, including methyltrie-thoxysilane, H_2O, ammonia, and methyl alcohol [62]. With a 155° WCA and < 7° roll-off angle, the resultant coating was created. Hu *et al.* [64] came up with a novel electrodeposition-supported sol-gel method. Due to elevated pH in solution near the substrate (applied with cathodic potentials), the improved sol-gel condensation process resulted in rapid sol-gel coating's growth on the surface of mild steel. The resulting superhydrophobic coating's corrosion resistance was corroborated by the EIS results, which showed a substantially higher impedance modulus. Inorganic/organic sol-gel surfaces are more thermally and UV-stable than organic surfaces because of the high strength of the Si-O linkage.

4.8. Layer-by-Layer Self-Assembly

As a simple and low-cost approach for creating thin-film surfaces, this assembly involves depositing oppositely charged coats with intermediate washing phases. Electrostatic contact and covalent bonds combine to create multilayer grafts. Due to the films' linear growth in response to the number of bi-layers, this approach has a high level of molecular control on the film thickness.

4.9. Nanocomposite Coatings

This manner can be used to create anticorrosive superhydrophobic coatings using a bottom-up method [66, 67]. In this manner, surface roughness is created by NPs, whereas coating's non-wettability is given by both NPs and organic binders, which can bind to the NPs. Even after a 250-minute scouring assessment with 10 m/s water stream [68], the coating's non-wettability can be significantly maintained. Different nanocomposite coatings can be developed by employing various structures of NPs.

4.10. Laser Surface Texturing

Due to its benefits in variable surface shape design, precise machining with case hardening, and quick patterning for various metallic surfaces, this approach has received a lot of interest in generating superhydrophobic surfaces with varied micro- and nanostructures [69]. The synergistic interactions between laser intensity, scanning speed, and repetition rate regulate the surface geometry and determine whether the laser-textured surface can absorb organic molecules. As a result, the surface is tuned from hydrophilicity to superhydrophobicity [70].

For showing superhydrophobic effects on metals using direct laser texturing, laser texturing patterns and wetting transition conditions are critical. Ma *et al.* created a highly durable superhydrophobic surface using nanosecond laser texturing [71]. The friction surface in their investigation was 400# sand paper, and the WCA was dropped from 162° to 156° after a 20-meter abrasion distance. For controlling the laser processing, the scanning speed was 500 mm/s with an average laser power of 40 W. The laser's incident angle was 90°, and the repetition rate is 400 kHz. Micro/nanoscale roughness and superhydrophobic structures can be fabricated using laser texturing techniques [72].

5. ANTICORROSIVE COATINGS

Organic/inorganic nanocomposites can develop anticorrosive superhydrophobic coatings [67]. Surface roughness is created by NPs, while organic binder polymers can bind to the NPs to provide the superhydrophobic character [68]. Another advantage of nanocomposite coatings is the variety of types, sizes, and shapes of the employed NPs. Abrasion resistance, chemical protection, and corrosion prevention are all features of anticorrosive coatings. Anticorrosion refers to the process of shielding metal surfaces against corrosive conditions. Organic-inorganic nanocomposite coatings are interesting mater due to their simplicity of use and inexpensive cost. These coatings are usually filmed in many layers. The coating serves as a protective shield between corrosive ions and the attached object [73, 74].

Coatings create a protective film to prevent corrosive ions' diffusion, allowing the anodic and cathodic corrosion inhibition. Anticorrosion coatings commonly passivate steel by creating a barrier between it and the corrosive conditions. (Fig. **4**) depicts a general representation of the iron steel's corrosion mechanism in the air. Some anticorrosive nanocomposite coatings work by passivating steel with zinc and corrosion-protection pigments. The types of the employed solvent and metallic surface have an effect on the coating's anticorrosion performance. The coating efficiency is further influenced by the exposed environment, the film thickness, and the application conditions [74, 75]. Environmental elements such as humidity %, oxygen concentration, UV radiation, temperature, and pollutants affect the coating's anticorrosion performance [76]. Nanostructured coatings with anticorrosive properties can minimize or prevent corrosion interactions between metallic surfaces and their environment [77]. The potential of inhibitors' incorporation into the coating is enticing [78 - 80]. However, simply mixing small organic molecules into the coating's matrix may impair its mechanical durability. Anticorrosive and self-healing surfaces can achieve long-term corrosion-protection for steel buildings [81]. Self-cleaning properties could be achieved by TiO$_2$ NPs dispersed in a water-repellent resin [82].

Polymers, as corrosion inhibitors, have recently gained much interest. The polymers' functional units can form complexes with metal ions and cover the metallic surface to form a barrier against corrosive chemicals [83]. The heteroatoms such as oxygen and nitrogen, cyclic rings, and active centres of adsorption can structurally influence the polymers' inhibitive efficiency. Advanced coatings, especially graphene-based nanocomposites, have much attention for being used as anticorrosive coatings. Some metallic surfaces can be protected from oxidation by graphene coatings [84].

Graphene-based coatings improve the steel's anticorrosion performance by up to 74%, allowing it to withstand or minimize corrosion. There has been a slew of modern publications on the subject of superhydrophobicity [85, 86]. The water droplets roll off on the superhydrophobic coatings with bouncing a modest applied force when released from a specific height. SFE and surface morphology can influence the surface's degree of superhydrophobicity. Hydrophobicity increases as the SFE decreases. The SFE and non-wettability are influenced by the coatings' chemical compositions [87].

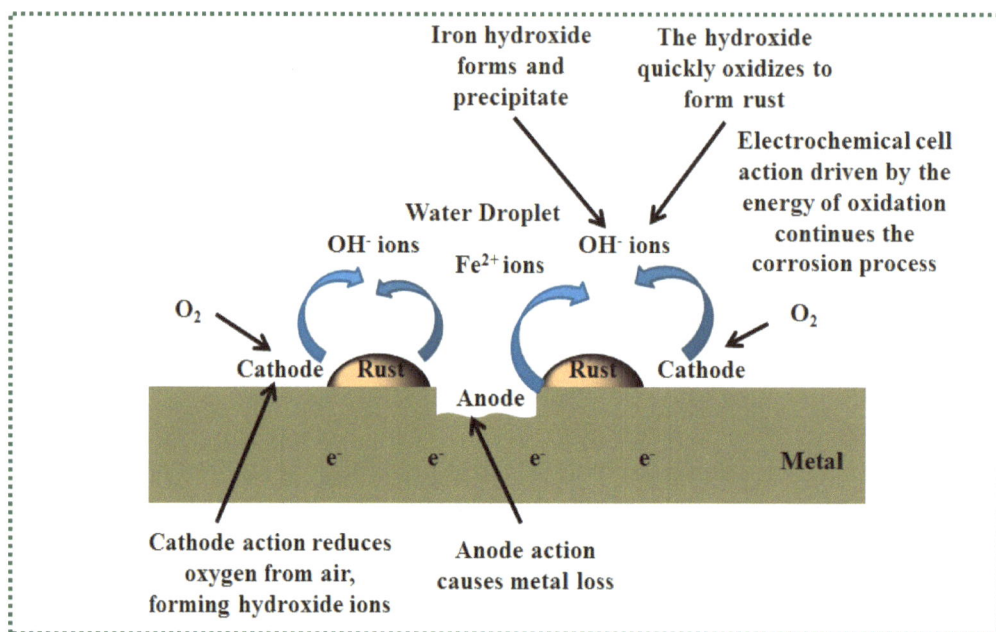

Fig. (4). Steel corrosion diagram owing to ambient moisture condensation in a drop electrolyte cross-section.

6. ARTIFICIAL SUPERHYDROPHOBIC COATINGS FOR STEEL

This section seeks to highlight well-known structured materials utilized in superhydrophobic coating manufacturing. These materials have a considerable impact on the self-cleaning capabilities of surfaces [88]. The employed superhydrophobic anticorrosive materials were discussed in this chapter. These materials are divided into three categories: metallic, non-metallic, and hybrid nanostructures.

6.1. Metallic Materials

Metals and their compounds are widely employed in the development of anticorrosive self-cleaning surfaces. Zinc-based coatings have been widely used to impart corrosion resistance to steel materials [89]. The electrodeposition approach was employed to generate a rough Zn-film on the steel surface in the presence of stearic acid. The zinc stearate complex could form a coating film with low SFE and anticorrosive features. Aluminium is an anticorrosion metallic material by nature, thanks to an air-formed metal oxide coating that forms over the surface.

A major concern is directed toward the mechanical and chemical fragility of the air-sprayed coating. Chemical modification after sandblasting was performed using perfluorooctanoic acid to develop a direct approach for generating a thermally stable and mechanically durable superhydrophobic aluminium surface [90]. The superior qualities of the superhydrophobic surface are determined by the fluorination degree and the surface's hierarchical structured surface. This research has introduced a superhydrophobic Al surface with high mechanical durability and corrosion resistance. Also, metal oxides have the potential to boost the corrosion-protection of superhydrophobic coatings. A TiO_2 non-wettable surface that has been enhanced with Polyvinylidene fluoride on a Cu surface is created *via* dip-coating. Chemical treating of TiO_2 by adding $-CF_2$ and $-CF_3$ groups could improve the sliding angle to 5.5° and WCA to 160° [91]. Electrochemical impedance and potentiodynamic polarisation were used to demonstrate the improved anticorrosion features of Cu substrate in 3.5 wt.% NaCl solution.

Enhanced corrosion-protection was achieved by combining high surface roughness with micro/nanostructured topology and low SFE assembly of polyvinylidene fluoride. The capacity of this material to be converted from non-wettable to wettable surface and back again using UV irradiation and heat is a fascinating property. Employing ZnO nanorods/actyflon-G502 coating could develop superhydrophobic and low SFE film on a metallic Zn surface [92].

The protruding structures of the loosely stacked ZnO nanorods could trap air, reducing the metal-corrosive media contact area. In addition, Laplace pressure pushes the corrosive ions out of the air layer. In 2018, a new hydrothermal method for manufacturing micro/nano rough superhydrophobic iron oxide was developed. Octadecylamine and dopamine were added to create micro/nano-scale coating. After modification, it had a 158.3° WCA and 3.3° sliding angle [93]. The coverage of the macro/nanoscaled surface determined the superhydrophobicity extent.

The developed surface has a 32-fold higher impedance than virgin-N80 steel after being treated with 5 wt.% aqueous NaCl solution. Corrosion media cannot access the metallic surface because of the enormous volume of air-trapped within micro/nanostructures. The addition of octadecylamine in the non-wettable coating also helped to increase its corrosion-protection capabilities. Thermal stability, UV resistance, and durability were all excellent on superhydrophobic surfaces.

Yin and his coworkers [94] used the solvothermal method to create a robust superhydrophobic Ni_3S_2 coating, which was then modified with myristic acid. They studied the superhydrophobic capability coated surface. The results exhibited high resistance against UV irradiation, robustness, and thermal steadiness. An electrodeposition approach was used to create a superhydrophobic covering of MoS_2 NPs dispersed into a Cu-based resin [95]. N80 carbon steel and

AISI 316L stainless steel were employed as substrates. Potentiodynamic polarization measurements in H_2SO_4, NaCl, and NaOH mediums were employed to investigate the coatings' corrosion properties. A MoS_2 filmed layer is used as a reference electrode. The formed $Cu-MoS_2$ surface could provide anticorrosive and superhydrophobic properties. Metallic materials and their superhydrophobic surfaces were approved to be important for preventing steel corrosion.

6.2. Non-Metallic Materials

Non-metals including carbonaceous materials, silica, polymeric materials, and other non-metallic materials have also been discovered to improve the anticorrosive properties of superhydrophobic coatings.

6.2.1. Silica-based

Silica-based coatings are commonly employed in bioinspired superhydrophobic surfaces because of their reduced bulk density, excellent optical quality, elevated rough topology, and low bulk density.

Mahadhik *et al*. [96] presented a facile sol-gel spray coating process for synthesizing silica-derived self-cleaning coatings. The developed coatings were applied to the pre-treated aluminum, steel, copper, and germanium surfaces. The manufactured superhydrophobic surface has long-term mechanical durability, thermal resistance, and good anticorrosive properties. For superhydrophobic materials' design, the preparation techniques and characteristics were investigated using different SiO_2 types [97]. A spraying technique for creating a silica-based superhydrophobic film on the AZ31B Mg alloy surface was described by Qian *et al*. [98]. Binary structures were built using nano- and micro-sized silica particles. The coating's corrosion current density was 3-times lesser than that of the substrate. After a single-step silica deposition and oil infusion, the durability and anticorrosion features of the SiO_2 NPs-based superhydrophobic surface sprayed onto Al alloy substrates were studied [99]. Dual scale roughness came from the existence of micro protuberances and NPs is resulted in increasing the superhydrophobic property. Because oil diffusion into the pores forms a permanent oil film, the SiO_2-based surface-displayed anticorrosion durability. As water and oil are immiscible, the oil film may endure even the most severe corrosive conditions, providing long-term stability.

6.2.2. Carbon-based Materials

The use of carbon-derived nanomaterials, including graphene, carbon nanofibers, multi-walled carbon nanotubes (MWCNTs), and fullerenes in the manufacturing of corrosion-resistant superhydrophobic surfaces has gained popularity [100]. Through electrodeposition, Zhu *et al.* [101] devised a unique approach for fabricating MWCNTs and NPs on carbon-derived films. Carbon-derived coating offers high non-wettability, anticorrosion performance, and superhydrophobic qualities. Co NPs and MWCNTs are anticorrosive agents, preventing chloride ions from penetrating into the NaCl solution. MWCNTs-cobalt-a-C:H based film demonstrated a decreased corrosion rate as compared with a virgin film of diamond-like carbon. A superhydrophobic carbon-derived surface with fluorine-free merit was developed using B-doped carbon NPs and a two-stage hydrothermal process [102]. The prepared superhydrophobic carbon-based materials exhibited exceptional corrosion resistance for the maritime industry. Shi *et al.* [103] used a spray-drying process to create a non-wettable nanostructured paper made of SiO_2 and MWCNTs. Superhydrophobic, chemically stable, protection against corrosion are among the advantages of the demonstrated modified nano paper. PDMS-filled with graphene composites were created as robust superhydrophobic coating materials (Fig. **5**) [104]. A hybrid PDMS nanocomposite filled with diatomaceous earth (DE), TiO_2 NPs, crushed GO exhibited an anticorrosive and robust surface with WCA of 172° and resistance to abrasion and scratching. The coatings were sprayed using an air-spraying approach, and the results confirmed the development of a superhydrophobic self-cleaning surface. The potentiodynamic polarization and the Tafel tests showed superior anticorrosion capabilities with a 96.78% inhibition efficiency.

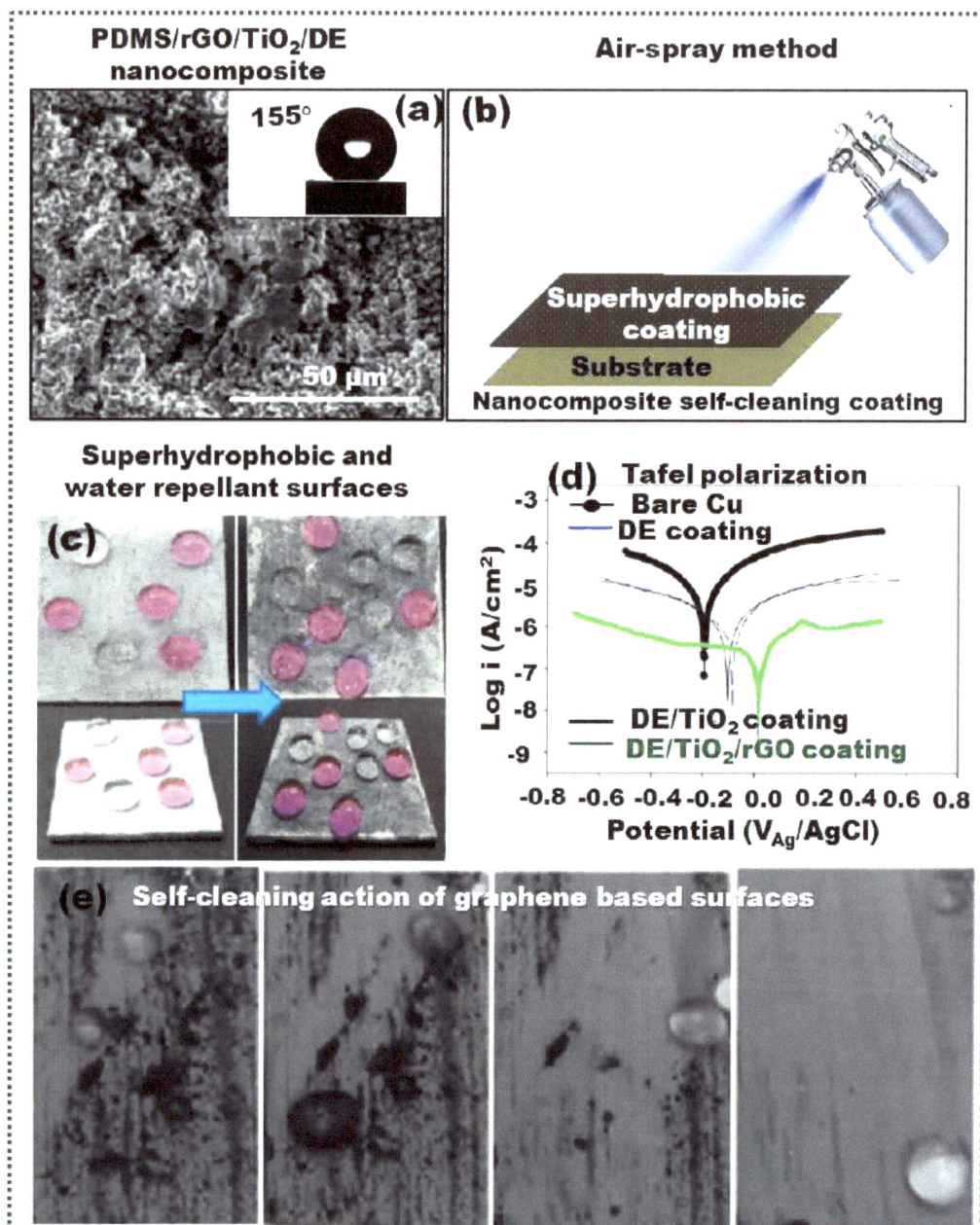

Fig. (5). **(a)** SEM capture of silicone/rGO/TiO$_2$/DE coating filmed with superhydrophobic character (155° WCA), **(b)** Illustration of a nanocomposite coating sprayed by a gun on steel surface, **(c)** Superhydrophobic activity against abrasion examination for the silicone/DE/TiO$_2$/rGO surface, **(d)** corrosion's electrochemical testing known as Tafel polarization for the prepared virgin and PDMS/rGO/TiO$_2$/DE composites compared to bare DE and Cu, and **(e)** Illustrates the self-cleaning action of graphene-based surfaces. Copyright 2015, used with the American Chemical Society's permission [104].

6.2.3. Polymer-based Materials

Due to the exceptional polymeric properties, such as cost-effectiveness and elasticity, various mechanically durable polymer material-based superhydrophobic surfaces have been produced [104]. Das *et al.* [105], for example, discussed the latest advances and uses of superhydrophobic polymer coatings. A spraying technique was used to make an organosilicon-epoxy-derived PVF nanocomposite coating for corrosion-protection purposes [106]. Liu *et al.* [107] used original polycondensation and photo-polymerization techniques to generate a superhydrophobic polymer material. Roughness regeneration is a feature of the produced mechanically durable polymer superhydrophobic material. Ghose *et al.* [108] used $Cu(OH)_2$ nanowires to fabricate various superhydrophobic surfaces based on polycaprolactone and PDMS. The ready-to-use surfaces might be found in a variety of fields, including surface and medical applications. A superhydrophobic biodegradable polymer-chitosan was produced by a facile preparation technique [109]. Meena *et al.* [110] used a spin-coating approach to create a non-wettable polyurethane/SiO_2/hexadecyltrimethoxysilane nanocomposite.

6.2.4. Hybrid Materials

The superhydrophobic activity of polymeric resin enriched with graphene-MWCNTs, organo-metallic, and metal-metalloid hybrid materials have been studied. Ding *et al.* [111] used an electrodeposition approach and myristic acid modification to create a non-wettable hybrid composite of the graphene-nickel surface with WCA 160.4° and sliding angle of 4°. In the 3.5% NaCl solution, such a superhydrophobic surface demonstrated outstanding durable and anticorrosive composite. The hierarchical surface's gaps could trap air and form a barrier film that protects the mild steel substrate from chloride ion attack. Wang *et al.* [112] used a non-solvent induced phase separation approach to produce a non-wettable porous monolith of polycarbonate/metal nanocomposite. Jena *et al.* [113] described an electrodeposition approach for coating carbon steel substrate with the non-wettable surface of Ni-rGO modified with myristic acid. Zhang *et al.* [114] used a hydrothermal approach to create an advanced polypropylene/$Mg(OH)_2$ superhydrophobic surface over an AZ31 Mg alloy substrate. The polypropylene coating displayed good water repellency and corrosion resistance because of its low SFE and micro/nano-rough topology. Selim *et al.* [115] coated steel with a superhydrophobic anticorrosive ternary film of silicone/GO decorated with ZnO nanocomposite (Fig. **6**). The hydrophobicity and anticorrosive performance of different GO-ZnO filling concentrations distributed in the silicone resin were tested. The developed silicone/GO-ZnO (1 wt.%) nanocomposite coating exhibited a WCA of 151°. For producing superhydrophobic surfaces, many methods are presented. All of these techniques aimed to reduce SFE and roughen the surface on the micro/nanoscales. The chapter's next sector provides a complete overview of effective surface corrosion prevention mechanisms.

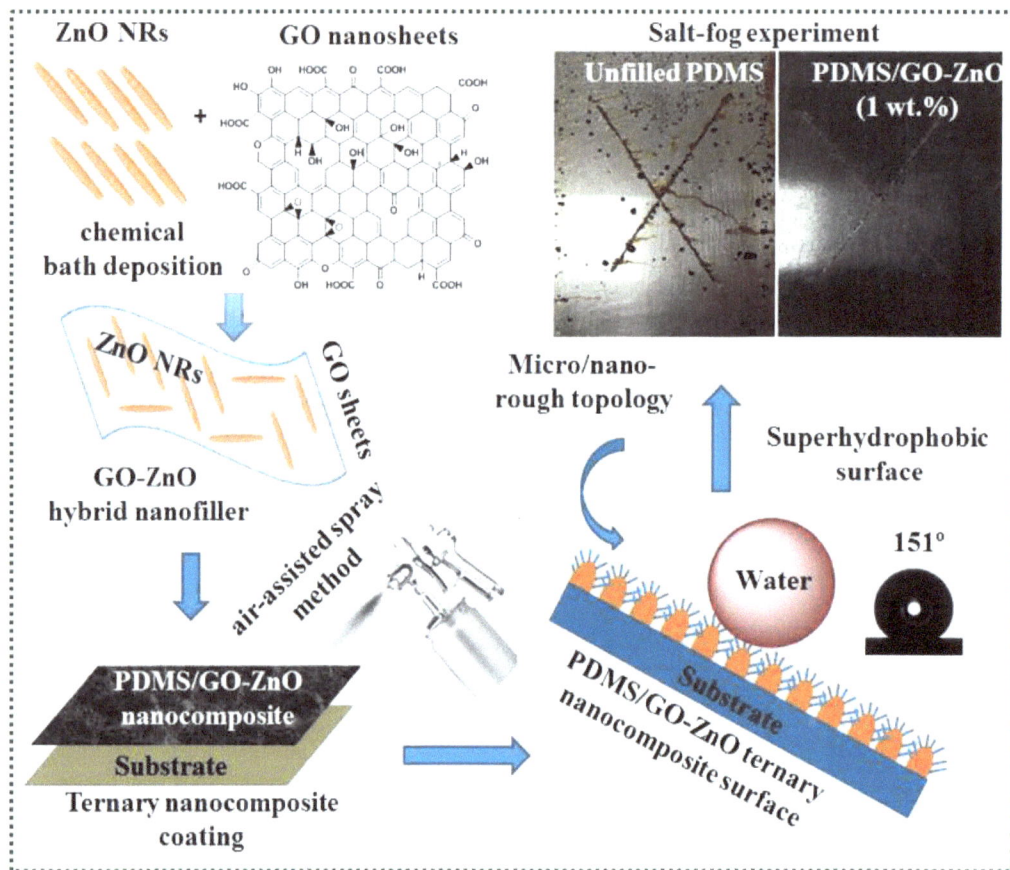

Fig. (6) . Single-step chemical deposition and air-supported spray techniques were used to make a hybrid coating of PDMS/GO-ZnO nanocomposite which exhibited micro/nano-rough topology for the homogenously distributed nanosurface. Copyright 2021, reproduced with permission from Elsevier [115].

7. SUPERHYDROPHOBIC SURFACES' CORROSION PREVENTION MECHANISM

Previously, we reported on numerous methods to develop anticorrosive superhydrophobic nanocomposite coatings [116]. The corrosion prevention of superhydrophobic coatings is not merely connected to their water repellency but can be satisfactorily explained by:

i. Rough surfaces' air pockets can provide a steady air layer, successfully minimizing the direct contact area at the metal-corrosive medium interface.

ii. Air pockets in the hierarchical structures formed on roughened surfaces generate a stable air film and actively reduce the direct contact area at the steel-corrosive conditions interface.

iii. On superhydrophobic surfaces with WCAs of more than 150°, the corrosive medium is repelled from the surface's grooves by capillary pressure, and therefore the surface is protected from deterioration. It describes how water-repellency contributes to anticorrosive action. Heterogeneous wetting reduces the amount of corrosive sites on the metal by limiting interaction between the metal and the corrosive environment. Despite the fact that the interfacial area of metal-electrolyte is limited in the early stages of metal-corrosive solutions interaction, OCP heterogeneity aids the formation of anodic and cathodic locations over time. Uniform wetting avoids the formation of cathodic/ionic positions over the metal surfaces, but non-uniform wetting causes hydrophobic molecules to be desorbed.

iv. Hydrophobic molecules deposited on the metallic surface work as corrosion inhibitors by occupying the corrosion active sites. The anodic reaction is efficiently inhibited by chemisorbed hydrophobic compounds at metal's active corrosion sites. These sites prevent aggressive halide ions from entering. Because of the reduced adsorption of molecular oxygen, the cathodic reaction is stopped.

v. The metallic surface's positive charges in electrolytes attracts the halide ions' negative charges electrostatically in neutral solution. In this solution and at the pH of 2-4, hydrophobic compounds own negative charges, causing ion dispersion in the electrical double layer. This results in depletion of the halide ion concentration at the metals' vicinity, which slows the corrosion.

The corrosion process is disrupted by the hydrophobic molecules' desorption from the metallic surface and the corrosion is prevented. Superhydrophobic coatings have two characteristics; (1) micro and/or nano-roughness and (2) low SFE. There are ups and downs on a rough surface. The depressions on the non-homogeneous surface allow air to be trapped easily. As a result, limited air effactually hinders corrosive compounds like chloride ions, preventing metal corrosion [117]. By functioning as a passive layer, the trapped air effectively protects the metal. To generate a superhydrophobic surface on tin plates, Zhu *et al.* [118] sprayed polyurethane, mesoporous silica, and benzoxazine. The superhydrophobic coating's WCA was greater than 150° when made from various polyurethane and benzoxazine mixtures.

The superhydrophobic surface of this system protects the substrate from corrosion. This influence was enhanced by enhancing the polyurethane proportion in the mix. This polymeric material formed a cross-linked structure and enhanced the supplementary reaction between the stearate salt film and the porous structure's air pockets. This prevents corrosive ions from penetrating the air layer that forms over the liquid-solid interface.

The trapped-air in the surface grooves between rough surfaces inhibits the penetration of corrosive ions. Because the WCA is more than 150° [119], water can easily move against gravity in porous materials. The capillarity effect on superhydrophobic anticorrosion coating was hypothesized by Liu *et al.* [117]. Superhydrophobic coatings are formed in a horizontal tube-shaped submerged in liquid. The water column's height (h) within the cylindrical tube is calculated as follows (Eq. 7).

$$h = \frac{2\gamma cos\theta}{\rho gr} \tag{7}$$

where, γ, θ, ρ, g, and r represent the surface tension, contact angle, liquid density, gravity-induced acceleration, and the radius of the cylindrical tube, respectively.

An *et al.* [120] described a simple and successful method for making PDMS/cerium dioxide nanocomposite as a non-wettable coating. Surface durability, non-wettability, and resistance against UV radiation were all taken into consideration to develop an advanced coating. The fluoroal- kylsilanes-PDM--CeO$_2$ nanocoating demonstrated good anticorrosion capabilities, thanks to the synergistic action of CeO$_2$ and air cushion, according to electrochemical research.

Corrosion inhibitors are now used in conjunction with superhydrophobic materials to boost corrosion-protection performance. The role of contact angle in the nanoporous films' corrosion inhibition was previously studied [121]. The 0° WCA surface is hydrophilic, enabling water infiltration, whereas the 160° WCA surface does not. Filmed coating with a 100° WCA had less infiltration, whereas a 134° surface had less corrosion. The water infiltration decreases and thus the film can work as a corrosion-resistant material at WCAs of 100° - 134°. The recognized processes of corrosion prevention by superhydrophobic surfaces are depicted in Fig. (7) [122].

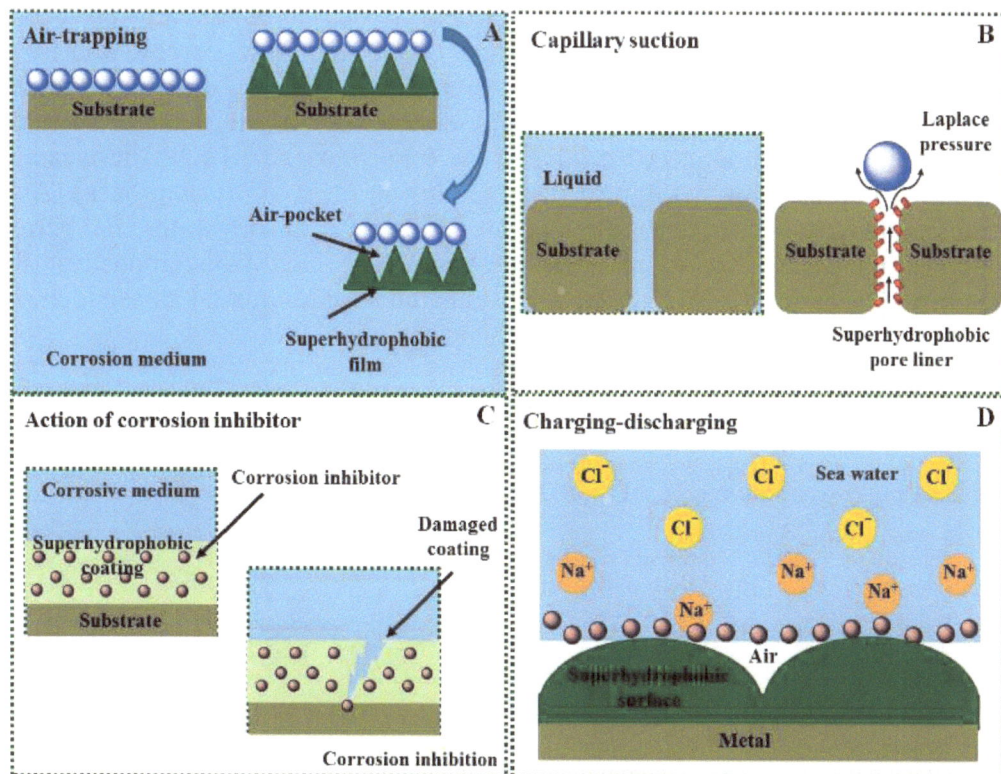

Fig. (7). Corrosion-resistance by superhydrophobic coatings has been established by **(A)** Air-trapping, **(B)** Capillary suction, **(C)** Action of corrosion inhibitor, and **(D)** Charging-discharging [122]. Copyright 2021, reproduced with permission from Elsevier Ltd.

CONCLUSION

In this chapter, we discussed how inorganic nanostructured materials can be used to develop advanced corrosion-resistant nanocomposite polymer coatings. Nanofillers have several advantages over microfillers, such as higher strength to weight ratio, less filler percentage in the polymeric resin, improved polymer-nanofiller interfacial interactions, higher tortuous path for the diffusion of corrosive species within the polymer coating, and increased anticorrosive properties. In a variety of industrial applications, especially material's corrosion protection, nanocomposite coatings can be customized to perform specific functions. Nanocomposite coatings can be used as a corrosion barrier layers. Several nanostructured fillers, such as NPs, nanowires, nanotubes, nanosheets, and nanocomposites, have been produced and evaluated for anticorrosive applications. The nanofillers' dispersion quality in the coatings matrix is critical for producing nanocomposites with improved corrosion-protection. The nanofillers' weight% in the coated material, their chemical structure, functionality, and the preparation method are all critical issues to achieving uniformly distributed polymer/nanofillers composites. To provide more durable and long-lasting protection in a variety of situations, hybrid organic-inorganic nanocomposite coatings with a combination of one or more nanostructured components can be produced. Because of the nanofillers' small size, these can be integrated into a variety of hybrid coatings during the hydrolysis process. As a result, superhydrophobic surfaces have low SFE, do not absorb water, and permit water droplets to easily roll-off. Organic-inorganic nanocomposites for anticorrosive paints have long-term detrimental environmental consequences, which are huge worries for future advancements.

REFERENCES

[1] Mittal, G.; Dhand, V.; Rhee, K.Y.; Park, S.J.; Lee, W.R. A review on carbon nanotubes and graphene as fillers in reinforced polymer nanocomposites. *J. Ind. Eng. Chem.,* **2015**, *21*, 11-25.
[http://dx.doi.org/10.1016/j.jiec.2014.03.022]

[2] Tsang, C.H.A.; Kwok, H.Y.H.; Cheng, Z.; Leung, D.Y.C. The applications of graphene-based materials in pollutant control and disinfection. *Prog. Solid State Chem.,* **2017**, *45-46*, 1-8.
[http://dx.doi.org/10.1016/j.progsolidstchem.2017.02.001]

[3] Habib, S.; Shakoor, R.A.; Kahraman, R. A focused review on smart carriers tailored for corrosion protection: Developments, applications, and challenges. *Prog. Org. Coat.,* **2021**, *154*, 106218.
[http://dx.doi.org/10.1016/j.porgcoat.2021.106218]

[4] Elbourne, A.; Crawford, R.J.; Ivanova, E.P. Nano-structured antimicrobial surfaces: From nature to synthetic analogues. *J. Colloid Interface Sci.,* **2017**, *508*, 603-616.
[http://dx.doi.org/10.1016/j.jcis.2017.07.021] [PMID: 28728752]

[5] Barthlott, W.; Neinhuis, C. Purity of the sacred lotus, or escape from contamination in biological surfaces. *Planta,* **1997**, *202*(1), 1-8.
[http://dx.doi.org/10.1007/s004250050096]

[6] Mohamed, A.M.A.; Abdullah, A.M.; Younan, N.A. Corrosion behavior of superhydrophobic surfaces: A review. *Arab. J. Chem.,* **2015**, *8*, 749-765.
[http://dx.doi.org/10.1016/j.arabjc.2014.03.006]

[7] Ma, M.; Hill, R.M. Superhydrophobic surfaces. *Curr. Opin. Colloid Interface Sci.,* **2006**, *11*, 193-202.
[http://dx.doi.org/10.1016/j.cocis.2006.06.002]

[8] Beitollahpoor, M.; Farzam, M.; Pesika, N.S. Determination of the sliding angle of water drops on surfaces from friction force measurements. *Langmuir,* **2022**, *38*(6), 2132-2136.
[http://dx.doi.org/10.1021/acs.langmuir.1c03206] [PMID: 35104147]

[9] Nyankson, E.; Agbe, H.; Takyi, G.K.S.; Bensah, Y.D.; Sarkar, D.K. Recent advances in nanostructured superhydrophobic surfaces: fabrication and long-term durability challenges. *Curr. Opin. Chem. Eng.,* **2022**, *36*, 100790.
[http://dx.doi.org/10.1016/j.coche.2021.100790]

[10] Pierce, E.; Carmona, F.J.; Amirfazli, A. Understanding of sliding and contact angle results in tilted plate experiments. *Colloids Surf. A Physicochem. Eng. Asp.,* **2008**, *323*(1–3), 73-82.
[http://dx.doi.org/10.1016/j.colsurfa.2007.09.032]

[11] Martines, E.; Seunarine, K.; Morgan, H.; Gadegaard, N.; Wilkinson, C.D.W.; Riehle, M.O. Superhydrophobicity and superhydrophilicity of regular nanopatterns. *Nano Lett.,* **2005**, *5*(10), 2097-2103.
[http://dx.doi.org/10.1021/nl051435t] [PMID: 16218745]

[12]   Nakajima, A.; Hashimoto, K.; Watanabe, T. Recent studies on superhydrophobic films. *Monatsh. Chem.,* **2001**, *132*, 31-41.
       [http://dx.doi.org/10.1007/s007060170142]

[13]   Deyab, M.A. Anticorrosion properties of nanocomposites coatings: A critical review. *J. Mol. Liq.,* **2020**, *313*, 113533.
       [http://dx.doi.org/10.1016/j.molliq.2020.113533]

[14]   Selim, M.S.; Shenashen, M.A.; El-Safty, S.A.; Sakai, M.; Higazy, S.A.; Selim, M.M.; Isago, H.; Elmarakbi, A. Recent progress in marine foul-release polymeric nanocomposite coatings. *Prog. Mater. Sci.,* **2017**, *87*, 1-32.
       [http://dx.doi.org/10.1016/j.pmatsci.2017.02.001]

[15]   Dong, C.; Gu, Y.; Zhong, M.; Li, L.; Sezer, K.; Ma, M.; Liu, W. Fabrication of superhydrophobic Cu surfaces with tunable regular micro and random nano-scale structures by hybrid laser texture and chemical etching. *J. Mater. Process. Technol.,* **2011**, *211*(7), 1234-1240.
       [http://dx.doi.org/10.1016/j.jmatprotec.2011.02.007]

[16]   Gao, X.; Guo, Z. Biomimetic superhydrophobic surfaces with transition metals and their oxides: A review. *J. Bionics Eng.,* **2017**, *14*, 401-439.
       [http://dx.doi.org/10.1016/S1672-6529(16)60408-0]

[17]   Woodward, J.W.; Gwin, H.; Schwartz, D.K. Contact angles on surfaces with mesoscopic chemical heterogeneity. *Langmuir,* **2000**, *16*(6), 2957-2961.
       [http://dx.doi.org/10.1021/la991068z]

[18]   Ma, M.; Mao, Y.; Gupta, Y.; Gleason, K.K.; Rutledge, G.C. Superhydrophobic fabrics produced by electrospinning and chemical vapor deposition. *Macromolecules,* **2005**, *38*(23), 9742-9748.
       [http://dx.doi.org/10.1021/ma0511189]

[19]   Sun, T.; Feng, L.; Gao, X.; Jiang, L. Bioinspired surfaces with special wettability. *Acc. Chem. Res.,* **2005**, *38*(8), 644-652.
       [http://dx.doi.org/10.1021/ar040224c] [PMID: 16104687]

[20]   Selim, M.S.; El-Safty, S.A.; Shenashen, M.A. Superhydrophobic foul resistant and self-cleaning polymer coating.*Superhydrophobic Polymer Coatings*; Samal, S.K.; Mohanty, S.; Nayak, S.K., Eds.; Elsevier Scientific Publisher Company: New York, **2019**, pp. 181-203.
       [http://dx.doi.org/10.1016/B978-0-12-816671-0.00009-6]

[21]   aSelim, M.S.; Yang, H.; Wang, F.Q.; Fatthallah, N.A.; Li, X.; Li, Y.; Huang, Y. Superhydrophobic silicone/SiC nanowire composite as a fouling release coating material. *J. Coat. Technol. Res.,* **2019**, *16*, 1165-1180.
       [http://dx.doi.org/10.1007/s11998-019-00192-8] bSelim, M.S.; El-Safty, S.A.; Azzam, A.M.; Shenashen, M.A.; El-Sockary, M.A.; Abo Elenien, O.M. Superhydrophobic Silicone/TiO$_2$-SiO$_2$ nanorod-like composites for marine fouling release coatings. *ChemistrySelect,* **2019**, *4*, 3395-3407.
       [http://dx.doi.org/10.1002/slct.201803314] cSelim, M.S.; Shenashen, M.A.; Fatthallah, N.A.; Elmarakbi, A.; El-Safty, S.A. *In situ* fabrication of onedimensional-based lotus-like silicone/γ-Al$_2$O$_3$ nanocomposites for marine fouling release coatings. *ChemistrySelect,* **2017**, *2*(30), 9691-9700.
       [http://dx.doi.org/10.1002/slct.201701235]

[22]   aSelim, M.S.; Yang, H.; Wang, F.Q.; Fatthallah, N.A.; Huang, Y.; Kuga, S. Silicone/ZnO nanorod composite coating as a marine antifouling surface. *Appl. Surf. Sci.,* **2019**, *466*, 40-50.
       [http://dx.doi.org/10.1016/j.apsusc.2018.10.004] bSelim, M.S.; El-Safty, S.A.; Fatthallah, N.A.; Shenashen, M.A. Silicone/graphene oxide sheet- alumina nanorod ternary super hydrophobic antifouling coating. *Prog. Org. Coat.,* **2018**, *121*, 160-172.
       [http://dx.doi.org/10.1016/j.porgcoat.2018.04.021]

[23]   Selim, M.S.; El-Safty, S.A.; Shenashen, M.A.; Higazy, S.A.; Elmarakbi, A. Progress in biomimetic leverages for marine antifouling using nanocomposite coatings. *J. Mater. Chem. B Mater. Biol. Med.,* **2020**, *8*(17), 3701-3732.

[http://dx.doi.org/10.1039/C9TB02119A] [PMID: 32141469]

[24]   Faustino, C.M.C.; Lemos, S.M.C.; Monge, N.; Ribeiro, I.A.C. A scope at antifouling strategies to prevent catheter-associated infections. *Adv. Colloid Interface Sci.,* **2020**, *284*, 102230.
[http://dx.doi.org/10.1016/j.cis.2020.102230] [PMID: 32961420]

[25]   Gao, L.; McCarthy, T.J. Contact angle hysteresis explained. *Langmuir,* **2006**, *22*(14), 6234-6237.
[http://dx.doi.org/10.1021/la060254j] [PMID: 16800680]

[26]   Ensikat, H.J.; Ditsche-Kuru, P.; Neinhuis, C.; Barthlott, W. Superhydrophobicity in perfection: the outstanding properties of the lotus leaf. *Beilstein J. Nanotechnol.,* **2011**, *2*, 152-161.
[http://dx.doi.org/10.3762/bjnano.2.19] [PMID: 21977427]

[27]   Goodwyn, P.P.; Maezono, Y.; Hosoda, N.; Fujisaki, K. Waterproof and translucent wings at the same time: problems and solutions in butterflies. *Naturwissenschaften,* **2009**, *96*(7), 781-787.
[http://dx.doi.org/10.1007/s00114-009-0531-z] [PMID: 19322552]

[28]   Wang, G.; Guo, Z.; Liu, W. Interfacial effects of superhydrophobic plant surfaces: A review. *J. Bionics Eng.,* **2014**, *11*, 325-345.
[http://dx.doi.org/10.1016/S1672-6529(14)60047-0]

[29]   Zheng, Y.; Gao, X.; Jiang, L. Directional adhesion of superhydrophobic butterfly wings. *Soft Matter,* **2007**, *3*(2), 178-182.
[http://dx.doi.org/10.1039/B612667G] [PMID: 32680261]

[30]   Gao, X.; Jiang, L. Biophysics: water-repellent legs of water striders. *Nature,* **2004**, *432*(7013), 36.
[http://dx.doi.org/10.1038/432036a] [PMID: 15525973]

[31]   Bechert, D.; Bruse, M.; Hage, W. Experiments with three dimensional riblets as an idealized model of shark skin. *Exp. Fluids,* **2000**, *28*(5), 403-412.
[http://dx.doi.org/10.1007/s003480050400]

[32]   Neinhuis, C.; Barthlott, W. Characterization and distribution of water-repellent, self-cleaning plant surfaces. *Ann. Bot.,* **1997**, *79*(6), 667-677.
[http://dx.doi.org/10.1006/anbo.1997.0400]

[33]   Otten, A.; Herminghaus, S. How plants keep dry: a physicist's point of view. *Langmuir,* **2004**, *20*(6), 2405-2408.
[http://dx.doi.org/10.1021/la034961d] [PMID: 15835702]

[34]   Srivastav, A.D.; Singh, V.; Singh, D.; Giri, B.S.; Singh, D. Analysis of natural wax from *Nelumbo nucifera* leaves by using polar and non-polar organic solvents. *Process Biochem.,* **2021**, *106*, 96-102.
[http://dx.doi.org/10.1016/j.procbio.2021.04.007]

[35]   Ma JW. Impact of surface topography on colloidal and bacterial adhesion, Master thesis, Rice University, Houston, Texas, USA, 2011.

[36]   Feng, L.; Zhang, Y.; Xi, J.; Zhu, Y.; Wang, N.; Xia, F.; Jiang, L. Petal effect: a superhydrophobic state with high adhesive force. *Langmuir,* **2008**, *24*(8), 4114-4119.
[http://dx.doi.org/10.1021/la703821h] [PMID: 18312016]

[37]   Bhushan, B.; Jung, Y.C. Natural and biomimetic artificial surfaces for superhydrophobicity, self-cleaning, low adhesion, and drag reduction. *Prog. Mater. Sci.,* **2011**, *56*(1), 1-108.
[http://dx.doi.org/10.1016/j.pmatsci.2010.04.003]

[38]   Nosonovsky, M.; Bhushan, B. Superhydrophobic surfaces and emerging applications: non-adhesion, energy, green engineering. *Curr. Opin. Colloid Interface Sci.,* **2009**, *14*(4), 270-280.
[http://dx.doi.org/10.1016/j.cocis.2009.05.004]

[39]   Patankar, N.A. Mimicking the lotus effect: influence of double roughness structures and slender pillars. *Langmuir,* **2004**, *20*(19), 8209-8213.
[http://dx.doi.org/10.1021/la048629t] [PMID: 15350093]

[40]   Wenzel, R.N. Resistance of solid surfaces to wetting by water. *Ind. Eng. Chem.,* **1936**, *28*(8), 988-994.

[http://dx.doi.org/10.1021/ie50320a024]

[41]   Cassie, A.B.D.; Baxter, S. Wettability of porous surfaces. *Trans. Faraday Soc.,* **1944**, *40*, 546-551.
       [http://dx.doi.org/10.1039/tf9444000546]

[42]   Wenzel, R.N. Surface roughness and contact angle. *J. Phys. Chem.,* **1949**, *53*(9), 1466-1467.
       [http://dx.doi.org/10.1021/j150474a015]

[43]   Jung, Y.C.; Bhushan, B. Contact angle, adhesion and friction properties of micro- and nanopatterned
       polymers for superhydrophobicity. *Nanotechnology,* **2006**, *17*(19), 4970-4980.
       [http://dx.doi.org/10.1088/0957-4484/17/19/033]

[44]   Koishi, T.; Yasuoka, K.; Fujikawa, S.; Zeng, X.C. Measurement of contact-angle hysteresis for
       droplets on nanopillared surface and in the Cassie and Wenzel states: a molecular dynamics simulation
       study. *ACS Nano,* **2011**, *5*(9), 6834-6842.
       [http://dx.doi.org/10.1021/nn2005393] [PMID: 21838303]

[45]   Lafuma, A.; Quéré, D. Superhydrophobic states. *Nat. Mater.,* **2003**, *2*(7), 457-460.
       [http://dx.doi.org/10.1038/nmat924] [PMID: 12819775]

[46]   Yeganeh, M.; Mohammadi, N. Superhydrophobic surface of Mg alloys: A review. *J Magnes Alloys,*
       2018, *6*(1), 59-70.
       [http://dx.doi.org/10.1016/j.jma.2018.02.001]

[47]   Liu, T.; Yin, Y.; Chen, S.; Chang, X.; Cheng, S. Super-hydrophobic surfaces improve corrosion
       resistance of copper in seawater. *Electrochim. Acta,* **2007**, *52*(11), 3709-3713.
       [http://dx.doi.org/10.1016/j.electacta.2006.10.059]

[48]   Liang, J.; Hu, Y.; Fan, Y.; Chen, H. Formation of superhydrophobic cerium oxide surfaces on
       aluminum substrate and its corrosion resistance properties. *Surf. Interface Anal.,* **2013**, *45*(8), 1211-
       1216.
       [http://dx.doi.org/10.1002/sia.5255]

[49]   Ganesh, V.A.; Raut, H.K.; Nair, A.S.; Ramakrishna, S. A review on self-cleaning coatings. *J. Mater.
       Chem.,* **2011**, *21*(41), 16304-16322.
       [http://dx.doi.org/10.1039/c1jm12523k]

[50]   Chen, X.; Yuan, J.; Huang, J.; Ren, K.; Liu, Y.; Lu, S.; Li, H. Large-scale fabrication of
       superhydrophobic polyurethane/nano-Al_2O_3 coatings by suspension flame spraying for anti-corrosion
       applications. *Appl. Surf. Sci.,* **2014**, *311*, 864-869.
       [http://dx.doi.org/10.1016/j.apsusc.2014.05.186]

[51]   Qi, Y.; Cui, Z.; Liang, B.; Parnas, R.S.; Lu, H. A fast method to fabricate superhydrophobic surfaces
       on zinc substrate with ion assisted chemical etching. *Appl. Surf. Sci.,* **2014**, *305*, 716-724.
       [http://dx.doi.org/10.1016/j.apsusc.2014.03.183]

[52]   Liu, K.; Zhang, M.; Zhai, J.; Wang, J.; Jiang, L. Bioinspired construction of Mg–Li alloys surfaces
       with stable superhydrophobicity and improved corrosion resistance. *Appl. Phys. Lett.,* **2008**, *92*(18),
       183103.
       [http://dx.doi.org/10.1063/1.2917463]

[53]   Wang, N.; Xiong, D.; Deng, Y.; Shi, Y.; Wang, K. Mechanically robust superhydrophobic steel
       surface with anti-icing, UV-durability, and corrosion resistance properties. *ACS Appl. Mater.
       Interfaces,* **2015**, *7*(11), 6260-6272.
       [http://dx.doi.org/10.1021/acsami.5b00558] [PMID: 25749123]

[54]   Xue, C-H.; Zhang, P.; Ma, J.Z.; Ji, P.T.; Li, Y.R.; Jia, S.T. Long-lived superhydrophobic colorful
       surfaces. *Chem. Commun. (Camb.),* **2013**, *49*(34), 3588-3590.
       [http://dx.doi.org/10.1039/c3cc40895g] [PMID: 23525214]

[55]   Guo, X.; Xu, S.; Zhao, L.; Lu, W.; Zhang, F.; Evans, D.G.; Duan, X. One-step hydrothermal
       crystallization of a layered double hydroxide/alumina bilayer film on aluminum and its corrosion
       resistance properties. *Langmuir,* **2009**, *25*(17), 9894-9897.

[http://dx.doi.org/10.1021/la901012w] [PMID: 19441823]

[56]   Ou, J.; Hu, W.; Xue, M.; Wang, F.; Li, W. Superhydrophobic surfaces on light alloy substrates fabricated by a versatile process and their corrosion protection. *ACS Appl. Mater. Interfaces,* **2013**, *5*(8), 3101-3107.
       [http://dx.doi.org/10.1021/am4000134] [PMID: 23496751]

[57]   Su, F.; Yao, K.; Liu, C.; Huang, P. Rapid fabrication of corrosion resistant and superhydrophobic cobalt coating by a one-step electrodeposition. *J. Electrochem. Soc.,* **2013**, *160*(11), D593-D599.
       [http://dx.doi.org/10.1149/2.047311jes]

[58]   Liu, Y.; Li, S.; Zhang, J.; Liu, J.; Han, Z.; Ren, L. Corrosion inhibition of biomimetic superhydrophobic electrodeposition coatings on copper substrate. *Corros. Sci.,* **2015**, *94*, 190-196.
       [http://dx.doi.org/10.1016/j.corsci.2015.02.009]

[59]   He, T.; Wang, Y.; Zhang, Y.; Xu, T.; Liu, T. Super-hydrophobic surface treatment as corrosion protection for aluminum in seawater. *Corros. Sci.,* **2009**, *51*(8), 1757-1761.
       [http://dx.doi.org/10.1016/j.corsci.2009.04.027]

[60]   Xiao, F.; Yuan, S.; Liang, B.; Li, G.; Pehkonen, S.O.; Zhang, T. Superhydrophobic CuO nanoneedlecovered copper surfaces for anti-corrosion. *J. Mater. Chem. A Mater. Energy Sustain.,* **2015**, *3*(8), 4378-4388.
       [http://dx.doi.org/10.1039/C4TA05730A]

[61]   Darmanin, T.; Taffin de Givenchy, E.; Amigoni, S.; Guittard, F. Superhydrophobic surfaces by electrochemical processes. *Adv. Mater.,* **2013**, *25*(10), 1378-1394.
       [http://dx.doi.org/10.1002/adma.201204300] [PMID: 23381950]

[62]   Rao, A.V.; Latthe, S.S.; Mahadik, S.A.; Kappenstein, C. Mechanically stable and corrosion resistant superhydrophobic sol-gel coatings on copper substrate. *Appl. Surf. Sci.,* **2011**, *257*(13), 5772-5776.
       [http://dx.doi.org/10.1016/j.apsusc.2011.01.099]

[63]   Wang, S.; Guo, X.; Xie, Y.; Liu, L.; Yang, H.; Zhu, R.; Gong, J.; Peng, L.; Ding, W. Preparation of superhydrophobic silica film on Mg–Nd–Zn–Zr magnesium alloy with enhanced corrosion resistance by combining micro-arc oxidation and sol-gel method. *Surf. Coat. Tech.,* **2012**, *213*, 192-201.
       [http://dx.doi.org/10.1016/j.surfcoat.2012.10.046]

[64]   Wu, L-K.; Zhang, X-F.; Hu, J-M. Corrosion protection of mild steel by one-step electrodeposition of superhydrophobic silica film. *Corros. Sci.,* **2014**, *85*, 482-487.
       [http://dx.doi.org/10.1016/j.corsci.2014.04.026]

[65]   Figueira, R.; Silva, C.; Pereira, E. Organic–inorganic hybrid sol–gel coatings for metal corrosion protection: a review of recent progress. *J. Coat. Technol. Res.,* **2014**, *12*(1), 1-35.
       [http://dx.doi.org/10.1007/s11998-014-9595-6]

[66]   Xu, X.; Zhang, Z.; Guo, F.; Yang, J.; Zhu, X. Fabrication of superhydrophobic binary nanoparticles/PMMA composite coating with reversible switching of adhesion and anticorrosive property. *Appl. Surf. Sci.,* **2011**, *257*(16), 7054-7060.
       [http://dx.doi.org/10.1016/j.apsusc.2011.02.136]

[67]   Weng, C.J.; Peng, C.W.; Chang, C.H.; Chang, Y.H.; Yeh, J.M. Corrosion resistance conferred by superhydrophobic fluorinated polyacrylate–silica composite coatings on cold-rolled steel. *J. Appl. Polym. Sci.,* **2012**, *126*(S2), E48-E55.
       [http://dx.doi.org/10.1002/app.36380]

[68]   Cui, Z.; Yin, L.; Wang, Q.; Ding, J.; Chen, Q. A facile dip-coating process for preparing highly durable superhydrophobic surface with multi-scale structures on paint films. *J. Colloid Interface Sci.,* **2009**, *337*(2), 531-537.
       [http://dx.doi.org/10.1016/j.jcis.2009.05.061] [PMID: 19552913]

[69]   Tang, X.; Tian, Y.; Tian, X.; Li, W.; Han, X.; Kong, T.; Wang, L. Design of multi-scale textured surfaces for unconventional liquid harnessing. *Mater. Today,* **2021**, *43*, 62-83.

[http://dx.doi.org/10.1016/j.mattod.2020.08.013]

[70]   Tong, W.; Xiong, D. Direct laser texturing technique for metal surfaces to achieve superhydrophobicity. *Mater Tod Phys,* **2022**, *23*, 100651.
[http://dx.doi.org/10.1016/j.mtphys.2022.100651]

[71]   Ma, Q.; Tong, Z.; Wang, W.; Dong, G. Fabricating robust and repairable superhydrophobic surface on carbon steel by nanosecond laser texturing for corrosion protection. *Appl. Surf. Sci.,* **2018**, *455*, 748-757.
[http://dx.doi.org/10.1016/j.apsusc.2018.06.033]

[72]   Liu, W.; Cai, M.; Luo, X.; Chen, C.; Pan, R.; Zhang, H.; Zhong, M. Wettability transition modes of aluminum surfaces with various micro/nanostructures produced by a femtosecond laser. *J. Laser Appl.,* **2019**, *31*, 022503.
[http://dx.doi.org/10.2351/1.5096076]

[73]   Selim, M.S.; Yang, H.; Li, Y.; Wang, F.Q.; Li, X.; Huang, Y. Ceramic hyperbranched alkyd/γ-Al$_2$O$_3$ nanorods composite as a surface coating. *Prog. Org. Coat.,* **2018**, *120*, 217-227.
[http://dx.doi.org/10.1016/j.porgcoat.2018.04.002]

[74]   Selim, MS; Hao, Z; Mo, P; Yi, J; Ou, H . Biobased alkyd/graphene oxide decorated with β–MnO$_2$ nanorods as a robust ternary nanocomposite for surface coating. *Colloid Surf A Physicochem Engin Asp,* **2020**, *601*, 125057.
[http://dx.doi.org/10.1016/j.colsurfa.2020.125057]

[75]   Guzmán, A.; Ocampo, L. Evaluation of the corrosion resistance of systems epoxy zinc rich primer/polysiloxane topcoat by means of electrochemical impedance spectroscopy. *Dyna (Medellin),* **2011**, *167*, 87-95.

[76]   Sharafudeen, R. Smart superhydrophobic anticorrosive coatings, Editor(s): Abdel SH Makhlouf, NY Abu-Thabit, Advances in smart coatings and thin films for future industrial and biomedical engineering applications, *Elsevier,* **2020**, 515-534.

[77]   Hu, K.; Zhuang, J.; Zheng, C.; Ma, Z.; Yan, L.; Gu, H.; Zeng, X.; Ding, J. Effect of novel cytosine-l-alanine derivative based corrosion inhibitor on steel surface in acidic solution. *Mol. Liq.,* **2016**, *222*, 109-117.
[http://dx.doi.org/10.1016/j.molliq.2016.07.008]

[78]   Kermannezhad, K.; Chermahini, A.N.; Momeni, M.M.; Rezaei, B. Application of amine-functionalized MCM-41 as pH-sensitive nano container for controlled release of 2-mercaptobenzoxazole corrosion inhibitor. *Chem. Eng. J.,* **2016**, *306*, 849-857.
[http://dx.doi.org/10.1016/j.cej.2016.08.004]

[79]   Gao, Q.; Wang, S.; Luo, W.J.; Feng, Y.Q. Facile synthesis of magnetic mesoporous titania and its application in selective and rapid enrichment of phosphopeptides. *Mater. Lett.,* **2013**, *107*, 202-205.
[http://dx.doi.org/10.1016/j.matlet.2013.06.017]

[80]   Raja, P.B.; Sethuraman, M.G. Natural products as corrosion inhibitor for metals in corrosive media — A review. *Mater. Lett.,* **2008**, *62*, 113-116.
[http://dx.doi.org/10.1016/j.matlet.2007.04.079]

[81]   Zhang, F.; Qian, H.; Wang, L.; Wang, Z.; Du, C.; Li, X.; Zhang, D. Superhydrophobic carbon nanotubes/epoxy nanocomposite coating by facile one-step spraying. *Surf. Coat. Tech.,* **2018**, *341*, 15-23.
[http://dx.doi.org/10.1016/j.surfcoat.2018.01.045]

[82]   Selim, M.S.; El-Safty, S.A.; El-Sockary, M.A.; Hashem, A.I.; Abo Elenien, O.M. EL-Saeed AM, Fatthallah NA. Smart photo-induced silicone/TiO$_2$ nanocomposites with dominant [110] exposed surfaces for self-cleaning foul-release coatings of ship hulls. *Mater. Des.,* **2016**, *101*, 218-225.
[http://dx.doi.org/10.1016/j.matdes.2016.03.124]

[83]   Rajendran, S.; Sridevi, S.P.; Anthony, N.; Amalraj, A.J.; Sundearavadivelu, M. Corrosion behaviour of

carbon steel in polyvinyl alcohol. *Anti-Corros. Methods Mater.,* **2005**, *52*(2), 102-107.
[http://dx.doi.org/10.1108/00035590510584816]

[84] Chen, S.; Brown, L.; Levendorf, M.; Cai, W.; Ju, S.Y.; Edgeworth, J.; Li, X.; Magnuson, C.W.; Velamakanni, A.; Piner, R.D.; Kang, J.; Park, J.; Ruoff, R.S. Oxidation resistance of graphene-coated Cu and Cu/Ni alloy. *ACS Nano,* **2011**, *5*(2), 1321-1327.
[http://dx.doi.org/10.1021/nn103028d] [PMID: 21275384]

[85] George, J.S.; Vijayan, P.P.; Hoang, A.T.; Kalarikkal, N.; Nguyen-Tri, P.; Thomas, S. Recent advances in bio-inspired multifunctional coatings for corrosion protection. *Prog. Org. Coat.,* **2022**, *168*, 106858.
[http://dx.doi.org/10.1016/j.porgcoat.2022.106858]

[86] Ma, M.L.; Hill, R.M. Superhydrophobic surfaces. *J. Colloid Interface Sci.,* **2006**, *11*(4), 193-202.

[87] Li, W.; Zhan, Y.; Amirfazli, A.; Siddiqui, A.R.; Yu, S. Recent progress in stimulus-responsive superhydrophobic surfaces. *Prog. Org. Coat.,* **2022**, *168*, 106877.
[http://dx.doi.org/10.1016/j.porgcoat.2022.106877]

[88] Parvate, S.; Dixit, P.; Chattopadhyay, S. Superhydrophobic Surfaces: Insights from Theory and Experiment. *J. Phys. Chem. B,* **2020**, *124*(8), 1323-1360.
[http://dx.doi.org/10.1021/acs.jpcb.9b08567] [PMID: 31931574]

[89] Hu, C.; Xie, X.; Zheng, H.; Qing, Y.; Ren, K. Facile fabrication of superhydrophobic zinc coatings with corrosion resistance *via* an electrodeposition process. *New J. Chem.,* **2020**, *44*, 8890-8901.
[http://dx.doi.org/10.1039/D0NJ00561D]

[90] Sun, R.; Zhao, J.; Li, Z.; Mo, J.; Pan, Y.; Luo, D. Preparation of mechanically durable superhydrophobic aluminum surface by sandblasting and chemical modification. *Prog. Org. Coat.,* **2019**, *133*, 77-84.
[http://dx.doi.org/10.1016/j.porgcoat.2019.04.020]

[91] Qing, Y.; Yang, C.; Yu, N.; Shang, Y.; Sun, Y.; Wang, L.; Liu, C. Superhydrophobic TiO_2 /polyvinylidene fluoride composite surface with reversible wettability switching and corrosion resistance. *Chem. Eng. J.,* **2016**, *290*, 37-44.
[http://dx.doi.org/10.1016/j.cej.2016.01.013]

[92] Li, L.; Zhang, Y.; Lei, J.; He, J.; Lv, R.; Li, N.; Pan, F. A facile approach to fabricate superhydrophobic Zn surface and its effect on corrosion resistance. *Corros. Sci.,* **2014**, *85*, 174-182.
[http://dx.doi.org/10.1016/j.corsci.2014.04.011]

[93] He, S.; Wang, Z.; Hu, J.; Zhu, J.; Wei, L.; Chen, Z. Formation of superhydrophobic micro-nanostructured iron oxide for corrosion protection of N80 steel. *Mater. Des.,* **2018**, *160*, 84-94.
[http://dx.doi.org/10.1016/j.matdes.2018.09.002]

[94] Yin, X.; Yu, S.; Bi, X.; Liu, E.; Zhao, Y. Robust superhydrophobic 1D Ni_3S_2 nanorods coating for self-cleaning and anti-scaling. *Ceram. Int.,* **2019**, *45*, 24618-24624.
[http://dx.doi.org/10.1016/j.ceramint.2019.08.192]

[95] Prado, L.H.; Virtanen, S. Cu–MoS_2 superhydrophobic coating by composite electrodeposition. *Coatings,* **2020**, *10*, 238.
[http://dx.doi.org/10.3390/coatings10030238]

[96] Mahadik, S.A.; Pedraza, F.; Vhatkar, R.S. Silica based superhydrophobic coating for long-term industrial and domestic applications. *J. Alloys Compd.,* **2016**, *663*, 487-493.
[http://dx.doi.org/10.1016/j.jallcom.2015.12.016]

[97] Tian, P.; Guo, Z. Bioinspired silica-based superhydrophobic materials. *Appl. Surf. Sci.,* **2017**, *426*, 1-18.
[http://dx.doi.org/10.1016/j.apsusc.2017.07.134]

[98] Qian, Z.; Wang, S.; Ye, X.; Liu, Z.; Wu, Z. Corrosion resistance and wetting properties of silica-based superhydrophobic coatings on AZ31B Mg alloy surfaces. *Appl. Surf. Sci.,* **2018**, *453*, 1-10.
[http://dx.doi.org/10.1016/j.apsusc.2018.05.086]

[99]   Qin, Y.; Li, Y.; Zhang, D.; Zhu, X. Stability and corrosion property of oil-infused hydrophobic silica nanoparticle coating. *Surf. Eng.,* **2021**, *37*(2), 206-211.
[http://dx.doi.org/10.1080/02670844.2019.1697040]

[100]   Si, Y.; Guo, Z. Superhydrophobic nanocoatings: from materials to fabrications and to applications. *Nanoscale,* **2015**, *7*(14), 5922-5946.
[http://dx.doi.org/10.1039/C4NR07554D] [PMID: 25766486]

[101]   Zhu, X.; Zhou, S.; Yan, Q.; Wang, S. Multi-walled carbon nanotubes enhanced superhydrophobic MWCNTs-Co/a-C:H carbon-based film for excellent self-cleaning and corrosion resistance. *Diamond Related Materials,* **2018**, *86*, 87-97.
[http://dx.doi.org/10.1016/j.diamond.2018.04.021]

[102]   Li, Y.; Gou, L.; Wang, H.; Wang, Y.; Zhang, J.; Li, N.; Hu, S.; Yang, J. Fluorine-free superhydrophobic carbon-based coatings on the concrete. *Mater. Lett.,* **2019**, *244*, 31-34.
[http://dx.doi.org/10.1016/j.matlet.2019.01.149]

[103]   Shi, C; Wu, Z; Xu, J; Wu, Q; Li, D; Chen, G; He, M; Tian, J. Fabrication of transparent and superhydrophobic nanopaper *via* coating hybrid SiO$_2$/MWCNTs composite. *Carbohydr Polym,* **2019**, *225*(2019), 115229.
[http://dx.doi.org/10.1016/j.carbpol.2019.115229]

[104]   Nine, M.J.; Cole, M.A.; Johnson, L.; Tran, D.N.H.; Losic, D. Robust superhydrophobic graphene-based composite coatings with self-cleaning and corrosion barrier properties. *ACS Appl. Mater. Interfaces,* **2015**, *7*(51), 28482-28493.
[http://dx.doi.org/10.1021/acsami.5b09611] [PMID: 26632960]

[105]   Das, S.; Kumar, S.; Samal, S.K.; Mohanty, S.; Nayak, S.K. A review on superhydrophobic polymer nanocoatings: Recent development and applications. *Ind. Eng. Chem. Res.,* **2018**, *57*, 2727-2745.
[http://dx.doi.org/10.1021/acs.iecr.7b04887]

[106]   Chen, X.; Wang, H.; Wang, C.; Zhang, W.; Lv, C.; Zhu, Y. A novel antiscaling and anti-corrosive polymer-based functional coating. *J. Taiwan Inst. Chem. Eng.,* **2019**, *97*, 397-405.
[http://dx.doi.org/10.1016/j.jtice.2019.01.016]

[107]   Liu, M.; Luo, Y.; Jia, D. Synthesis of mechanically durable superhydrophobic polymer materials with roughness-regeneration performance. *Compos Part Appl Sci Manuf,* **2020**, *133*, 105861.
[http://dx.doi.org/10.1016/j.compositesa.2020.105861]

[108]   Ghose, A.; Kumar, A.; Raj, S.; Modak, C.D.; Tripathy, A.; Sen, P. Fabrication of polymer-based water-repellent surfaces of complex shapes by physical transfer of nanostructures. *ISSS J. Micro Smart Syst.,* **2020**, *9*, 69-78.
[http://dx.doi.org/10.1007/s41683-020-00049-y]

[109]   Jana, N.; Parbat, D.; Mondal, B.; Das, S.; Manna, U. A biodegradable polymer-based common chemical avenue for optimizing switchable, chemically reactive and tunable adhesive superhydrophobicity. *J. Mater. Chem. A Mater. Energy Sustain.,* **2019**, *7*, 9120-9129.
[http://dx.doi.org/10.1039/C9TA01423C]

[110]   Meena, M.K.; Tudu, B.K.; Kumar, A.; Bhushan, B. Development of polyurethane-based superhydrophobic coatings on steel surfaces. *Philos. Trans.- Royal Soc., Math. Phys. Eng. Sci.,* **2020**, *378*(2167), 20190446.
[http://dx.doi.org/10.1098/rsta.2019.0446] [PMID: 32008453]

[111]   Ding, S.; Xiang, T.; Li, C.; Zheng, S.; Wang, J.; Zhang, M.; Dong, C.; Chan, W. Fabrication of self-cleaning super-hydrophobic nickel/graphene hybrid film with improved corrosion resistance on mild steel. *Mater. Des.,* **2017**, *117*, 280-288.
[http://dx.doi.org/10.1016/j.matdes.2016.12.084]

[112]   Wang, Y.; Yan, J.; Wang, J.; Zhang, X.; Wei, L.; Du, Y.; Yu, B.; Ye, S. Superhydrophobic metal organic framework doped polycarbonate porous monolith for efficient selective removal oil from

water. *Chemosphere,* **2020**, *260*, 127583.
[http://dx.doi.org/10.1016/j.chemosphere.2020.127583] [PMID: 32698115]

[113]   Jena, G.; Thinaharan, C.; George, R.P.; Philip, J. Robust nickel-reduced graphene oxide-myristic acid superhydrophobic coating on carbon steel using electrochemical codeposition and its corrosion resistance. *Surf. Coat. Tech.,* **2020**, *397*, 125942.
[http://dx.doi.org/10.1016/j.surfcoat.2020.125942]

[114]   Zhang, Z.Q.; Zeng, R.C.; Yan, W.; Lin, C.G.; Wang, L.; Wang, Z.L.; Chen, D.C. Corrosion resistance of one-step superhydrophobic polypropylene coating on magnesium hydroxide-pretreated magnesium alloy AZ31. *J. Alloys Compd.,* **2020**, *821*, 153515.
[http://dx.doi.org/10.1016/j.jallcom.2019.153515]

[115]   Selim, M.S.; El-Safty, S.A.; Abbas, M.A.; Shenashen, M.A. Facile design of graphene oxide-ZnO nanorod-based ternary nanocomposite as a superhydrophobic and corrosion-barrier coating. *Colloids Surf Physicochem Eng Asp,* **2020**, *611*, 125793.
[http://dx.doi.org/10.1016/j.colsurfa.2020.125793]

[116]   Boinovich, L.B.; Emelyanenko, A.M.; Modestov, A.D.; Domantovsky, A.G.; Emelyanenko, K.A. Not simply repel water: the diversified nature of corrosion protection by superhydrophobic coatings. *Mendeleev Commun.,* **2017**, *27*, 254-256.
[http://dx.doi.org/10.1016/j.mencom.2017.05.012]

[117]   Liu, T.; Chen, S.; Cheng, S.; Tian, J.; Chang, X.; Yin, Y. Corrosion behavior of super-hydrophobic surface on copper in seawater. *Electrochim. Acta,* **2007**, *52*, 8003-8007.
[http://dx.doi.org/10.1016/j.electacta.2007.06.072]

[118]   Zhu, H.; Hu, W.; Zhao, S.; Zhang, X.; Pei, L.; Zhao, G.; Wang, Z. Flexible and thermally stable superhydrophobic surface with excellent anti-corrosion behaviour. *J. Mater. Sci.,* **2020**, *55*, 2215-2225.
[http://dx.doi.org/10.1007/s10853-019-04050-1]

[119]   Wang, G.; Liu, S.; Wei, S.; Liu, Y.; Lian, J.; Jiang, Q. Robust superhydrophobic surface on Al substrate with durability, corrosion resistance and ice-phobicity. *Sci. Rep.,* **2016**, *6*, 20933.
[http://dx.doi.org/10.1038/srep20933] [PMID: 26853810]

[120]   An, K.; Long, C.; Sui, Y.; Qing, Y.; Zhao, G.; An, Z.; Wang, L.; Liu, C. Large-scale preparation of superhydrophobic cerium dioxide nanocomposite coating with UV resistance, mechanical robustness, and anti-corrosion properties. *Surf. Coat. Tech.,* **2020**, *384*, 125312.
[http://dx.doi.org/10.1016/j.surfcoat.2019.125312]

[121]   Doshi, D.A.; Shah, P.B.; Singh, S.; Branson, E.D.; Malanoski, A.P.; Watkins, E.B.; Majewski, J.; van Swol, F.; Brinker, C.J. Investigating the interface of superhydrophobic surfaces in contact with water. *Langmuir,* **2005**, *21*(17), 7805-7811.
[http://dx.doi.org/10.1021/la050750s] [PMID: 16089386]

[122]   Krishnan, A.; Krishnan, A.V.; Ajith, A.; Shibli, S.M.A. Influence of materials and fabrication strategies in tailoring the anticorrosive property of superhydrophobic coatings. *Surf Inter,* **2021**, *25*, 101238.
[http://dx.doi.org/10.1016/j.surfin.2021.101238]

CHAPTER 7

Morphologies and Properties of Virgin and Waste PP Nanocomposites

Khmais Zdiri[1,2,3,*], A. Elamri[2], O. Harzallah[1] and M. Hamdaoui[1]

[1] *Université de Haute Alsace, ENSISA, Laboratoire de Physique et Mécanique Textiles, EA 4365, 68100, Mulhouse, France*

[2] *Université de Monastir, ENIM, Unité de Recherche Matériaux et Procédés Textiles, UR17ES33, 5019, Monastir, Tunisie*

[3] *Université de Lille, ENSAIT, Laboratoire de Génie et Matériaux Textiles, EA 2461, 59056, Roubaix, France*

Abstract: Polypropylene is a semicrystalline thermoplastic addition polymer used in several applications, such as plastic parts for many industries, consumer product packaging, special devices like living hinges, and textiles. Thanks to its excellent properties, it could be used only or in a nanocomposite system.

This article presents the different types of nanoparticles used for the enhancement of thermo-mechanical and physical behaviors of PP nanocomposites. The analysis of morphologies and thermo-mechanical behaviors of virgin PP nanocomposites are described. Moreover, this paper also discusses the improvements of properties of waste PP by nanoparticle incorporation. Finally, the last section of the paper covers a case study about influence of clay nanoparticles on waste PP based nanocomposites.

Keywords: Morphology, Nanocomposites, Polypropylene, Virgin, Waste.

1. INTRODUCTION

Production and consumption of plastic materials have augmented rapidly over the past decades, due to their excellent properties such as versatile, lightweight, relatively inexpensive and moisture resistant. They can be easily used to replace traditional materials, such as wood, paper, glass and metals [1]. Polypropylene (PP) is considered an essential polymer in the modern society. Indeed, it is very

---

* **Corresponding author Khmais Zdiri:** ENSAIT- Laboratoire de Génie et Matériaux Textiles (GEMTEX), EA 2461, 2 allée Louise et Victor Champier BP 30329, 59056 Roubaix, France; Email: khmaiszdiri@gmail.com

Shazia Anjum (Ed.)

durable and commonly used in harsh environments such as fuel containers, automotive battery cases, *etc* [2].

Each year, large quantities of plastic materials like PP are thrown in nature causing environmental threats. In order to address this problem, considerable efforts have been achieved to re-use the waste PP in the manufacture of useful products [3, 4].

The recycling and reuse of waste PP not only reduces the environmental impacts, but also decreases the costs of the production. Therefore, it is more economic and ecologic to recycle PP [5]. Nevertheless, waste PP tends to have an inferior performance during the manufacturing process when compared with virgin materials. The reinforcement of waste polymers by inorganic and organic fillers is one of the techniques to modify used polymers for technological utilizations.

In this article, a comprehensive study on the incorporation of nanoparticles in virgin and waste polypropylene is investigated, which can help industrials and researchers in their future work.

2. POLYPROPYLENE

Polypropylene is one of the most common semi-crystalline thermoplastics used in industry. Roughly, 30 million tons of PP were consumed in the world during 2015. As we can see on Fig (**1**), it occupied the second range regarding mondial consumption of polymers after polyethylene (PE).

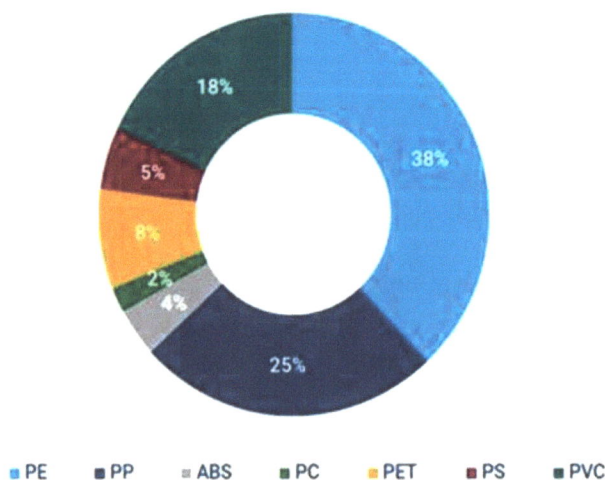

Fig. (1). Thermoplastic polymers consumption in 2015 year [6].

PP has very interesting thermo-mechanical and physical properties for use at room temperature. In fact, PP has high melting point, low density (0.9 to 0.91 g/cm^3), high stiffness and impact resistance. All these characteristic properties permit its use in the manufacture of several industrial materials. The areas of application are numerous and include textile fibers, geotextile, packaging, aerospace and automobile applications.

PP is obtained by metallocene catalysis or by Ziegler-Natta polymerization of propylene monomers. PP is a vinyl polymer with a chemical structure similar to that of PE [7]. Compared to the synthesis of PE, the particular characteristic of PP polymerization is the symmetry of monomer addition in the growing polymer chains due to the existence of -CH3 in the PP monomer (Fig. **2**) [8].

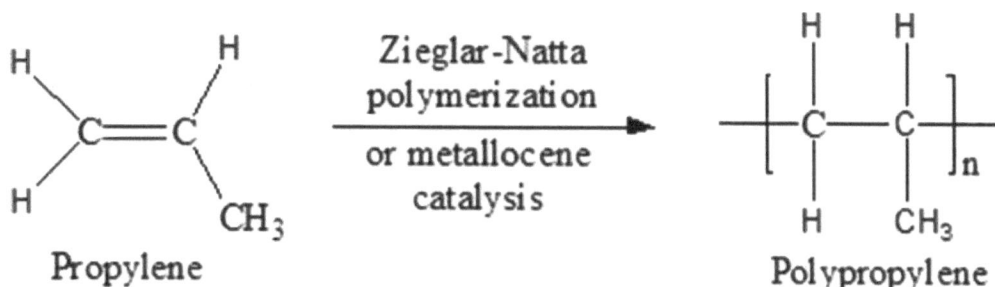

Fig. (2). Polypropylene PP Polymerization.

This polymer has a glass transition temperature (T_g) of about -18°C and a melting temperature (T_f) of about 165°C. Another important feature of polypropylene is its viscosity which has an important effect on molten state processing by extrusion or injection. Its hardness and mechanical strength are relatively interesting and its crystallinity is between 65 and 75%. The elongation limit, yield strength, tensile strength and elasticity modulus of PP are 100-600%, 31-37.2 MPa, 31-41.4 MPa and 1.14-1.55 GPa, respectively [9]. The main characteristics of PP (homopolymers, block and static copolymers) are summarized on the next table (Table **1**).

Table 1. Main characteristics of polypropylene [10].

Properties	Homo-polymers	Block Copolymers	Statistical Copolymers
Density (kg / m³)	900-910	890-905	890-905
MFI (g / 10 min)	1-55	0.8-100	1.7-40
Water Absorption (%)	<0.05	<0.05	<0.05

(Table 1) cont....

Melting temperature (°C)	160-170	160-168	130-164
Glass transition temperature (°C)	-25 to -15	-25 to -15	-25 to -15
Thermal conductivity (W / (m.k))	0.22	0.22	0.22
Heat capacity (J / (kg.k))	1700	1700	1700
Strength at break (MPa)	31-42	28-38	28-38
Elongation at break (%)	100-600	800-900	450-900
Bending strength (MPa)	42-58	35-49	40-45
Tensile Young Modulus (MPa)	1200-1700	1100-1500	900-1300
Izod impact (Kj / m²)	2.5-6	7-50	5-20

PP polymer has a spherulitic structure, when solidified from the pure melt. This spherulitic structure exhibits crystalline lamellae composed of folded chain crystallites [11, 12]. The amorphous regions are found in the interlamellar regions in the form of the linker chains. The molecular structure of PP is presented in Fig (**3**) [12 - 14].

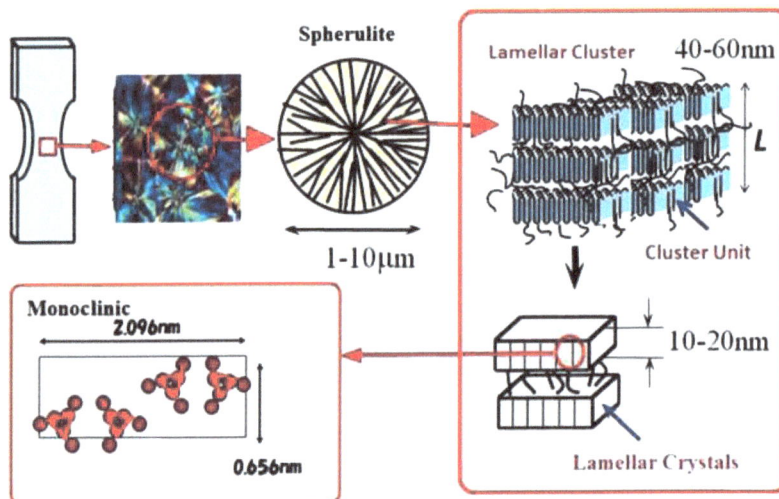

Fig. (3). Crystalline structure of polypropylene.

3. NANOPARTICLES USED IN POLYPROPYLENE NANOCOMPOSITES

As is well known, the incorporation of nanoparticles into the PP matrix can significantly improve the optical, barrier and thermo-mechanical behaviors [15, 16]. This incorporation of nanoparticles allows the increase of the superficial area/volume ratio and the interphase surface of the components, which leads to an enhancement of the overall performances [17]. Contrary to the conventional particles, a small content of nanoparticles is adequate to attain the greatest enhancement of the polymer behaviors without growing the cost or the density [18].

Different types of nanofillers could be used to increase PP properties. On Table **2** are reported the main nanoparticles used to tailor PP nanocomposites with corresponding improved properties.

Table 2. Nanofillers used to improve the properties of PP nanocomposites.

Nanofillers	Enhanced Properties	References
Clays	Physical, thermo-mechanical, morphological, and optical	[19, 20]
Montmorillonite (MMT)	Thermo-mechanical and rheological	[21, 22]
Silica (SiO$_2$)	Thermal and rheological	[23, 24]
Calcium carbonate (CaCO$_3$)	Thermal, Morphology and mechanical	[25, 26]
Carbon nanotubes (CNTs)	Thermal, Morphology and mechanical	[27, 28]
Graphene (GN)	Mechanical, Thermal, and Rheological	[29, 30]
Zinc Oxide (ZnO)	Thermal stability, photo-oxidation and mechanical	[31, 32]

4. MORPHOLOGY AND THERMO-MECHANICAL BEHAVIOR OF VIRGIN PP NANOCOMPOSITES

Several nano-composites with excellent behaviors have been generated by the integration of nanoparticles into virgin polypropylene (vPP).

C.K. Hong *et al.* [33] studied organo-montmorillonite (O-MMT) nanoparticles' performances in improving morphological properties of vPP. According to transmission electron microscopy (TEM), these authors indicated that clay nanoparticles was partly exfoliated and intercalated (Fig. **4**).

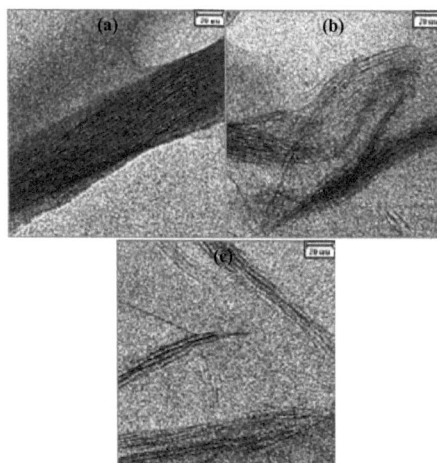

Fig. (4). TEM micrographs of PP/O-MMT5 **(a)**, PP/O-MMT5/C8 (mixing route 1) **(b)** and PP/O-MMT5/C8 (mixing route 2) **(c)** [33].

They showed that the incorporation of organo-montmorillonite enhanced the elastic modulus and the tensile strength of PP nanocomposites (Fig. **5**).

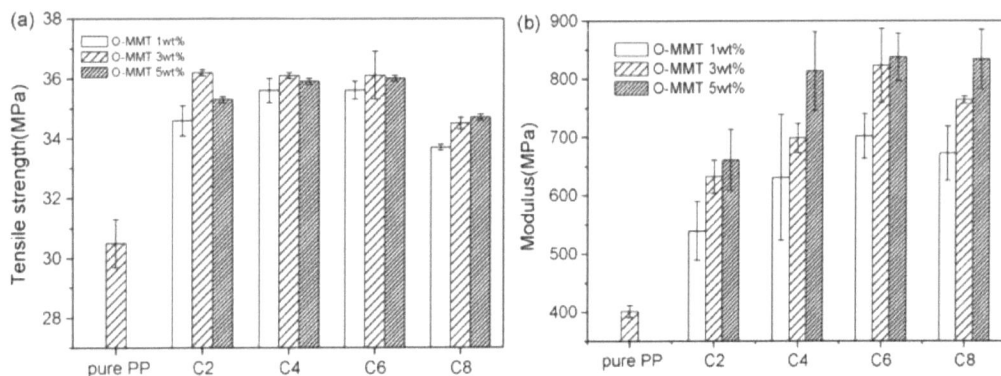

Fig. (5). Influences of O-MMT and PP-g-(MA/ST) amount on the tensile strength **(a)** and the modulus **(b)** of PP nano-composites [33].

Rahman *et al.* [34] showed that 0.5% (by weight) loading of nano-clay enhanced the ductile characteristic (Fig. **6**), which allows the decrease of the virgin PP void fractions. In fact, when the holes were filled with clay nano-fillers, the void fraction was marginally reduced.

Fig. (6). Surface morphologies of **(a)** polypropylene and **(b)**PP reinforced with 0.5 wt% nano-clay [34].

Using thermogravimetric analysis, V. Selvakumar *et al.* [35] reported that the incorporation of montmorillonite (MMT) into PP matrix improved the thermal stability (Fig. **7**). This enhancement in the thermal stability can be attributed to the effect of these nanoparticles on the nucleation stage. Nanocomposite containing layered fillers such as clays are able to create tortuous paths, allowing for increased barrier properties.

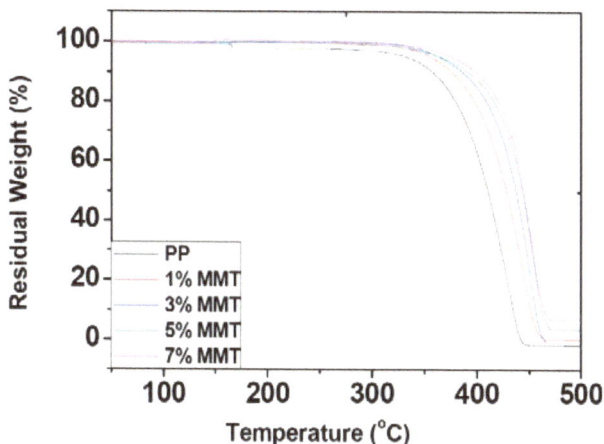

Fig. (7). TGA Thermogram of PP and PP/MMT nanocomposites [35].

In order to analyze the influence of nanofillers on melting and crystallization temperature of polypropylene, Wenjing Yuan *et al.* [36] added Silica nanoparticles to PP matrix (Fig. **8**). They have noticed that the addition of SiO_2 nanoparticles allows to have an increase in the crystallization temperature (Tc) of PP polymer. This effect can be explained by the nucleating abilities of SiO_2 nanoparticles. Whereas, the increase of the melting temperature (Tm) by the incorporation of SiO_2 nanoparticles was not significant.

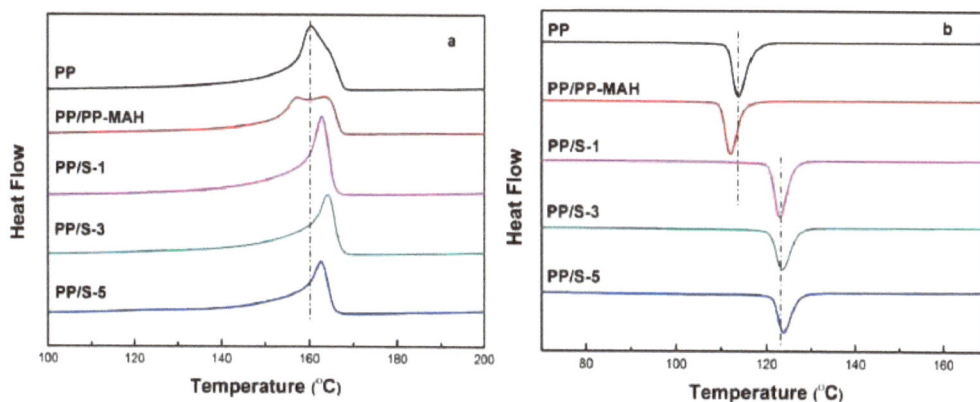

Fig. (8). Melting (**a**) and crystallization (**b**) curves for raw PP and PP nano-composites [36].

Kun Yang *et al.* [37] evaluated the influence of calcium carbonate (CC0.07) on the mechanical behaviors of polypropylene (Table **3**). They indicated that the modulus and flexural strength of PP nanocomposites increased with increasing calcium carbonate content. Their results show that the addition of low amounts of CC0.07 increased slowly the impact strength, flexural modulus and flexural strength of PP nano-composites. Also, these authors found that the yield strength of PP nano-composites increased with increasing CC0.07 content. Whereas, with further increase in CC0.07 amount from at 20 wt%, a decrease in the yield strength was observed.

Table 3. The Mechanical Behaviors of PP/CaCO$_3$ Nano-Composites [37].

Compositions		Yield Strength (MPa)	Flexural Strength (MPa)	Flexural Modulus (MPa)	Izod Impact Strength (KJ/m^2)
PP	CC0.07				
100	0	32.4	35.4	1386.7	2.5
98	2	34.8	35.4	1420.8	2.5
96	4	34.2	35.6	1451.4	2.5
94	6	33.6	35.8	1469.4	2.8
92	8	33.2	36.1	1485.1	2.9
90	10	33.1	36.4	1536.3	3.2
80	20	26.7	37.5	1722.4	4.7
70	30	26.0	38.3	1868.5	5.7
60	40	24.2	39.2	2085.9	6.0

Yanhui Liu *et al.* [38] examined the change in Young's modules, compact strength, ultimate strain and tensile strength *versus* CNTs concentration (Fig. **9**). These authors have shown that incorporation of 1–4 wt.% of CNTs into PP polymer increased the tensile strength and Young's modules. They concluded that these improvements in mechanical properties were significant at low concentrations of CNT. Indeed, the addition of 3 wt% of CNTs enhanced the Young's modules and the tensile strength by 37% and 72%, respectively, compared to virgin PP. This improvement in tensile properties can be attributed to the intercalation of CNTs nanoparticles between PP molecular chains, which allows for more efficient charge transfer from the PP matrix to the nanofillers.

El Achaby *et al.* [39] have investigated the impact of Graphene (GN) nanoparticles on the rheological properties of PP (Fig. **10**). Their results show that GN had a dramatic effect on the rheological behaviors of the PP nano-composites. In fact, storage modulus and complex viscosity of PP nanocomposites increased with increasing GNs loading.

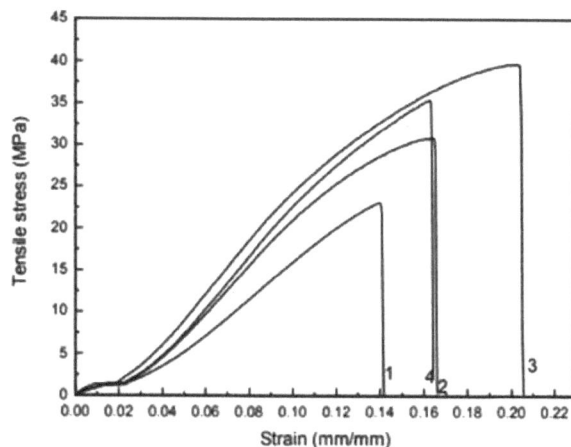

Fig. (9). Impact of CNTs on the mechanical properties of PP nanocomposites [37].

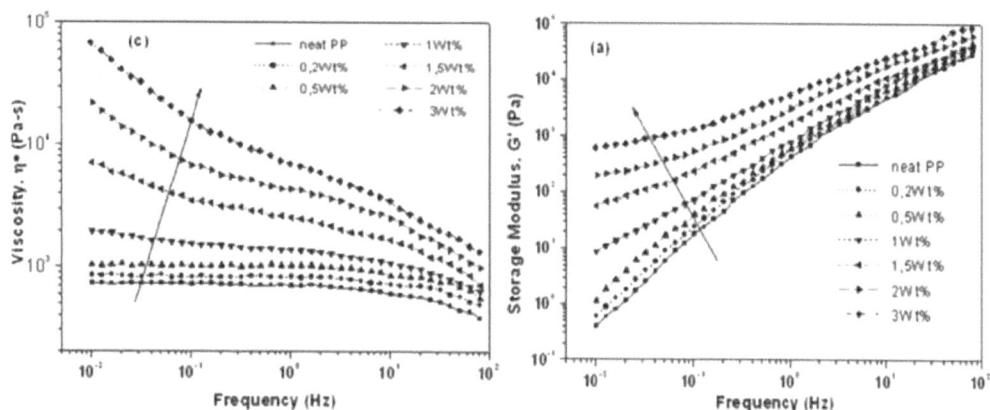

Fig. (10). Complex viscosity (η^*) and storage modulus (G') of PP and its nano-composites [39].

5. IMPROVEMENTS OF PROPERTIES OF WASTE PP BY NANOPARTICLES INCORPORATION

The the elaboration process, nanofiller content and the type of surfactant are characteristic parameters that have a very important influence on the properties of waste polypropylene nano-composites. An overview of the results of several works on improving the properties of waste polypropylene (WPP) will be reviewed in this section.

5.1. Calcium Carbonate (CaCO₃)

Elloumi *et al.* [40] investigated the influence of $CaCO_3$ nano-fillers on the thermal properties of virgin and waste PP (WPP). These authors showed that there are no differences of Tc and Tm between virgin and waste polypropylene compounded with calcium carbonate nanofillers. In contrast, they found that the incorporation of $CaCO_3$ nanofillers into WPP has a considerable impact on the increase of crystal domains. A little reduction in the crystal loading of nanocomposites based on WPP, compared to nanocomposites based on vPP, is noticed up to 10wt% of nanofillers loading. Whereas, the addition of 20wt% of filler increased the crystallinity ratio of the WPP, as compared to vPP nano-composite (Table **4**). This effect can be explained by the interaction between $CaCO_3$ nano-filler and WPP matrix.

Table 4. Thermal Properties of VPP and WPP Nano-Composites [40].

Specimens	Tm₂ (°C)	Tc (°C)	ΔHc (J/g)	ΔHm₂ (J/g)	Xc (%)
vPP	166.4	121.8	69.1	69.3	33
vPP/3%NCC	167.4	121.9	73.8	75.1	36
vPP/10%NCC	167.4	122.2	78.4	82.1	39
vPP/20%NCC	167.7	122.4	83.5	84.2	40
WPP	166.2	122.6	77.5	78.1	37
WPP/3%NCC	166.2	122.7	77.2	66.5	32
WPP/10%NCC	167.1	121.3	69.2	72.2	34
WPP/20%NCC	165.8	122.7	87.5	86.5	41

In another work, Jawad *et al.* [41] studied the impact of $CaCO_3$ nanoparticles on the thermal conductivity of PP polymers. Their results showed that the thermal conductivity increased with increasing the nanofiller concentration. The virgin and waste PP polymers show a considerable difference in thermal conductivity up to a concentration of 3% (by weight) of calcium carbonate nanofillers. The convergence appears progressively up to a concentration of 10wt% of nanofillers loading (Fig. **11**). The observed effect could be justified by the large surface area of $CaCO_3$ nanofillers and their intrinsic thermal conductivity. This effect becomes stronger at a very high concentration of $CaCO_3$ nanofillers.

5.2. Carbon Nanotubes (CNTs)

Gao *et al.* [42] were elaborated WPP/CNT nano-composites using the technique of the melting blend. They have shown that the thermal degradation of WPP nano-composite increased with increasing CNT content. They also observed that when the temperature of the WPP/CNTs nano-composite was increased from 700°C to 1000°C, there was no additional mass loss.

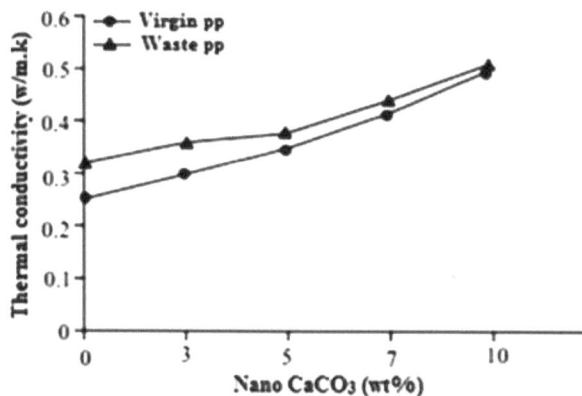

Fig. (11). The evolution of the thermal conductivity versus CaCO$_3$ concentrations [41].

According to tensile tests (Fig. **12**), these authors indicated that the increase in the CNT content up to 3 wt% by weight improves the elongation at break and the tensile strength of the elaborated nano-composites, as compared to virgin WPP. This improvement can be attributed to the nano-filler used that allow the WPP crystallization.

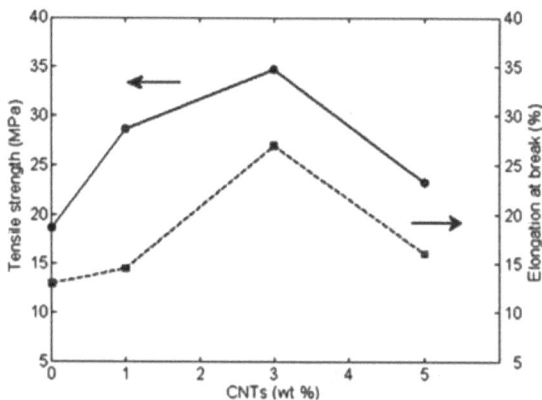

Fig. (12). Mechanical behaviors of WPP/CNTs nano-composite [42].

A decrease in the mechanical properties was observed with further increase in CNT content from at 3 wt% by weight. This effect can be explained by the phenomenon of agglomeration of nanoparticles at high concentrations.

In the same focus, Zhang *et al.* [43] analysed the impact of recycling on the behaviors of polypropylene filled with CNT. The results showed that the recycling increases the toughness of PP reinforced with CNT (Fig. **13**). This can be explained by the fact that recycling allows for changes in the crystallization properties of PP/CNT nano-composites. Furthermore, these authors concluded that the resistance to degradation of PP polymer was increased after the addition of CNT nanoparticles. Also, these authors concluded that the addition of CNT to PP matrix increases the resistance to degradation of the PP. Consequently, to improve the recycling resistance of PP, we can reinforce our polymer with the CNT nanoparticles.

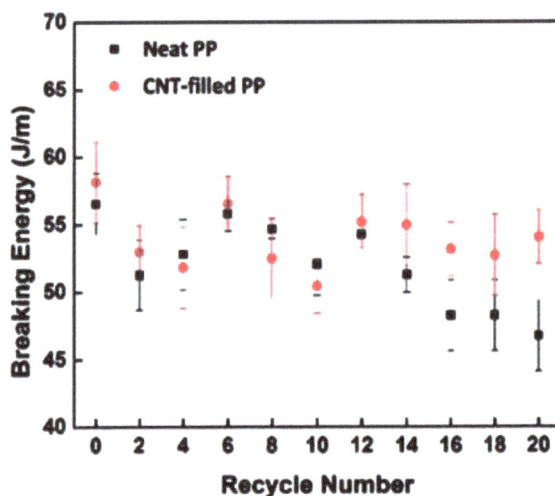

Fig. (13). The izod impact resistance of waste virgin PP and PP/CNT nano-composites [43].

5.3. Zinc Oxide (ZnO)

In their works, Hadi *et al.* [44] evaluated the effects of ZnO nano-fillers on the viscosity and the melt flow rate (MFR) of WPP. These authors found that the addition of ZnO nano-particles increases the viscosity of WPP. This can be explained by the intercalation of ZnO nano-fillers between the waste PP chains. The observed effect was verified by the reduction of the MFR (Fig. **14**).

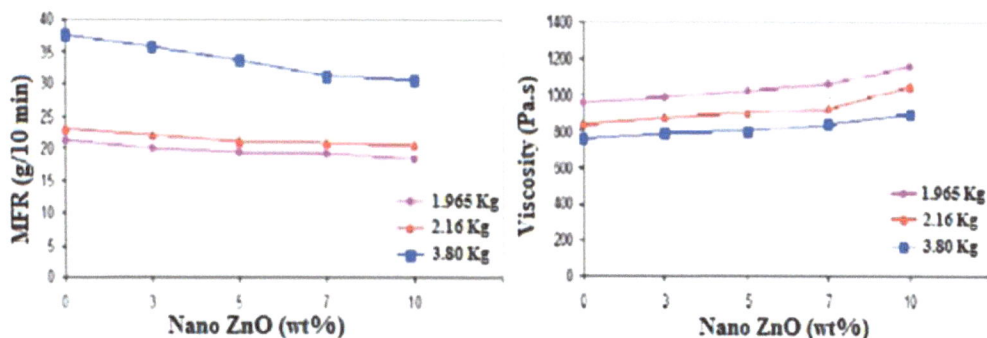

Fig. (14). 14 Impact of ZnO on the MFR and the viscosity of WPP nano-composites [44].

In another work, the impacts of ZnO nano-fillers on the thermal behavior of WPP were investigated by Jawad *et al.* [41]. The reinforcement of WPP with ZnO nanoparticles was performed in two steps. First, ZnO nano-fillers were mixed with acetone solvents using an ultrasonic device. Then, a twin screw extruder was used to produce WPP/ZnO nano-composite. Their results showed that the thermal conductivity of waste PP was improved with increasing nano-particle amount (Fig. **15**). In fact, the large specific surface area and the intrinsic thermal conductivity of ZnO nano-particles are responsible for this improvement in thermal conductivity.

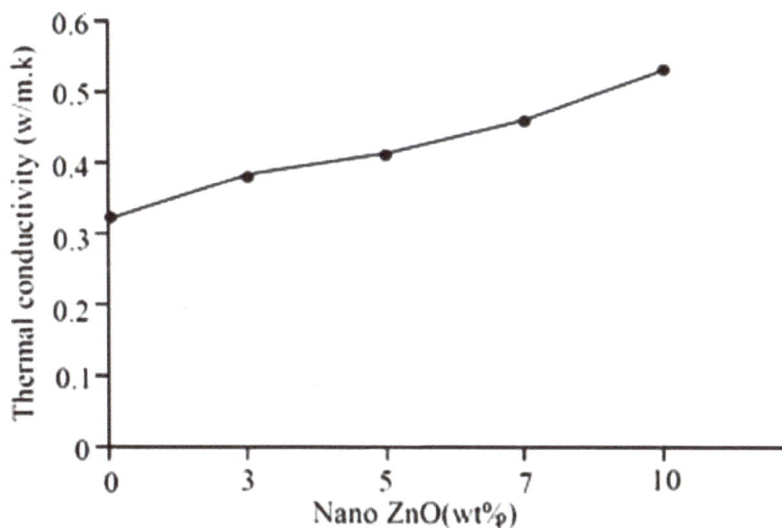

Fig. (15). The thermal conductivity of WPP nano-composites [41].

5.4. Silica (SiO$_2$)

In order to study the impact of silica nanoparticles on the rheological properties of waste PP, Hadi NJ *et al.* [45] have examined the MFR and the viscosity of neat WPP and WPP/SiO$_2$ nano-composites. It was found that the MFR of WPP enhanced with the addition of the SiO$_2$ nano-fillers (Fig. **16**). This can be explained by that the interaction between silica nanoparticles and WPP allows the increase of the chains movement and the free volume. Also, their results showed that the addition of SiO$_2$ nano-fillers decreases the viscosity of WPP. The observed effect can be attributed to the scission of molecular chains throughout the melt mixing process.

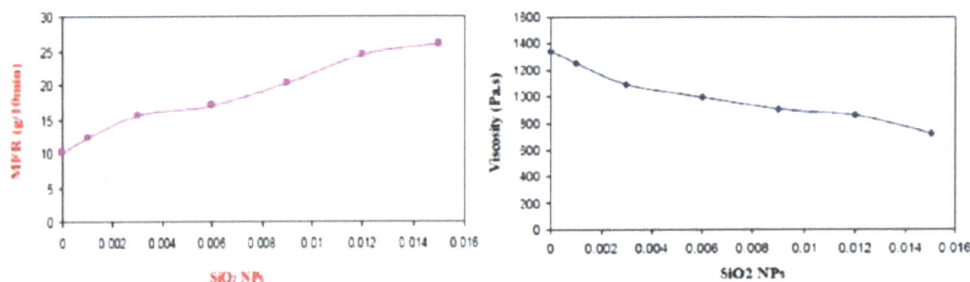

Fig. (16). The rheological properties of neat PP and WPP/SiO$_2$ nano-composites [45].

Dorigato A *et al.* [46] analysed the impact of time on thermal and mechanical behaviors of PP polymer reinforced with 2wt% of both hydrophobic and hydrophilic fumed silica nanofillers. These authors showed that the chain scission reactions caused by the thermal treatment generated a considerable reduction of the melt viscosity of neat polypropylene polymer and its nano-composites. However, this effect was remarkably hindered by the incorporation of SiO$_2$ nanofillers into PP polymer. Also, they found that the addition of hydrophobic SiO$_2$ nanoparticle reduces the size of nanoparticle aggregates with increasing mixing time. It was observed that the melting temperature (Tm) did not appear to be affected by the processing time.

5.5. Graphene (GN)

Husin *et al.* [47] examined the influence of graphene (GN) nano-particles incorporation on the mechanical properties of waste polypropylene with cross linked polyaniline (PANI) elaborated *via* the technique of ultrasound extrusion. These authors deduced from the tensile tests that the enhancement in elastic modulus and tensile strength (Fig. **17**) was achieved in the case of waste PP reinforced with 1.5wt% and 2wt% of graphene content. These enhancements can be explained by the effective transfer of the stress between WPP polymer and GN nanofiller and also to the stiffness of the used particles.

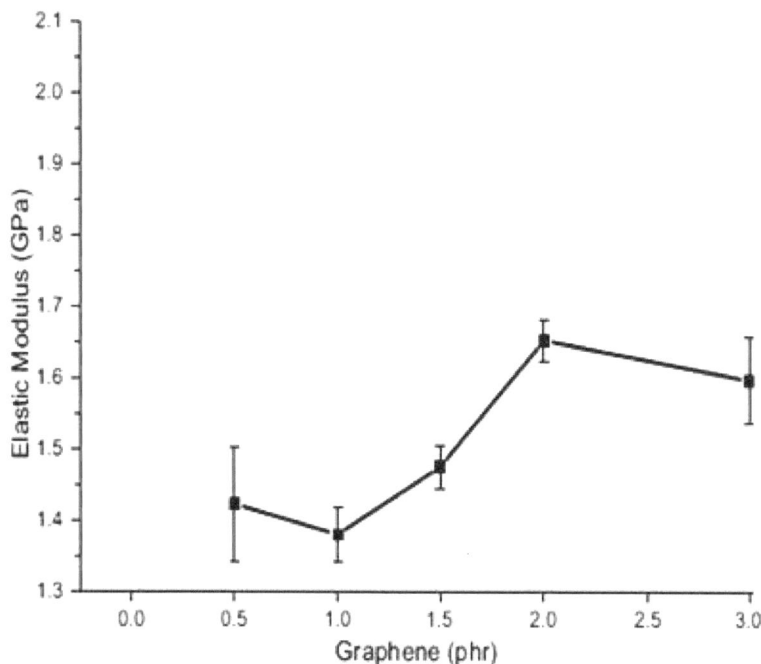

Fig. (17). Elastic modulus of WPP/GN nanocomposites [47].

In another work, Marianna Triantou *et al.* [48] investigated the impact of graphene on re-extruded and aged PP nanocomposites. Their tensile tests indicated that the tensile behaviors of the re-extruded nano-composites and aged nano-composites were very similar to those of the unaged PP. While the tensile strength and strain of aged PP were significantly decreased, compared to unaged PP.

According to FTIR analysis, they found that graphene nanoplatelets protected the PP matrix from the ageing. In fact, the peak of the degradation was hardly obvious in the infrared spectra of the aged nano-composites. Through the oxidation induction time (OIT) measurement, the OIT was longer, as compared to the aged unreinforced PP (Fig. **18**). Whereas, aging process has resulted in a deterioration of the mechanical properties of the virgin PP.

In conclusion, we can be said that the effect of anti-aging of graphene could have an extension of the life of the PP polymers.

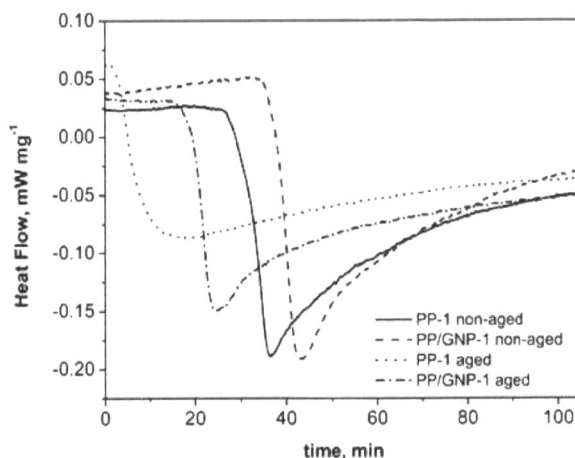

Fig. (18). OIT measurement for aged PP, non-aged PP and PP/GNP nano-composite [48].

6. INFLUENCE OF CLAY NANOPARTICLES ON WASTE PP BASED NANOCOMPOSITES

In this section, we studied the impact of Tunisian clay nano-filler on the mineralogical and thermo-mechanical properties of waste polypropylene (WPP). The amounts of Tunisian clays used to reinforce the waste PP *via* a twin screw extruder are 1%, 3%, 5% and 7%.

6.1. Mineralogical Behaviors

We have used X-ray diffraction (XRD) to analyze the mineralogical behaviors of waste PP and its nano-composites.

(Fig. **19**) shows that the crystallographic nature of the WPP polymer is not affected by the addition of Tunisian clay. The appearance of the characteristic peaks of the PP structure is observed in all samples. These peaks appear at 21.9°, 21.2°, 18.6°, 16.8° and 14.0°, which correspond to the planes of (041), (111), (130), (040) and (110), respectively [49, 50].

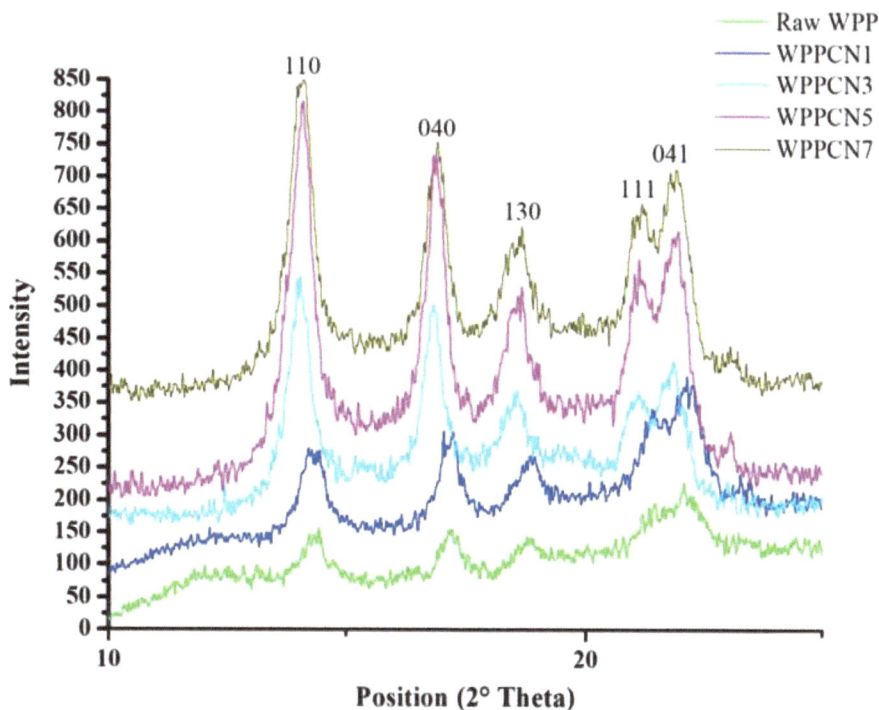

Fig. (19). Mineralogical properties of WPP and WPP filled with Tunisian clay.

It was observed that all samples show the same number of X-ray diffraction peaks in all diffractograms with a difference in the intensity of these peaks that increases with increasing clay content. This effect confirms that the incorporation of nano-particles does not significantly change the crystal structures of the waste PP polymer. The intensity of reflections of elaborated nano-composites is significant increased, particularly in the case of WPP reinforced with 5 wt% of clay nanofiller (WPPCN5).

6.2. Mechanical Behavior

The influences of Tunisian clay and polypropylene-graft-maleic anhydride (PP--MA) on the mechanical properties of the WPP were studied in this section. It may be observed from the stress strain curves (Fig. **20**) of WPP and WPP/clay nano-composites that:

- The increase in the Tunisian clay particles concentration leads to enhance the tensile strength and the Young's modulus of the elaborated nanocomposites.

Indeed, the incorporation of 5% of nano-clay by weight increases the tensile strength and the Young's modulus by approximately 86% and 70%, respectively. This increase in tensile modulus may be justified by the intercalation of the polymer molecules inside the Tunisian clay layers [51].

- A clear decrease in the tensile strength and the Young's modulus of the WPPCN nano-composites is obtained in the case of a WPP filled with 7% by weight of clay content. In fact, the tensile strength of WPPCN7 is equal to 51.53 MPa, whereas it is equal to 55.86 MPa for WPPCN5 and only 30.04 MPa for pure WPP. The observed effect could be justified by the agglomeration phenomenon at high nanoclay concentrations [52].

- The incorporation of Tunisian clay nanoparticles into WPP increases the elongation at break. The maximum value of elongation at break (19.64%) was obtained when WPP was reinforced with 5 wt% of clay nano-filler. The observed effect can be explained by the plasticizing effect of PP-g-MA (used as a compatibilizer) [53].

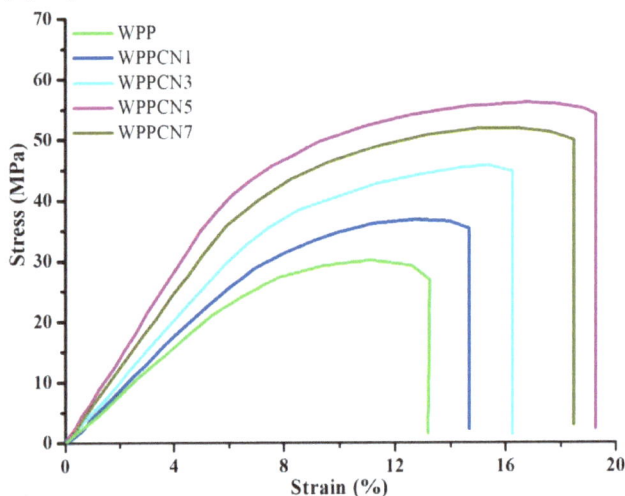

Fig. (20). The mechanical properties of WPP and WPP/clay nano-composites.

6.3. Thermal Analysis

Thermogravimetric analysis (TGA) and differential thermal analysis (DTA) were investigated to evaluate the thermal stability of raw WPP and WPP filled with Tunisian clay.

Fig. (**21a**) shows that with the increase of the temperature decreases the weight of all samples, confirming the thermal degradation of specimens. The temperature of the thermal degradation for composites reinforced with nanoclay is higher compared with that of the unfilled WPP. In fact, the temperature of the thermal degradation of the unfilled WPP is 245°C, whereas it is 280°C for nano-composites containing 7% by weight of clay filler. The enhancement in the thermal stability of WPP might be attributed to the effect of Tunisian clay fillers, which acts as nucleation agent [54].

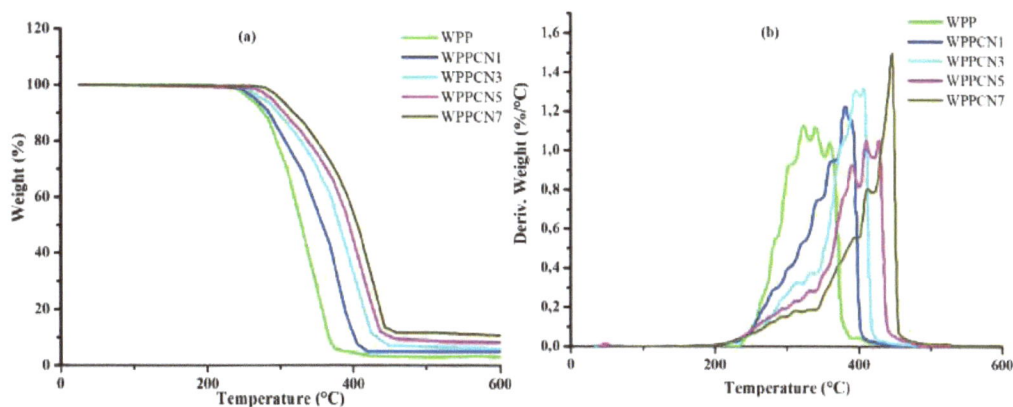

Fig. (21).   TGA **(a)** and DTA **(b)** curves of neat WPP and WPP nano-composites.

As observed in the (Fig. **21b**), the incorporation of the Tunisian clay fillers into WPP reduces the intensity of the released energy amount. This can be attributed to the development of the coal layer on the external surface of the polymer matrix by nanoclay particles [55].

CONCLUSION

The present study indicates that the incorporation of low amount of nano-particles in both virgin and waste PP is a successful way to improve their performances. Various nano-fillers such as $CaCO_3$, nano-clay, CNTs, GN, SiO_2, ZnO and MMT have been used for PP nano-composites elaboration. When incorporated in PP matrix, these nanofillers gave considerable improvements in physical, mechanical and thermal properties.

The last section covered our laboratory case study about influence of nanoclay on waste PP based nanocomposites. We found that good dispersion of clay nanoparticles enables upgrading of waste PP thermal and mechanical behaviors.

REFERENCES

[1]   Zdiri, K.; Elamri, A.; Hamdaoui, M.; Harzallah, O.; Khenoussi, N.; Brendlé, J. Reinforcement of recycled PP polymers by nanoparticles incorporation. *Green Chem. Lett. Rev.,* **2018**, *11*(3), 296-311.
       [http://dx.doi.org/10.1080/17518253.2018.1491645]

[2]   Zdiri, K.; Elamri, A.; Hamdaoui, M.; Harzallah, O.; Khenoussi, N.; Brendlé, J. Elaboration and Characterization of Recycled PP/Clay Nanocomposites. *J. Mater. Environ. Sci.,* **2018**, *9*, 2370-2378.

[3]   Ziaei Tabari, H.; Khademieslam, H. A Study on Nanocomposite Properties Made of Polypropylene/Nanoclay and Wood Flour. *World Appl. Sci. J.,* **2012**, *16*, 275-279.

[4]   Zdiri, K.; Elamri, A.; Hamdaoui, M.; Harzallah, O.; Khenoussi, N.; Brendlé, J. Valorization of Post-consumer PP by (Un)modified Tunisian Clay Nanoparticles Incorporation. *Waste Biomass Valoriz.,* **2020**, *11*(5), 2285-2296.
[http://dx.doi.org/10.1007/s12649-018-0427-2]

[5]   Arjmandi, R.; Hassan, A.; Othman, N.; Mohamad, Z. *Characterizations of Carbon-Based Polypropylene Nanocomposites. Carbon-Based Polymer Nanocomposites for Environmental and Energy Applications*; Elsevier, **2018**, pp. 57-78.

[6]   Ansari, M. Manufacture and Comparison of Mechanical Properties of Reinforced Polypropylene Nanocomposite with Carbon Fibers and Calcium Carbonate Nano-particles. Iran. J. Mech.Eng. **2019**, *6*(5), 1-12.

[7]   Zahedi, R.; Afshar Taromi, F.; Mirjahanmardi, S.H.; Nekoomanesh Haghighi, M.; Jadidi, K.; Jamjah, R. Propylene Polymerization over $MgCl_2$ -Supported Ziegler-Natta Catalysts Containing Tri-Ether as the Internal Donor. *Adv. Polym. Technol.,* **2018**, *37*(1), 144-153.
[http://dx.doi.org/10.1002/adv.21651]

[8]   Greene, J.P. Commodity Plastics. In *Automotive Plastics and Composites*; William Andrew Publishing. **2021**, 83-105.

[9]   Santos, A.M. *Estudo de compósitos híbridos polipropileno/fibras de vidro e coco para aplicações em Engenharia*; Federal University of Paraná, **2006**, pp. 1-90.

[10]   Duval, C. Duval, C. Plastiques et automobile : d'hier à aujourd'hui. Techniques de l'ingénieur. **2007**.

[11]   Fiebig, J.; Gahleitner, M.; Paulik, C.; Wolfschwenger, J. Ageing of polypropylene: processes and consequences. *Polym. Test.,* **1999**, *18*(4), 257-266.
[http://dx.doi.org/10.1016/S0142-9418(98)00023-3]

[12]   Galeski, A. Strength and toughness of crystalline polymer systems. *Prog. Polym. Sci.,* **2003**, *28*(12), 1643-1699.
[http://dx.doi.org/10.1016/j.progpolymsci.2003.09.003]

[13]   Pawlak, A.; Galeski, A. Cavitation during tensile deformation of polypropylene. *Macromolecules,* **2008**, *41*(8), 2839-2851.
[http://dx.doi.org/10.1021/ma0715122]

[14]   Nitta, K-H.; Yamana, M. *Poisson's Ratio and Mechanical Nonlinearity Under Tensile Deformation in Crystalline Polymers*; Division of Material Sciences, Graduate School of Natural Science and Technology: Japan, **2014**, pp. 114-132.

[15]   Tscharnuter, D.; Jerabek, M.; Major, Z.; Pinter, G. Irreversible deformation of isotactic polypropylene in the pre-yield regime. *Eur. Polym. J.,* **2011**, *47*(5), 989-996.
[http://dx.doi.org/10.1016/j.eurpolymj.2011.03.003]

[16]   Zdiri, K.; Elamri, A.; Hamdaoui, M.; Khenoussi, N.; Harzallah, O.; Brendle, J. Impact of Tunisian clay nanofillers on structure and properties of post-consumer polypropylene-based nanocomposites. *J. Thermoplast. Compos. Mater.,,* **2019**, *32*(9), 1159-1175.
[http://dx.doi.org/10.1177/0892705718792377]

[17]   Zare, Y.; Garmabi, H. A developed model to assume the interphase properties in a ternary polymer nanocomposite reinforced with two nanofillers. *Compos., Part B Eng.,* **2015**, *75*, 29-35.
[http://dx.doi.org/10.1016/j.compositesb.2015.01.031]

[18]   Jafari, S.H.; Kalati-vahid, A.; Khonakdar, H.A.; Asadinezhad, A.; Wagenknecht, U.; Jehnichen, D. Crystallization and melting behavior of nanoclay-containing polypropylene/poly(trimethylene

terephthalate) blends. *Express Polym. Lett.,* **2012**, *6*(2), 148-158.
[http://dx.doi.org/10.3144/expresspolymlett.2012.16]

[19]   Bunekar, N.; Tsai, T-Y.; Huang, J-Y.; Chen, S-J. Investigation of thermal, mechanical and gas barrier properties of polypropylene-modified clay nanocomposites by micro-compounding process. *J. Taiwan Inst. Chem. Eng.,* **2018**, *88*, 252-260.
[http://dx.doi.org/10.1016/j.jtice.2018.04.016]

[20]   Raji, M.; Mekhzoum, M.E.M.; Rodrigue, D.; Qaiss, A.; Bouhfid, R. Effect of silane functionalization on properties of polypropylene/clay nanocomposites. *Compos., Part B Eng.,* **2018**, *146*, 106-115.
[http://dx.doi.org/10.1016/j.compositesb.2018.04.013]

[21]   Fitaroni, L.B.; de Lima, J.A.; Cruz, S.A.; Waldman, W.R. Thermal stability of polypropylene–montmorillonite clay nanocomposites: Limitation of the thermogravimetric analysis. *Polym. Degrad. Stabil.,* **2015**, *111*, 102-108.
[http://dx.doi.org/10.1016/j.polymdegradstab.2014.10.016]

[22]   Majka, T.M.; Bartyzel, O.; Raftopoulos, K.N.; Pagacz, J.; Leszczyńska, A.; Pielichowski, K. Recycling of polypropylene/montmorillonite nanocomposites by pyrolysis. *J. Anal. Appl. Pyrolysis,* **2016**, *119*, 1-7.
[http://dx.doi.org/10.1016/j.jaap.2016.04.005]

[23]   Mallakpour, S.; Naghdi, M. Polymer/SiO_2 nanocomposites: Production and applications. *Prog. Mater. Sci.,* **2018**, *97*, 409-447.
[http://dx.doi.org/10.1016/j.pmatsci.2018.04.002]

[24]   Oseh, J.O.; Mohd Norddin, M.N.A.; Ismail, I.; Gbadamosi, A.O.; Agi, A.; Mohammed, H.N. A novel approach to enhance rheological and filtration properties of water–based mud using polypropylene–silica nanocomposite. *J. Petrol. Sci. Eng.,* **2019**, *181*106264
[http://dx.doi.org/10.1016/j.petrol.2019.106264]

[25]   Chan, C.M.; Wu, J.; Li, J.X.; Cheung, Y.K. Polypropylene/calcium carbonate nanocomposites. *Polymer (Guildf.),* **2002**, *43*(10), 2981-2992.
[http://dx.doi.org/10.1016/S0032-3861(02)00120-9]

[26]   Lam, T.D.; Hoang, T.V.; Quang, D.T.; Kim, J.S. Effect of nanosized and surface-modified precipitated calcium carbonate on properties of $CaCO_3$/polypropylene nanocomposites. *Mater. Sci. Eng. A,* **2009**, *501*(1-2), 87-93.
[http://dx.doi.org/10.1016/j.msea.2008.09.060]

[27]   Ghoshal, S.; Wang, P.H.; Gulgunje, P.; Verghese, N.; Kumar, S. High impact strength polypropylene containing carbon nanotubes. *Polymer (Guildf.),* **2016**, *100*, 259-274.
[http://dx.doi.org/10.1016/j.polymer.2016.07.069]

[28]   Al-Saleh, M.H. Electrically conductive carbon nanotube/polypropylene nanocomposite with improved mechanical properties. *Mater. Des.,* **2015**, *85*, 76-81.
[http://dx.doi.org/10.1016/j.matdes.2015.06.162]

[29]   Bafana, A.P.; Yan, X.; Wei, X.; Patel, M.; Guo, Z.; Wei, S.; Wujcik, E.K. Polypropylene nanocomposites reinforced with low weight percent graphene nanoplatelets. *Compos., Part B Eng.,* **2017**, *109*, 101-107.
[http://dx.doi.org/10.1016/j.compositesb.2016.10.048]

[30]   Ajorloo, M.; Fasihi, M.; Ohshima, M.; Taki, K. How are the thermal properties of polypropylene/graphene nanoplatelet composites affected by polymer chain configuration and size of nanofiller? *Mater. Des.,* **2019**, *181*108068
[http://dx.doi.org/10.1016/j.matdes.2019.108068]

[31]   Zhao, H.; Li, R.K.Y. A study on the photo-degradation of zinc oxide (ZnO) filled polypropylene nanocomposites. *Polymer (Guildf.),* **2006**, *47*(9), 3207-3217.
[http://dx.doi.org/10.1016/j.polymer.2006.02.089]

[32]   Bustos-Torres, K.A.; Vazquez-Rodriguez, S.; la Cruz, A.M.; Sepulveda-Guzman, S.; Benavides, R.;
        Lopez-Gonzalez, R.; Torres-Martínez, L.M. Influence of the morphology of ZnO nanomaterials on
        photooxidation of polypropylene/ZnO composites. *Mater. Sci. Semicond. Process.,* **2017**, *68*, 217-225.
        [http://dx.doi.org/10.1016/j.mssp.2017.06.023]

[33]   Hong, C.K.; Kim, M.J.; Oh, S.H.; Lee, Y.S.; Nah, C. Effects of polypropylene-g-(maleic
        anhydride/styrene) compatibilizer on mechanical and rheological properties of polypropylene/clay
        nanocomposites. *J. Ind. Eng. Chem.,* **2008**, *14*(2), 236-242.
        [http://dx.doi.org/10.1016/j.jiec.2007.11.001]

[34]   Rahman, M.R.; Hamdan, S.; Hossen, M.F. *The effect of clay dispersion on polypropylene
        nanocomposites: Physico-mechanical, thermal, morphological, and optical properties. Silica and Clay
        Dispersed Polymer Nanocomposites*; Elsevier, **2018**, pp. 201-257.
        [http://dx.doi.org/10.1016/B978-0-08-102129-3.00011-7]

[35]   Selvakumar, V.; Manoharan, N. Thermal Properties of Polypropylene/ Montmorillonite
        Nanocomposites. *Indian J. Sci. Technol.,* **2014**, *7*(is7), 136-139.
        [http://dx.doi.org/10.17485/ijst/2014/v7sp7.3]

[36]   Yuan, W.; Wang, F.; Chen, Z.; Gao, C.; Liu, P.; Ding, Y.; Zhang, S.; Yang, M. Efficient grafting of
        polypropylene onto silica nanoparticles and the properties of PP/PP-g-SiO$_2$ nanocomposites. *Polymer
        (Guildf.),* **2018**, *151*, 242-249.
        [http://dx.doi.org/10.1016/j.polymer.2018.07.060]

[37]   Yang, K.; Yang, Q.; Li, G.; Sun, Y.; Feng, D. Morphology and mechanical properties of
        polypropylene/calcium carbonate nanocomposites. *Mater. Lett.,* **2006**, *60*(6), 805-809.
        [http://dx.doi.org/10.1016/j.matlet.2005.10.020]

[38]   Liu, Y.H.; Gao, J.L. Mechanical Properties and Wear Behavior of Polypropylene/Carbon Nanotube
        Nanocomposites. *Adv. Mat. Res.,* **2011**, *299-300*, 798-801.
        [http://dx.doi.org/10.4028/www.scientific.net/AMR.299-300.798]

[39]   El Achaby, M.; Arrakhiz, F.E.; Vaudreuil, S.; el Kacem Qaiss, A.; Bousmina, M.; Fassi-Fehri, O.
        Mechanical, thermal, and rheological properties of graphene-based polypropylene nanocomposites
        prepared by melt mixing. *Polym. Compos.,* **2012**, *33*(5), 733-744.
        [http://dx.doi.org/10.1002/pc.22198]

[40]   Elloumi, A.; Pimbert, S.; Bourmaud, A.; Bradai, C. Thermomechanical properties of virgin and
        recycled polypropylene impact copolymer/CaCO$_3$ nanocomposites. *Polym. Eng. Sci.,* **2010**, *50*(10),
        1904-1913.
        [http://dx.doi.org/10.1002/pen.21716]

[41]   Jawad Hadi, N.; Saad, N.; Jawad Mohamed, D. Thermal Behavior of Calcium Carbonate and Zinc
        Oxide Nanoparticles Filled Polypropylene by Melt Compounding. *Res. J. Appl. Sci. Eng. Technol.,*
        2016, *13*, 265-272.
        [http://dx.doi.org/10.19026/rjaset.13.2941]

[42]   Gao, J.L.; Liu, Y.H.; Li, D.M. Preparation and Properties of Recycled Polypropylene /Carbon
        Nanotube Composites. *Adv. Mat. Res.,* **2011**, *279*, 106-110.
        [http://dx.doi.org/10.4028/www.scientific.net/AMR.279.106]

[43]   Zhang, J.; Panwar, A.; Bello, D.; Jozokos, T.; Isaacs, J.A.; Barry, C.; Mead, J. The effects of recycling
        on the properties of carbon nanotube-filled polypropylene composites and worker exposures. *Environ.
        Sci. Nano,* **2016**, *3*(2), 409-417.
        [http://dx.doi.org/10.1039/C5EN00253B]

[44]   Hadi, N.J.; Saad, N.A.; Alkhfagy, D. *Rheological behavior of waste polypropylene reinforced with
        zinc oxide nanoparticles*; Ancona, Italy, **2016**, pp. 201-211.
        [http://dx.doi.org/10.2495/AFM160171]

[45]   Hadi, N.J.; Mohamed, D.J. Study the Relation between Flow, Thermal and Mechanical Properties of

Waste Polypropylene Filled Silica Nanoparticles. *Key Eng. Mater.,* **2016**, *724*, 28-38.
[http://dx.doi.org/10.4028/www.scientific.net/KEM.724.28]

[46]   Dorigato, A.; Pegoretti, A. Reprocessing effects on polypropylene/silica nanocomposites. *J. Appl. Polym. Sci.,* **2014**, *131*(10), n/a.
[http://dx.doi.org/10.1002/app.40242]

[47]   Hadi, N.J.; Mohamed, D.J. Study the Relation between Flow, Thermal and Mechanical Properties of Waste Polypropylene Filled Silica Nanoparticles. Key Engineering Materials. 2016, 724, 28–38.

[48]   Triantou, M.; Todorova, N.; Giannakopoulou, T.; Vaimakis, T.; Trapalis, C. Mechanical performance of re-extruded and aged graphene/polypropylene nanocomposites. *Polym. Int.,* **2017**, *66*(12), 1716-1724.
[http://dx.doi.org/10.1002/pi.5353]

[49]   Zheng, Q.; Shangguan, Y.; Yan, S.; Song, Y.; Peng, M.; Zhang, Q. Structure, morphology and non-isothermal crystallization behavior of polypropylene catalloys. *Polymer (Guildf.),* **2005**, *46*(9), 3163-3174.
[http://dx.doi.org/10.1016/j.polymer.2005.01.097]

[50]   Dong, Y.; Bhattacharyya, D. Effects of clay type, clay/compatibiliser content and matrix viscosity on the mechanical properties of polypropylene/organoclay nanocomposites. *Compos., Part A Appl. Sci. Manuf.,* **2008**, *39*(7), 1177-1191.
[http://dx.doi.org/10.1016/j.compositesa.2008.03.006]

[51]   Lin, Y.; Chen, H.; Chan, C.M.; Wu, J. High impact toughness polypropylene/CaCO$_3$ nanocomposites and the toughening mechanism. *Macromolecules,* **2008**, *41*(23), 9204-9213.
[http://dx.doi.org/10.1021/ma801095d]

[52]   Chandran, N.; Chandran, S.; Maria, H.J.; Thomas, S. Compatibilizing action and localization of clay in a polypropylene/natural rubber (PP/NR) blend. *RSC Advances,* **2015**, *5*(105), 86265-86273.
[http://dx.doi.org/10.1039/C5RA14352G]

[53]   Sharma, S.K.; Nayak, S.K. Surface modified clay/polypropylene (PP) nanocomposites: Effect on physico-mechanical, thermal and morphological properties. *Polym. Degrad. Stabil.,* **2009**, *94*(1), 132-138.
[http://dx.doi.org/10.1016/j.polymdegradstab.2008.09.004]

[54]   Borysiak, S.; Paukszta, D.; Helwig, M. Flammability of wood–polypropylene composites. *Polym. Degrad. Stabil.,* **2006**, *91*(12), 3339-3343.
[http://dx.doi.org/10.1016/j.polymdegradstab.2006.06.002]

[55]   Sun, Q.; Schork, F.; Deng, Y. Water-based polymer/clay nanocomposite suspension for improving water and moisture barrier in coating. *Compos. Sci. Technol.,* **2007**, *67*(9), 1823-1829.
[http://dx.doi.org/10.1016/j.compscitech.2006.10.022]

SUBJECT INDEX

A

Acetylation reactions 160
Acid(s) 10, 44, 51, 52, 62, 63, 64, 65, 92, 112,
 115, 117, 119, 120, 122, 126, 128, 132,
 135, 154, 172, 173, 176, 214, 217, 220,
 245, 250, 254
 aliphatic 126
 Alkylation of barbituric 120
 amino 51, 52
 aqueous citric 65
 carboxylic 172
 chiral phosphoric 62
 cyanuric 135
 deoxyribonucleic 220
 dialkylbarbituric 115, 128
 diethylbarbituric 115, 117
 dilituric 122
 dimethylbarbituric 117
 dimethyltiovioluric 117
 dimethylvioluric 117
 folic 214
 medium trifluoromethanesulfonic 119
 myristic 245, 250, 254
 nucleic 217
 perfluorooctanoic 250
 phosphoric 44
 ptoluenesulfonic 10
 purpuric 117
 sulfamic 63
 thiobarbituric 112, 117, 135
 tolylboronic 176
 triflic 92
 trifluoroacetic 63, 64
 tungstophosphoric 132
Aeromonas hydrophila 208
Antibacterial 85, 131, 190, 203, 217, 218, 222
 activity 85, 131, 217, 218, 222
 agents 190, 222
 effects 203
Antibreast cancer activities 20
Anticancer 1, 2, 5, 6, 7, 13, 33, 34, 35, 219,
 220
 action 2

activity 2, 5, 6, 7, 13, 33, 34, 220
 drugs 1, 35, 219
Antiepileptic drug 113, 114
Antifungal agents 2, 190
Antimicrobial 191, 194, 210, 212, 218, 219
 activity 191, 218, 219
 agents 194, 210, 212, 218
Antiproliferative activities 11, 21, 29, 34
Anti-tumor effects 5, 24
Aspergillus niger 219
Atomic force microscopy (AFM) 198, 203
Automotive battery cases 269

B

Bacillus subtilis 219
Bacterial infection 222
Ball milling processes 195
Band-gap energy 190, 209
Biphasic catalysis 152
Brønsted acid catalysts 44, 63, 87, 89, 92

C

Cancer 2, 7
 bone 7
 drug resistance 2
Chemical synthesis proceses 212
Chemical vapor 194, 195, 197
 deposition (CVD) 194
 deposition methods 195
 deposition processes 197
Chromenoacridine derivatives 86
Coatings 236, 238, 248, 249, 254, 255, 259
 air-sprayed 249
 anticorrosion 238, 248
 corrosion-resistant 238
 hybrid 255, 259
 polymeric 236
 superhydrophobic polymer 254
Corrosion 237
 metallic 237
 inhibition 257

www.ingramcontent.com/pod-product-compliance
Lightning Source LLC
Chambersburg PA
CBHW050811220326
41598CB00006B/178